39th International Microelectronics and Packaging Conference (IMAPS Poland 2015)

IOP Conference Series: Materials Science and Engineering
Volume 104

Gdansk, Poland
20 – 23 September 2015

Editors:

Piotr Jasinski
Krzysztof Gorecki
Robert Bogdanowicz

ISBN: 978-1-5108-1955-9
ISSN: 1757-8981

Printed from e-media with permission by:

Curran Associates, Inc.
57 Morehouse Lane
Red Hook, NY 12571

Some format issues inherent in the e-media version may also appear in this print version.

Printed by Curran Associates, Inc. (2016)

For permission requests, please contact the Institute of Physics
at the address below.

Institute of Physics
Dirac House, Temple Back
Bristol BS1 6BE UK

Phone: 44 1 17 929 7481
Fax: 44 1 17 920 0979

techtracking@iop.org

Additional copies of this publication are available from:

Curran Associates, Inc.
57 Morehouse Lane
Red Hook, NY 12571 USA
Phone: 845-758-0400
Fax: 845-758-2634
Email: curran@proceedings.com
Web: www.proceedings.com

Table of contents

Volume 104

39th International Microelectronics and Packaging IMAPS Poland 2015 Conference
20–23 September 2015, Gdansk, Poland

Accepted papers received: 27 November 2015
Published online: 21 January 2016

Preface

Papers

Proceeding of 39th International Microelectronics and Packaging IMAPS Poland Conference

Foreword

These proceedings are a collection of the selected articles presented at the 39th International Microelectronics and Packaging IMAPS Poland Conference, held in Gdansk, Poland on September 20-23, 2015 (IMAPS Poland 2015). The conference has been held under the scientific patronage of the International Microelectronics and Packaging Society Poland Chapter and the Committee of Electronics and Telecommunication, Polish Academy of Science and jointly hosted by the Gdansk University of Technology, Faculty of Electronics, Telecommunication and Informatics (GUT) and the Gdynia Maritime University, Faculty of Electrical Engineering (GMU). The IMAPS Poland conference series aims to advance interdisciplinary scientific information exchange and the discussion of the science and technology of advanced electronics.

The IMAPS Poland 2015 conference took place in the heart of Gdansk, two minutes walking distance from the beach. The surroundings and location of the venue guaranteed excellent working and leisure conditions. The three-day conference highlighted invited talks by outstanding scientists working in important areas of electronics and electronic material science.

The eight sessions covered areas in the fields of electronics packaging, interconnects on PCB, Low Temperature Co-fired Ceramic (LTCC), MEMS devices, transducers, sensors and modelling of electronic devices. The conference was attended by 99 participants from 11 countries. The conference schedule included 18 invited presentations and 78 poster presentations.

The scientific committee awarded presentations for the best posters. The ceremony for the granting of awards took place during final gala dinner. Prizes awarded were as following:

- First place to Marcin Myśliwiec, Arkadiusz Lewandowski, Wojciech Wiatr, Jerzy Weremczuk, Zbigniew Szczepański, Ryszard Kisiel, "*Challenges in packaging of IR detectors – technology of elastic electrical connections*";
- Second place to Katarzyna Dunst, Karolina Cysewska, Paweł Kalinowski, Piotr Jasiński, "*Polypyrrole based gas sensor for ammonia detection*";
- Third place to Dariusz Witek, Mariusz Wójcik, Tomasz Klej, Edward Ramotowski, "*Embedding components in voltage converter PCB for size reduction and heat management*".

Three additional prizes were awarded by IOP Publishing (Lubomir Skvarenina and Robert Macku "Noise and optical spectroscopy of single junction silicon solar cell") and the IMAPS Poland Chapter for Young Scientists (Kamil Janeczek, Aneta Araźna, Bartłomiej Salski, Krzysztof Lipiec, Małgorzata Jakubowska "Printed HF antennas for RFID on-metal transponders" and Tibor Rovensky, Alena Pietrikova, Igor Vehec, Martin Kmec "Influence of various multilayer LTCC systems on dielectric properties' stability in GHz frequency range") for the timely submission of manuscripts and the quality of submitted manuscripts.

The IMAPS Poland 2015 was accompanied by Summer School for PhD students "Materials, technologies and measurement strategies" supported by the project "The development of

IOP Conf. Series: Materials Science and Engineering **104** (2015) 011001 doi:10.1088/1757-899X/104/1/011001

interdisciplinary doctoral studies at the Gdansk University of Technology in modern technologies". Twelve PhD students from Gdansk University of Technology attended the school.

The General Assembly of the IMAPS Poland Chapter met during the conference and looked into various aspects of the IMAPS Poland Chapter and the future role of this conference for the scientific community. During this meeting the next location of the IMAPS Poland 2016 conference has been discussed, and it was announced that the 40th International Microelectronics and Packaging IMAPS Poland Conference will be organized by the Wroclaw University of Technology in September of 2016.

This special issue presents 38 papers that are selected via peer-review process from the papers submitted for the proceedings of the conference. The issue aims to be an essential reference work for the evolution and development of various aspects of electronics, but also for recent scientific advances in novel electronic materials and sensors, that were presented during the sessions of the conference. We finally want to acknowledge all the authors that have submitted their work to be included in this issue and a very special thank to all those collaborating persons who acted as reviewers of the papers included in this issue for their work.
We would like to acknowledge the rest of the Organizing Committee (Kalina Detka (GMU), Marcin Gnyba (GUT), Grzegorz Jasiński (GUT), Małgorzata Jędrzejewska-Szczerska (GUT), Tomasz Kocejko (GUT)) and the team of Gdansk University of Technology PhD students for their help and assistance before and during the conference.

The conference would not have been also possible without the generous sponsorship provided by the Chancellor of the Gdansk University of Technology, the Dean of the Faculty of Electronics, Telecommunication and Informatics (GUT), the Dean of the Faculty of Electrical Engineering (GMU), the project "The development of interdisciplinary doctoral studies at the Gdansk University of Technology in modern technologies", Port of Gdynia Authority S.A and Institute of Physics Publishing. The organizing committee is very grateful to these organizations for their support.

The proceeding editors:
Piotr Jasiński (GUT)
Krzysztof Górecki (GMU)
Robert Bogdanowicz (GUT)

IOP Conf. Series: Materials Science and Engineering **104** (2015) 011002 doi:10.1088/1757-899X/104/1/011002

Peer review statement

All papers published in this volume of *IOP Conference Series: Materials Science and Engineering* have been peer reviewed through processes administered by the proceedings Editors. Reviews were conducted by expert referees to the professional and scientific standards expected of a proceedings journal published by IOP Publishing.

IOP Conf. Series: Materials Science and Engineering **104** (2015) 012001 doi:10.1088/1757-899X/104/1/012001

LTCC based bioreactors for cell cultivation

H Bartsch, T Welker, K Welker, H Witte and J Müller

Institute for Micro- and Nanotechnologies IMN MacroNano®, Technische Universität Ilmenau, Gustav-Kirchhoff-Str. 7, 98693 Ilmenau, Germany

E-mail: heike.bartsch@tu-ilmenau.de

Abstract. LTCC multilayers offer a wide range of structural options and flexibility of connections not available in standard thin film technology. Therefore they are considered as material base for cell culture reactors. The integration of microfluidic handling systems and features for optical and electrical capturing of indicators for cell culture growth offers the platform for an open system concept. The present paper assesses different approaches for the creation of microfluidic channels in LTCC multilayers. Basic functions required for the fluid management in bioreactors include temperature and flow control. Both features can be realized with integrated heaters and temperature sensors in LTCC multilayers. Technological conditions for the integration of such elements into bioreactors are analysed. The temperature regulation for the system makes use of NTC thermistor sensors which serve as real value input for the control of the heater. It allows the adjustment of the fluid temperature with an accuracy of 0.2 K. The tempered fluid flows through the cell culture chamber. Inside of this chamber a thick film electrode array monitors the impedance as an indicator for the growth process of 3-dimensional cell cultures. At the system output a flow sensor is arranged to monitor the continual flow. For this purpose a calorimetric sensor is implemented, and its crucial design parameters are discussed. Thus, the work presented gives an overview on the current status of LTCC based fluid management for cell culture reactors, which provides a promising base for the automation of cell culture processes.

1. Introduction

The use of LTCC multilayer technology for the design of micro reactors provides the possibility of complex fluid handling in combination with electrical and electrochemical functions. These features of the technology brought forth a number of applications in the field of micro reactors [1], analytic devices [2-4] and PCR assays [5, 6].

First approaches in the field of LTCC based bio assays aim on the design of 2-dimensional electrode arrays [7]. The present work introduces a concept for the monitoring of 3D cell cultivation using an LTCC fluid system, which enables an optimized supply of the cell culture with oxygen and nutrients, guarantees stable temperature and flow conditions, and enables at the same time the monitoring of the cell growth by means of impedance analysis. The system consists of a temperature control unit, a cell culture chamber with impedance measurement system and a flow monitoring unit. At the current stage of development all components are realized as single modules linked via PEEK connectors [8]. This work introduces these single components and their features and thus defines the basis for a monolithic integrated bioreactor based on LTCC technology. The integration of fluidic channels into LTCC multilayer requires an advanced process schedule to avoid deformation of the hollows during lamination and sintering. Therefore, proven methods for channel forming are summarized to give an overview of the applied methods.

2. Methods for channel manufacturing in LTCC reactors

2.1 Sequential lamination

One method to avoid sagging and to achieve geometrical exact fluid channels in the multilayer is the use of sequential lamination of the different reactor parts and their uniaxial joining. It makes use of dependency of the shrinkage rate from the densification of the ceramic green body. Figure 1 illustrates the process. A typical lamination pressure of 20 MPa is applied uniaxially to the channel part. In parallel, top and bottom parts are uniaxially joined at a reduced pressure of 2 MPa. All parts subsequently are laminated using the same reduced pressure of approximately one tenth part of the lamination pressure applied to the channel part. Hence the shrinkage of the top and the bottom part is

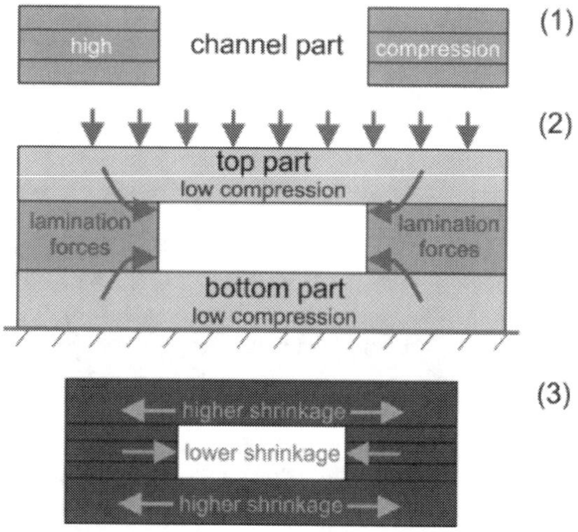

Figure 1. Process flow for the sequentially lamination method: (1) Uniaxial lamination of the channel parts at reduced pressure; (2) Force flow during uniaxial lamination of top and bottom part at reduced pressure; (3) Impact of the strain caused by different shrinkage rates during sintering.

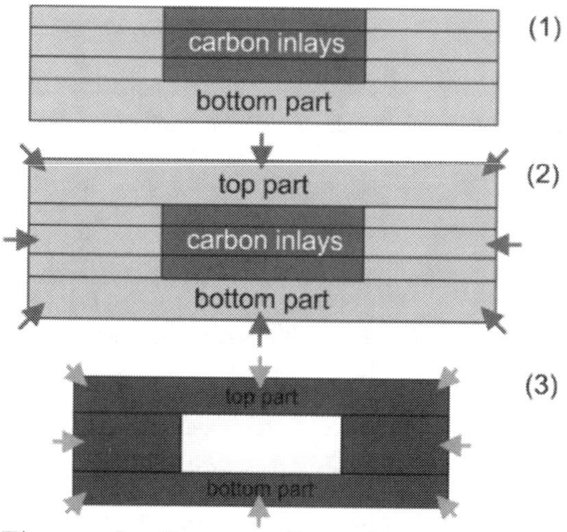

Figure 2. Process flow for the isostatic lamination method with sacrificial inlays: (1) Assembly of uniaxially pre-laminated green body; (2) Isostatic lamination at familiar conditions; (3) Sublimation of the carbon inlay and homogeneous shrinkage during firing.

Figure 3. Cross section of a fluid channel produced with the sequential lamination method.

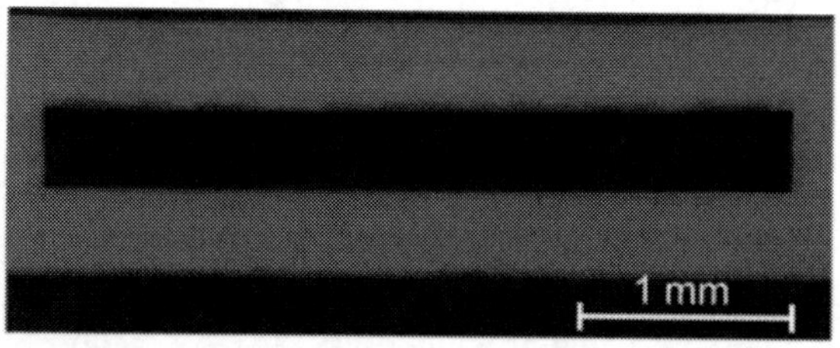

Figure 4. Cross section through a wide fluid chamber produced using carbon inlays.

IOP Conf. Series: Materials Science and Engineering **104** (2015) 012001 doi:10.1088/1757-899X/104/1/012001

higher than that of the channel part, and the generated tensile stress spans the cantilever structure over the channel. Figure 3 shows a cross section of a channel produced using this method. It can be applied for channels with a width up to 1 mm. However, delimitation can occur due to inadequate contact of the single sheets after lamination.
Additional metallization layers influence the strain and consequently the geometry. The impact of metallization structures and layout on the channel sagging can be predicted by DoE methods [9]. Accordingly, the lamination pressure for the system is fixed. One disadvantage of the method thus consists in the fact, that the correction of shrinkage tolerances by adjustment of lamination pressure - a widespread technique in commercial LTCC fabrication - is not applicable. Therefore, tape charge specific and layout induced shrinkage deviations, which can reach an order of 2 %, cannot be corrected; this circumstance heightens the effort to meet narrow tolerances. However, for many fluid systems these tolerances are acceptable. Thus, the method is applicable for a wide range of systems.

2.2 Use of sacrificial materials

An alternative way to generate hollows in the green body is the use of volatile inlays. Starch and polymer based materials serve this purpose [10]. Additionally, sublimating carbon inlays can be used to sustain the hollows and chambers during isostatic lamination. Figure 2 depicts the process flow for this manufacturing method. Channel part, bottom and top are prepared in parallel. A carbon tape is laser cut, matching channel geometry and inserted into the prepared fluid channel. Then all parts are joint first uniaxially, and then isostatically laminated under familiar conditions. During firing the temperature is hold at 600°C for a certain time to sublimate the carbon. The holding time strongly depends on the amount of inserted carbon. The higher effort is justified by the benefit of a uniform and controllable shrinkage. Even large hollows as shown in Figure 4 can be produced by the method [11].

2.3 Microchannels

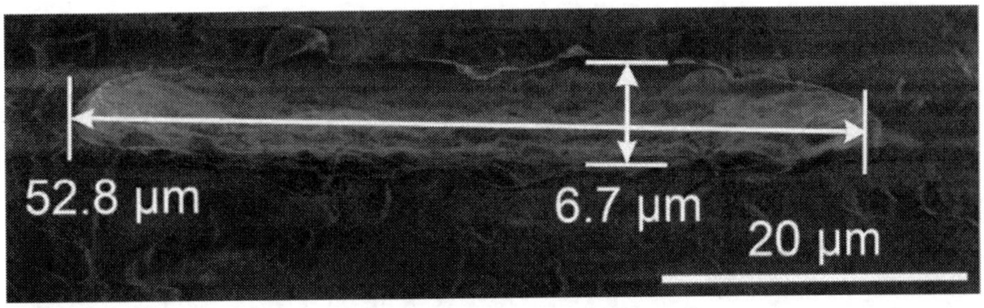

Figure 5. Micro channel produced by screen printing of fugitive carbon paste, isostatic lamination and subsequent firing. Typical dimensions are about 5-10 µm in height and more than 50 µm in width.

Figure 6. Embossed micro channels after uniaxial lamination at reduced pressure of 2 MPa (left) and after firing (right).

Fine channels are necessary to connect single scaffold tubes with the peripheral elements. Laser cut channels can be achieved with dimensions of 200 μm or larger. Channels with smaller dimensions can be manufactured using printed carbon layers or embossing. In the first case carbon paste is applied with stencils or standard screens. The channel height corresponds to the thickness of the printed paste. The tape is joined by means of standard isostatic lamination. The carbon sublimates during firing and lefts the channel behind. Figure 5 depicts the cross section of such a channel with typical dimensions. The method restricts the channel geometry to dimensions which have a height between 5 μm and 15 μm, a width larger than 50 μm and a characteristic shape resulting from screen printing.
An alternative method is the embossing of micro channels and subsequent uniaxial lamination at reduced pressure. Detailed descriptions of the embossing process are given in [12, 13]. The channels are moulded using silicon tools, for example. Exact geometries with rectangular cross section are achievable (see Figure 6). The finest achieved dimensions of fired channels are 35 μm x 35 μm. Larger geometries can be produced, free standing channels with a height of 50 μm and a width of 200 μm already were manufactured.The process requires an exact adjustment of the tool geometry and lamination process. Density gradients lead to shrinkage deviations, which are predictable but have to be adapted for new layouts. This effort is only justified for few applications with height requirements on the channel geometry.

3. Temperature control

3.1 NTC thermistors

The temperature control must maintain the temperature of the cell culture medium at 37 °C with an accuracy of 0.2 K. Temperature sensors must thus feature a high temperature coefficient, thus the material NTC-2114 (ESL ElectroScience) is chosen. Preliminary tests were carried out to determine sheet resistance and B parameter of the Steinhart–Hart equation (B parameter), and their respective variation as a function of the sensor geometry. A test structure with different length to width ratio is

Table 1. Resistance, sheet resistance and B parameter of the NTC-2114 composition (ESL ElectroScience) as a function of sensor geometry

		Resistance 30°C		Sheet resistance		B parameter	
Length (mm)	Width (mm)	X (kΩ)	SD (%)	X (kΩ/sq)	SD (%)	X	SD (%)
1.3	1	14.7	20	12.0	20	2149	3
3.3	1	55.7	10	18.1	10	2225	3
1.3	0.65	25.0	23	13.7	23	2260	1
3.3	0.65	88.0	7	18.2	7	2256	2
1.3	0.325	67.4	19	18.4	19	2309	2
3.3	0.325	204.5	8	23.5	8	2329	2
1.3	0.163	303.5	19	40.8	19	2461	4
3.3	0.163	829.5	8	39.2	8	2526	3

X : Mean value of the sample
SD: Standard deviation of the sample

Table 2. Resistance R and B parameter of the sensors integrated into the fluid channel.

R @ 30°C		R @ 40°C		B parameter	
X (kΩ)	SD (%)	X (kΩ)	SD (%)	X	SD (%)
4.8	6.68	3.9	6.34	2052	0.54

X : Mean value of the sample
SD: Standard deviation of the sample

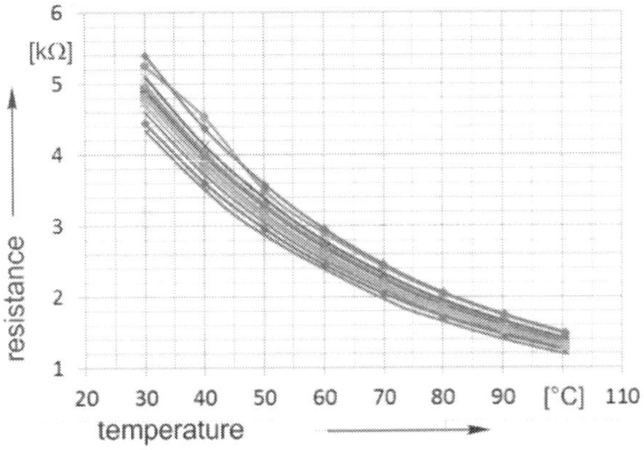

Figure 7. Temperature characteristic of the NTC sensors

used for this study. The paste is screen printed using a stainless steel fabric (325 mesh/inch and 0.024 wire thickness) with an emulsion thickness of 15 µm on DP 951 PX substrates (DuPont Nemours). Lamination and sintering is carried out under standard conditions. Two substrates with 40 test structures (5 of the respective geometry) are evaluated. The resistance of the test structures is measured by a multimeter (HP 34401) at 30 °C and 70 °C, respectively. The geometry data are captured by means of video measuring microscope (*UHL VMSergo,* Walter UHL GmbH), and the respective sheet resistance is calculated using these data. The temperature was adjusted with a hotplate and referred by the use of an external PT-100 platinum resistance thermometer. The data determined are gathered in table 1. The sheet resistance is in good agreement with the datasheet value of 10 kΩ for 1.3 x 1 mm². Other dimensions cause strong deviations because of substrate-paste interactions and substrate-termination interactions as well. The sensor is fitted into the channel and has a width of 4.5 mm and a length of 0.8 mm. Two sensors are situated directly above the channel at the inlet and outlet, respectively. The calibration is carried out in a convection oven, the sensors were put inside for 1 hour at each temperature point to achieve a homogenous temperature distribution over the whole module, and the resistance values are determined using a multimeter (HP 34401). Sensor temperature and the correlated resistance value are recorded in the interval from 30 °C to 100 °C in steps of 10 K, whereby a PT-100 platinum resistance thermometer serves as reference. The actually achieved values for 30 °C and 40 °C are highlighted in table 2. Figure 7 depicts the resultant temperature characteristic. The deviation of the resistance values is lower than 7 %, this fact attests a stable process. The logged characteristics are the base for temperature control unit parameters.

3.2 Heater control

The analogue voltage drop over the NTC sensors is transformed into a digital value with a resolution of 24 bit (AD7794, Analog Devices). The reference voltage amounts to 5 V. The resulting voltage resolution of 298 nV and corresponds to a resistance resolution of 1,42 mΩ. This value is interpreted by the microcontroller (Atmel ATmega32 8-bit AVR RISC), which here constitutes a PI controller. The control sets the electrical power of the heater element, which is arranged directly underneath the fluid channel. The silver composition DP6142D (DuPont Nemours) is used for the heater, the screen printing is carried out with a stainless steel fabric (325 mesh/inch and 0.024 wire thickness) with an emulsion thickness of 15 µm. Sensors and heater are covered with one sheet DP 951 C2. Figure 8 depicts an X-ray tomogram and a schematic cross section through the module. The channels have a cross section of 1 mm x 1mm. The fabrication follows the method described in section 0. The resistance of the heater amounts to 12.5 Ω at 35 °C and to 15 Ω at 50 °C. The power consumption in

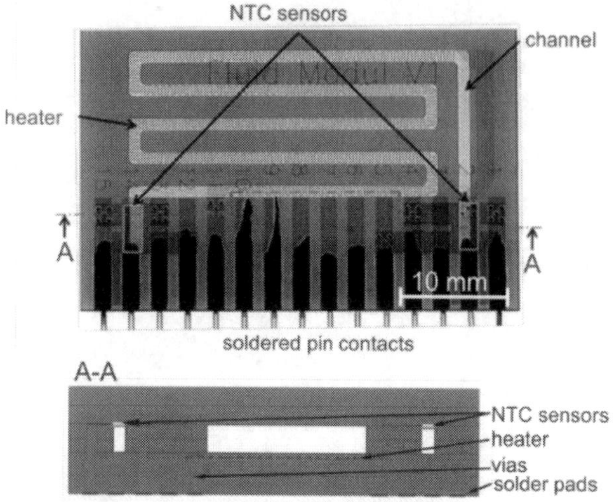

Figure 8. X-ray tomogram of the temperature conditioning module (above) and schematic cross section through the functional elements.

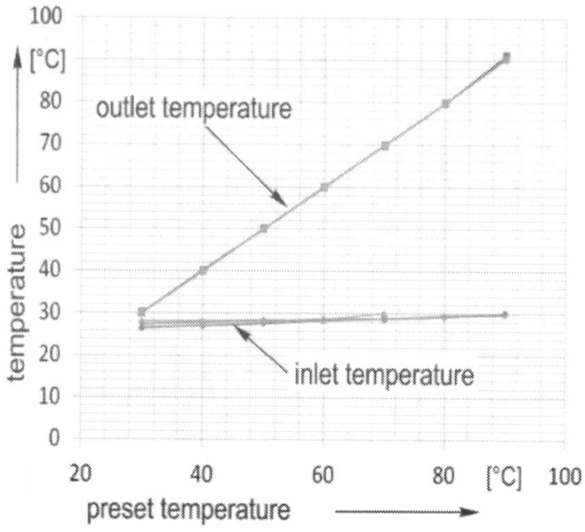

Figure 9. Temperature measured at the inlet and outlet of 3 modules.

this temperature range is less than 1 W. The time constant is determined analyzing the step response recorded at a temperature of 90 °C. It amounts to 24 s.

Three modules as depicted in Figure 8 are characterized in terms of their control accuracy. The actual temperature values at the inlet and the outlet are determined from the filed temperature characteristic of the respective NTC sensor and plotted in Figure 9. The temperature at the inlet varies between 26 °C and 30 °C. A warming up of approximately 2 K above ambient temperature is observed when the preset temperature is set to 90 °C. At the outlet the preset temperature is adjusted with high accuracy: the maximum deviation observed amounts to 0.2 K for preset temperatures up to 60 °C and to 2 K for higher preset temperatures.

4. Assay chamber

The core of the system is the cell culture chamber. The open cavity is formed with high accuracy by the use of silicon inlays during isostatic lamination [14]. The complex 3D fluid supply is embedded in a chamber module, which is schematically depicted in Figure 10. Inside of the chamber a scaffold with Schwarz P polymer structure serves as 3-dimensional framework for the cell growth. In the current case, a urethane dimethacrylate (UDMA) scaffold produced via two-photon polymerization [15, 16] is

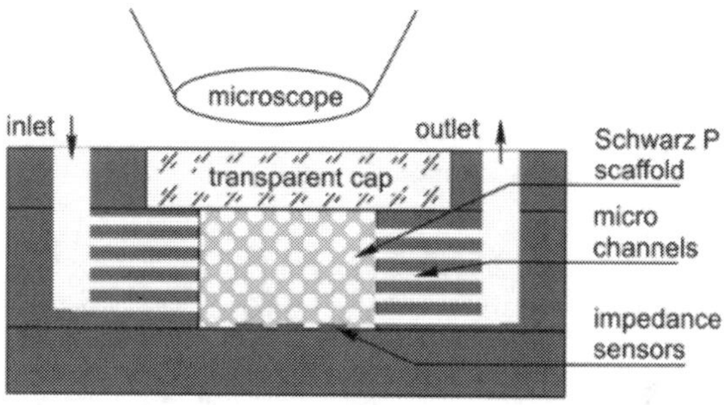

Figure 10. Schematic view of the cell culture chamber with fluid supply, Schwarz P scaffold and impedance sensors.

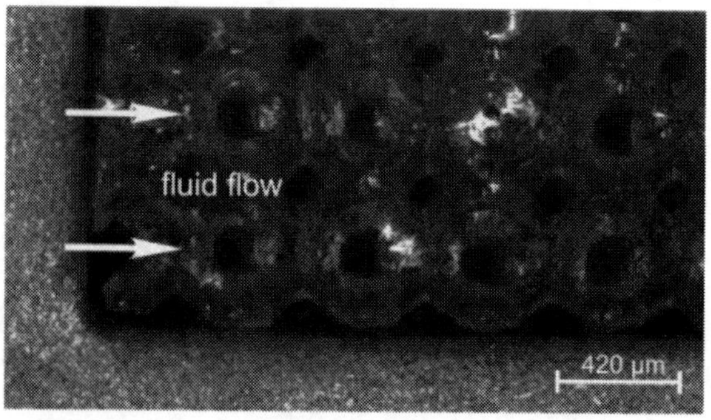

Figure 11. Detail of the cell culture chamber with UDMA scaffold and fluid supply.

inserted into the LTCC reaction camber as shown in Figure 11. The micro channels at the interface must match the scaffolds' pitch of 440 μm; the clear span of the scaffold channels amounts to 150 μm. In the first development stage the micro channels are laser cut in DP 951 PX, and the chamber body is laminated using carbon tape as sacrificial material (see section 0). The channel width amounts to 200 μm. The fluid concept intends to supply each second channel of the Schwarz P scaffold with nutrient solution in diagonal sequence and to use the other channels for the removal of metabolic products. The refinement of the system will apply embossing (see 0) for the channel forming to achieve finer dimensions and to enable addressing single channel for precise drug delivery in a defined region of the cell culture. Thus, e.g. the influence of pharmaceuticals on the differentiation of stem cells to osteoblasts and their mature to bone tissue can be investigated.
One method to gather information about the state of a cell culture is the impedance spectroscopy. It monitors the dielectric properties of biological cells and tissues. Conventionally, small AC voltages are applied, and the response of the system is measured [17]. The impedance sensors are electrodes, which are situated at the bottom of the assay chamber. A simple design, which is easily feasible in LTCC technology, is a round electrode. In the current design those electrodes have a diameter of 130 μm and are printed with the paste DP 5740A on gold vias (DP 5738). The impedance of these electrodes was measured in phosphate-buffered saline (PBS) as electrolyte at 1 kHz using an impedance measurement adapter described in [18]. The median impedance of the electrodes amounts to 24 kΩ.

5. Flow monitoring

5.1 Principle

The scaffold in the chamber is perfused permanently by the cell culture medium with a flow rate of 200 μl/min. The flow monitoring controls the steady transport of metabolic products. Flow control in LTCC fluid systems can be done optically [19] or thermally [20]. In the current application case a calorimetric principle with four temperature sensors allocated around a heater is applied. Two sensors are situated upstream and two downstream, as the X-ray tomogram of the module depicts in Figure 12. Figure 13 illustrates the temperature distribution over the sensor elements for a blocked channel (flow = 0) and a faultless one (flow > 0) schematically. The flow controller interprets the temperature deviation of upstream and downstream components based on the resistance change and detects a blockage.

5.2 Realization

All elements, heater and sensors, are situated on a free standing bridge, which is spanned in the middle of a 1 mm high fluid channel. The bridge consists of one layer DP 951 PX, heater and sensors are screen printed using a stainless steel fabric (325 mesh/inch and 0.024 wire thickness) with an emulsion thickness of 15 μm and covered with one layer DP 951 C2. The resulting thickness of the fired bridge amounts to 255 μm. The heater consists of the platinum thick film paste DP 9896. It is designed with a line width of 100 μm and a pitch of 200 μm and covers an area over the channel of 0.8 x 0.3 mm. The thermistor paste DP 5093 D is used for the temperature sensors. They are arranged symmetrically around the heater upstream and downstream. The respective distance amounts to 0.95 mm and 0.45 mm from the heater outline. They have a design length of 850 μm and thus span the whole fluid channel. The width was varied to secure higher resistance values upstream despite of manufacturing tolerances of the buried elements. The design values amount to 1060 Ω and 710 Ω for sensor 1 and 2 (upstream) and 530 Ω and 850 Ω for sensor 3 and 4 (downstream). The fabricated resistances feature a deviation between 10 % and 20 % from the design values. The bridge is laminated with a reduced pressure of 2 MPa and the fluid structure is laser cut subsequently. The fluid channels are punched with a 1 mm round tool and laminated at 20 MPa. The whole module is sequentially laminated applying a pressure of 2 MPa (see section 0).

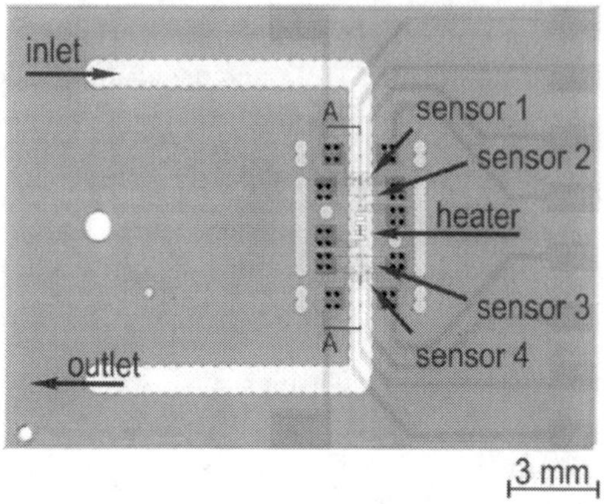

Figure 12. X-ray tomogram of the flow controller module with 4 sensor elements

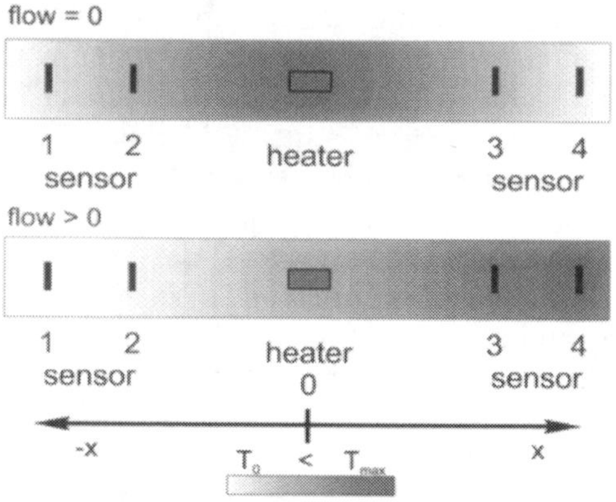

Figure 13. Temperature gradient around the heater along the cross section A-A (marked in Figure 12).

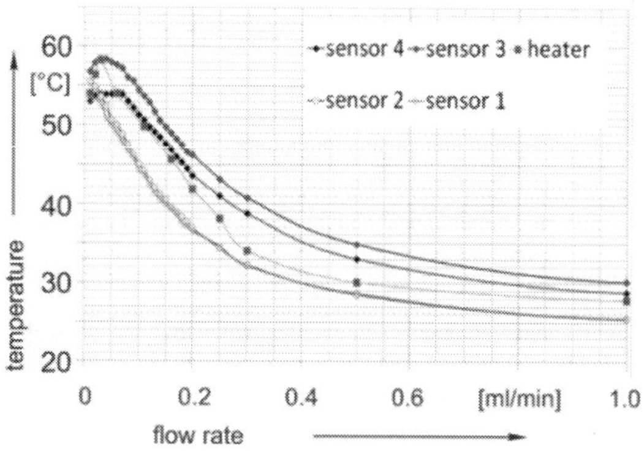

Figure 14. Temperature characteristic of the heater and upstream and downstream sensors.

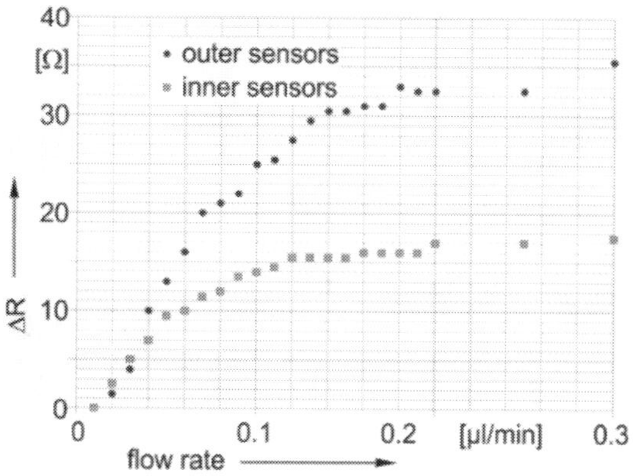

Figure 15. Resistance difference as a function of the flow rate for the outer and inner sensor pair

5.3 Characteristics

The temperature characteristic of the particular sensors is recorded with a PT-100 platinum resistance thermometer as reference. The mean of the temperature coefficient amounts to 2.1 Ω/K. However, the temperature characteristics of the paste show very strong deviation up to 40 %. Based on the particular data for each sensor, the real temperature distribution inside the system as a function of the flow rate is logged. For this purpose a constant flow is driven with high accuracy into the module by means of a High Pressure Liquid Chromatography (HPLC) analysis pump (LC-20AD, Shimadzu). The respective sensor temperature is determined from the actual resistance values measured with a multimeter while the heater is driven with a constant power of 0.45 W. Figure 14 depicts the plotted temperature data for the particular sensors and the heater element as well. The over temperature amounts to 40 °C at the heater element for the nominal flow rate of 200 µl/min. The resistance difference between upstream and downstream sensor amounts to 30 Ω at the outer sensors and to 15 Ω at the inner sensors. Figure 15 depicts the resistance difference between the upstream and downstream sensors as a function of the flow rate. The voltage drop over the upstream and downstream sensors is interpreted by means of a differential amplifier, which triggers an alert when the value falls below a critical resistance difference (e.g. 10 Ω). Each system must be attuned to compensate manufacturing tolerances of the sensors. For this purpose the heater is driven with a constant power and the resistance difference of upstream and downstream sensor is compensated with a potentiometer in series with the downstream sensor. A second potentiometer in the feedback path serves for the adjustment of the desired minimum flow rate. Then the flow monitor module is ready for use.

6. Summary

The current work describes three modules of a fluid handling system for cell cultures realized in LTCC multilayer technology. A temperature control based on buried NTC temperature sensors is described, which guarantees the comfort temperature for cell growth of 37°C with high accuracy of 0.2 K. The tempered cell culture medium enters subsequently into the cell culture chamber, which is equipped with a Schwarz P scaffold as support frame for the 3-dimensional cell growth. At the bottom of the chamber impedance sensors examine the cell growth. The steady perfuse of cell culture medium necessary for nutrient supply is monitored by means of a calorimetric sensor system, which gives an alert if the flow rate falls below a critical value. These modules build the basis for a cell monitoring system which allows the automated cell cultivation and analysis in bioreactors easy to use in bio labs.

7. References

[1] Groß G A, Thelemann T, Schneider S, Boskovic D and Körner J M 2008 Fabrication and fluidic characterization of static micromixers made of low temperature cofired ceramic (LTCC) *Chemical Engineering Science* **63** 2773–84

[2] Calvo-López A, Arasa-Puig E, Puyol M, Casalta J M and Alonso-Chamarro J 2013 Biparametric potentiometric analytical microsystem for nitrate and potassium monitoring in water recycling processes for manned space missions *Analytica chimica acta* **804** 190–6

[3] Almeida, S A A, Arasa E, Puyol M, Martinez-Cisneros C S, Alonso-Chamarro J, Montenegro, M C B S M and Sales, M G F 2011 Novel LTCC-potentiometric microfluidic device for biparametric analysis of organic compounds carrying plastic antibodies as ionophores: application to sulfamethoxazole and trimethoprim *Biosensors & bioelectronics* **30** 197–203

[4] Bustos López O N, Valdés Perezgasga F, Moreno Casillas H A, Chamarro J A and Coto Fuentes H 2014 Analytical Microsystem for the Monitoring and Analysis of Cobalt in Aqueous Solutions Using LTCC Technology *J. Mex. Chem. Soc.* **58** 411–5

[5] Walczak R, Bembnowicz P, Szczepańska P, Dziuban J A, Golonka L, Koszur J, Duong Bang D 2008 Miniaturized system for real-time PCR in low-cost disposable chip with integrated optical waveguide *Twelfth International Conference on Miniaturized Systems for Chemistry and Life Sciences (San Diego, California, USA, October 12 - 16)* pp 1078–80

[6] Bembnowicz P, Małodobra M, Kubicki W, Szczepańska P, Górecka-Drzazga A, Dziuban J, Jonkisz A, Karpiewska A, Dobosz T and Golonka L 2010 Preliminary studies on LTCC based PCR microreactor *Sensors and Actuators B: Chemical* **150** 715–21

[7] Ciosek P, Zawadzki K, Łopacińska J, Skolimowski M, Bembnowicz P, Golonka L J, Brzózka Z and Wróblewski W 2009 Monitoring of cell cultures with LTCC microelectrode array *Analytical and bioanalytical chemistry* **393** 2029–38

[8] Thelemann T, Fischer M, Groß G A and Müller J 2007 LTCC-based fluidic components for chemical applications *Journal of microelectronics and electronic packaging* **4** 167–72

[9] Bartsch H, Rensch C, Fischer M, Schober A, Hoffmann M and Müller J 2010 Thick film flow sensor for biological microsystems *Sensors and Actuators A: Physical* **160** 109–15

[10] Horváth E and Harsányi G 2010 Optimization of fluidic microchannel manufacturing processes in low temperature co-fired ceramic substrates *Per. Pol. Elec. Eng.* **54** 79

[11] Welker T, Geiling T, Bartsch H and Müller J 2013 Design and Fabrication of Transparent and Gas-Tight Optical Windows in Low-Temperature Co-Fired Ceramics *International Journal of Applied Ceramic Technology* **10** 405–12

[12] Bartsch de Torres H, Gade R, Albrecht A and Hoffmann M 2008 Systematic characterisation of embossing processes for LTCC-tapes *Journal of microelectronics and electronic packaging* **5** 142-149

[13] Bartsch H, Albrecht A, Hoffmann M and Müller J 2012 Microforming process for embossing of LTCC tapes *Journal of Micromechanics and Microengineering* **22** 015004 8pp

[14] Schulz A, Welker T, Gutzeit N, Stöpel D, Wollenschläger F and Müller J 2012 Optimized cavities for microwave applications using the new low loss LTCC material Du Pont 9k7 *8th International Conference and Exhibition on Ceramic Interconnect and Ceramic Microsystems Technologies: CICMT 2012 CICMT (Erfurt, Germany, April 16-19)* 269-74

[15] Weiß T, Hildebrand G, Schade R and Liefeith K 2009 Two-Photon polymerization for microfabrication of three-dimensional scaffolds for tissue engineering application *Eng. Life Sci.* **9** 384–90

[16] Weiß T, Schade R, Laube T, Berg A, Hildebrand G, Wyrwa R, Schnabelrauch M and Liefeith K 2011 Two-Photon Polymerization of Biocompatible Photopolymers for Microstructured 3D Biointerfaces *Adv. Eng. Mater.* **13** B264

[17] Sun T and Morgan H 2008 Impedance Measurements of Cells *Encyclopedia of Microfluidics and Nanofluidics* ed D Li (Springer US) 809–11

[18] Bartsch H, Stöpel D, Himmerlich M, Baca M, Stadie P, Hyttinen J, Müller J, Schober A 2015 LTCC Based Multi-Electrode Arrays for In-Vitro Cell Culture *IMAPS/ACerS 11th International Conference and Exhibition on Ceramic Interconnect and Ceramic Microsystems Technologies CICMT (Dresden, April 20-23)*

[19] Pudas M, Viollet S, Ruffier F, Kruusing A, Amic S, Leppävuori S and Franceschini N 2007 A miniature bio-inspired optic flow sensor based on low temperature co-fired ceramics (LTCC) technology *Sensors and Actuators A: Physical* **133** 88–95

[20] Jurków D, Maeder T, Dąbrowski A, Zarnik M S, Belavič D, Bartsch H and Müller J 2015 Overview on Low Temperature Co-fired Ceramic Sensors *Sensors and Actuators A: Physical* **233** 125-146

Acknowledgments
The authors wish to thank Carl Zeiss Foundation for the financial support of the project "System for Automated Cell Cultivation and Analysis".

IOP Conf. Series: Materials Science and Engineering **104** (2015) 012002 doi:10.1088/1757-899X/104/1/012002

Resistor trimming geometry; past, present and future

M Alafogianni, R Penlington, M Birkett,

[1]Faculty or Engineering and Environment
Northumbria University, Newcatle upon Tyne, NE1 8ST, UK

E-mail: maria.alafogianni@northumbria.ac.uk

Abstract. This paper explores the key developments in thin film resistive trimming geometry for use in the fabrication of discrete precision resistors. Firstly an introduction to the laser trimming process is given with respect to well established trim geometries such as the plunge, 'L' and serpentine cuts. The effect of these trim patterns on key electrical properties of resistance tolerance and temperature co-efficient of resistance (TCR) of the thin films is then discussed before the performance of more recent geometries such as the three-contact and random trim approaches are reviewed. In addition to the properties of the standard trim patterns, the concept of the heat affected zone (HAZ) and ablation energy and the effect of introducing a 'fine' trim in areas of low current density to improve device performance are also studied. It is shown how trimming geometry and laser parameters can be systematically controlled to produce thin film resistors of the required properties for varying applications such as high precision, long term stability and high power pulse performance.

1. Introduction

For thin film resistors it is generally impossible to deposit batches of product with resistance tolerances better than about ±10% [1-3]. This is due to problems in attaining uniform sheet resistance, but mainly due to dimensional variation of the individual resistor elements in the batch, a problem which is amplified as the resistor size decreases [3, 4].

Thus, it is normal to fabricate the resistor film to a lower resistance value than required and then adjust it by removing or 'trimming' away sections of the film material to increase the resistance to its target value in order precision of less than 10% to be succeeded [5, 6].

There are many different trimming methods which can be used to adjust the value of the resistor such as anodisation, heat trimming, electrical trimming, mechanical trimming, chemical trimming and laser trimming [1, 2]. However, laser trimming is by far the most effective and popular method and is still a subject of continuing theoretical and experimental research and optimisation [7-24].

This paper discusses the laser trimming process of thin film resistors with regard to both conventional and well established trim patterns and the influence of their geometry on key electrical performance properties such as resistor tolerance and temperature co-efficient of resistance (TCR). The concept of the heat affected zone (HAZ) bordering the kerf is also introduced and some recent trim patterns design to reduce its effects are discussed.

IOP Conf. Series: Materials Science and Engineering **104** (2015) 012002 doi:10.1088/1757-899X/104/1/012002

2. Laser trimming

2.1. Laser trimming process

The laser uses a light beam of a few μm in diameter to remove the resistive film from the ceramic substrate in a very short time period of less than 1 ms. The high intensity light pulse is then absorbed by the material causing it to heat and vaporise. This process depends on the intensity of the laser pulse, or the power level, the focus of the pulse as well as the properties of the material being ablated [1].

The laser beam is scanned across the resistor to produce a continuous kerf, changing the resistance value of the film as it progresses. The accuracy of the adjustment of the resistance value is dependent on the properties of the laser beam itself, the shape of the cut and also the speed at which the measurement system can switch the laser beam off between pulses once the target resistance value is reached [1, 12-13].

There are two main types of laser used for the adjustment of resistive films; the pulsed carbon dioxide (CO2) and the neodymium: yttrium-aluminium-garnet (Nd: YAG) laser. As for the CO2, it has a long pulse width with high energy per pulse, which causes vaporisation of the film [1]. However, the long pulse width can also cause damage to the substrate and the material at the edge of the kerf, the heat affected zone (HAZ) [17]. On the other hand, the YAG laser uses an acousto-optic Q-switch for a two way optical switching of the laser beam. This system is able to produce short pulses of high peak power at a wavelength of 1064nm, to rapidly vaporise the film, whilst minimizing heat flow and damage to the material surrounding the kerf [1]. Additional reductions in the HAZ can be succeeded by lasers operating in the green region of the visible spectrum due to the decrease in laser spot size resulting from the shorter wavelength of 532nm [14].

2.2. Conventional trimming geometries

There are various different laser trim patterns which can be used for the adjustment of surface mount bar shape resistors as shown in Figure 1. The influence of the width and length of the cutting line, the number of cuts is really important concerning the device geometry [25]. The plunge cut (Figure 1a) is the simplest and economical cut consisting of a single kerf orthogonal to the current flow through the resistive element but its overall tolerance accuracy can be less than other methods [5,15-16].

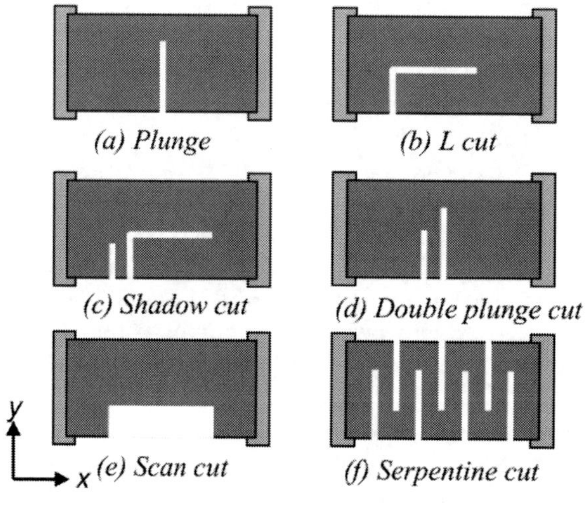

Figure 1. Commonly used laser trim kerf shapes [15]

The 'L' cut (Figure 1b) is perhaps the most frequently employed method due to its superior stability and tolerance accuracy. With this type of cut the resistance increases rapidly as the kerf is cut perpendicular to the current flow (y direction), which is called transverse cutting. Then, more gradually when it turns through 90° and cuts parallel to the current flow (x direction), which is called longitudinal cutting, until target value is reached i.e. in an area of equal current density, see Figure 2. [26, 27]

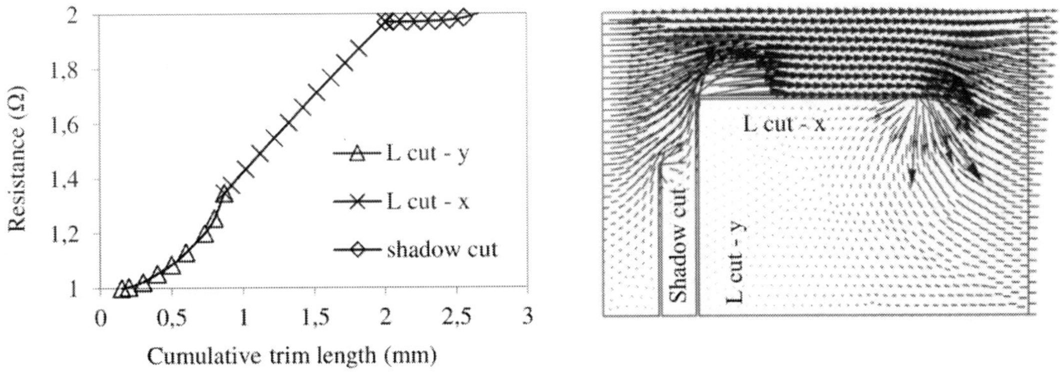

Figure 2. Typical plots of resistance increase with kerf length and corresponding model of current density for the 'L' cut and shadow cut [15].

The stability of the resistor can be maximised according to the variations on the x and y lengths. It is worth noting that when the y leg is as short as possible leaving maximum resistor line width remaining, the optimum performance is achieved. However, the trim time is increased and there is also the risk of trimming into the resistor termination with the extended x leg. Thus, with the 'L' cut, there is the opportunity to find a balance between trim speed and tolerance and stability accuracy but it is slightly more expensive than the plunge cut due to the additional time required to perform this cut [12-13, 23].

A shadow cut (Figure 1c) consists of an additional plunge to the side of an 'L' cut or plunge cut (double plunge (Figure 1d)). As a result, tighter resistance tolerances can be achieved as the kerf is cut in an area of low current density in the 'shadow' of the first cut, see Figure 3. The additional trimming time needed makes this type of cut quite expensive [5, 15].

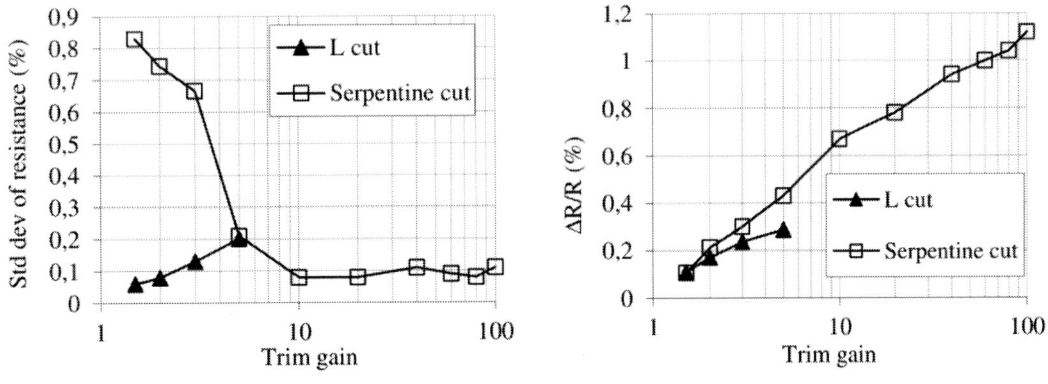

Figure 3. Variation in standard deviation of resistance and resistance stability with increasing trim gain for CuAlMo TFR's trimmed with both 'L' cut and serpentine cut [15].

IOP Conf. Series: Materials Science and Engineering **104** (2015) 012002 doi:10.1088/1757-899X/104/1/012002

A serpentine cut (Figure 1e) consists of multiple cuts made in areas of high current density which effectively increase the geometric length of the resistor and thus its resistance value. It has the ability to give a large resistance gain. As a result, the tolerance accuracy of the resistor is improved as shown in Figure 4. This type of cut can be employed when the pre-value resistance is much lower than the target value. However, a long trim length can lead to stability problems due to the large amount of HAZ surrounding it. In addition to this, when the resistance gain is very large, the TCR increases and causes instabilities due to the amount of material which has been raised to temperatures in excess of the zero TCR heat treatment temperature without being vaporised. [28, 29]

However, a solution to this issue would be a two stage serpentine trim to be employed when precision thin film resistors are required. During the first trim stage, the majority of the trimming is performed in order the resistor to be adjusted to around -1% from its target value. Then, the device is stabilised, typically overnight at around 200°C to relieve stresses built up in the HAZ [15]. Predicting the resistance change that occurs during this stabilisation operation is not possible. As a consequence, the second stage of the process, which is called fine trim, is required for the accurate adjustment of the resistor to target value. It usually consists of one or two plunges in the shadow of the first serpentine as shown in Figure 4. However, this type of cut can be very time consuming and expensive, but the increasing availability of auto substrate handling equipment can help reducing the labour cost involved [15, 16].

The scan cut (Figure 1f) is the most commonly used due to its stability [28] and especially when the device is required for high frequency applications to minimise creation of capacitive reactance components of an RC circuit. It also finds use in high voltage situations as the likelihood of voltage breakdown across the trim kerf is greatly reduced. This type of cut is very time consuming to perform and is not cost effective for general use [15, 28].

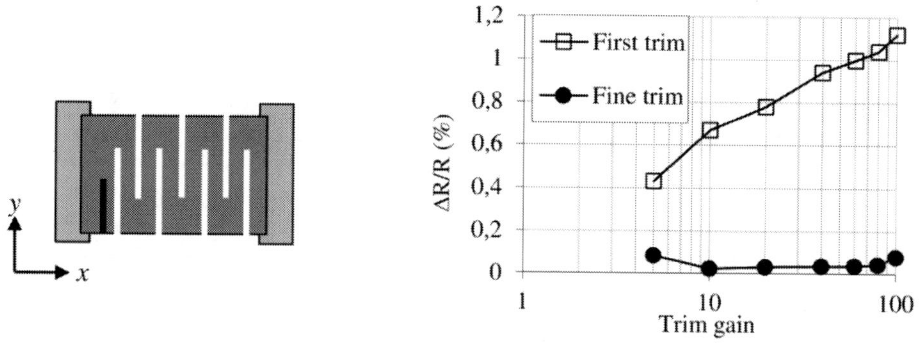

Figure 4. Two stage serpentine trim pattern and corresponding variation in resistance stability with increasing trim gain for CuAlMo TFR's trimmed with both first and fine trims [15].

2.3. The heat affected zone (HAZ)

A major cause of post-trim drift in laser trimmed resistors is due to the HAZ bordering the kerf [15, 29] and it would be beneficial to consider this area in more detail. It is worth commenting that the energy profile of the beam has a Gaussian distribution due to the fact that a Q-switched YAG laser is adjusted to operate in single traverse electromagnetic (TEM_{00}) mode. This profile can be translated into a Gaussian temperature profile as far as the thin films are concerned and it is shown in Figure 5.

It is noted that the central part of the profile has sufficient intensity to cause vaporisation of the thin film. As for the areas of film adjacent to the kerf, the energy absorbed from the laser beam is less than the threshold required for vaporisation. Thus, this region of the film becomes a heat affected zone and can considerably cause more changes and variations in the resistance than areas of film that are not irradiated by the laser beam [20, 21].

This aging effect is caused due to the fact that the film which was stabilised prior to the adjustment process and it has now been re-exposed along the edges of the laser kerf. The changes in the resistance are related to changes in the structural properties of the film material because of the rapid heating and cooling during the trimming process such as the sheet resistance and TCR [29-31].

It is worthwhile thinking that the length of the kerf and the effect of the HAZ in the trimmed resistors are important parameters that can determine the overall stability of the resistor [22]. These parameters can be optimised by varying the shape of the continuous trim and the properties of the laser beam itself.

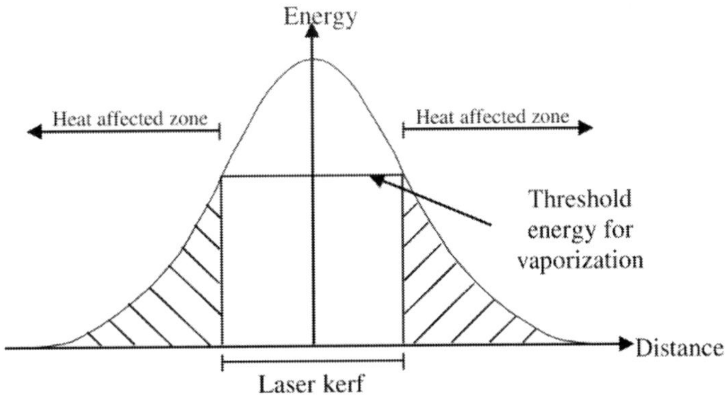

Figure 5. The fundamental mode laser beam profile as a Gaussian distribution [15, 29-31].

2.4. Alternative trimming geometries

An alternative trimming method in order to avoid the issues associated with the HAZ is link cutting [13]. This process can be applied by opening up shorting bars in loop and ladder type patterns to increase the resistance in discrete steps. Once the trimming bars are opened the current is redirected around the new longer path and any current crowding in the trimmed area or HAZ is eliminated. Although this method can provide a solution to the issues caused by the aging effects, a greater chip area is required compared to the conventional bar design. Moreover, as the resistance is adjusted in discrete steps a very large number of links may be required [13].

However, there have been methods applying the link cutting approach to bar resistors. The Swiss cheese pattern can be thought as the most popular cut using the benefits of the link cutting approach. This method forces current crowding only in non-heat affected areas [29-31].

Another study proposed a type of laser trimming performing an L-cut in a 'top hat' shape thin film resistor, avoiding excessive crowding of the current lines after trimming [27]. A more recent study has focused on the effect of replacing the two contact bar resistor with a three contact distributed structure and then trimming the resistor by narrow cuts in a variety of shapes around the additional contact and it can be shown in Figure 6(a) [17]. In addition to this, another approach proposed for designing resistors was the random trimming. This method is performed by cutting a hole in a random point that was combined with a single plunge cut as shown in Figure 6(b). First, the single plunge is applied in order to offer rough adjustment and wide trim range fast and then random trim spots and voids helped to get closer to the target resistance [5].

Both of these methods have shown encouraging results and it was found that the resistance could be increased more controllably than an L-cut [5,6], but the additional trimming time and materials required were negative factors concerning satisfying improvements in the production of thin film resistors.

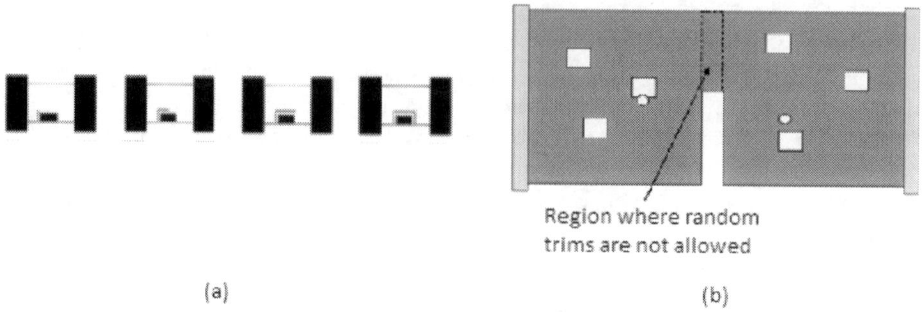

Figure 6. (a) Illustration of laser cut length used in three-contact resistors with rectangular additional contact [17], (b) Random trimming with single plunge cut [5].

3. Conclusions

This paper introduced the laser trimming process which can be thought as the most precise and reliable method in order to adjust the resistance value of bar-shaped thin film resistors. The advantages and disadvantages of common trim patterns were discussed and the effect of the HAZ on the performance of the resistors was also introduced. It is worth noting that trimming geometries play an important role on the characteristics of the resistors such as stability and tolerance accuracy. Moreover, trim pattern design has an effect on the resistance distribution and long term resistance performance of thin film resistors. It was also noted that the L-cut is the most commonly used trim pattern due to its expected tolerance which is less than 1% as well as its stability.

In addition to this, recent studies have focused on alternative trimming geometries and were presented in this paper such as the three-contact geometry and the random trim approach besides their positive results. However, these approaches, besides their promising results, did not appear to fulfil the improvements in the resistor performance reported due to the additional trimming time and materials needed. Therefore computer modelling of laser trim patterns could help in predicting and optimizing the resistor design process by designing new trim strategies. Within these limits, the maximum trim speed, overall performance and quality characteristics could be achieved. Thus, further research could focus on the possibility of improving the trimmed structures taking into account the factors that affect the performance of the resistors such as TCR and HAZ.

4. References

[1] Elshabini-Riad A and Barlow F 1998 *Thin Film Technology Handbook* (New York: McGraw-Hill) chapter 3 1-19

[2] Elshabini-Riad A and Bhutta I 1993 Lightly trimming the hybrids *Circuits and Devices* 30-4

[3] Fjeldsted K and Gottfried L 2004 Optimal designs for embedded passives laser trim *Circuitree* **7** 10-4

[4] Maissel L and Glang R 1970 *Handbook of Thin Film Technology* (New York: McGraw-Hill) 18-24.

[5] Sandborn P and Sandborn P 2008 A random trimming approach for obtaining high-precision embedded resistors *IEEE T. Adv. Packaging* **31** 76-81

[6] Sandborn P and Sandborn P 2006 Using embedded resistor emulation and trimming to demonstrate measurement methods and associated engineering model development *Int. J. Eng. Educ.* **22** 1-7

[7] Meunier M, Gagnon Y, Lacourse A, Ducharme M, Rioux S and Savaria Y 2007 Precision resistor laser trimming for analog microelectronics *IEEE Leos. Ann. Mtg.*1

[8] Meier M, Bertsch D, Onda N, Etter M, Gutsche M, Dommann A, Romano V and Nicolet N 2006 Laser trimming of amorphous Ta42Si13N45 thin films with ultrashort pulses *Microelectron Eng.* **83** 2234-7

[9] Antonov Y 2006 Use of laser trimming for the development of a project for accuracy-normalization of the resistance of resistors in hybrid integrated circuits *J. Commun. Technol. El+* **51** 1351-5

[10] Lullo G, Castiglia A, Carini G and Arnone C 2004 Chalcogenide thin films for direct resistors fabrication and trimming. *Mat. Sci. Semicon. Proc.*7 337-41

[11] Schultze V and Fischer A 1989 Laser trimming of NiCr thin film resistors I: Thin film resistors without a protective layer Thin Solid Films 182 23-33

[12] Schmidt H and Couch B 1989 Predicting the effect of high speed laser trimming on resistor stability *Electronic Components and Technology Conf.* (Houston:IEEE)

[13] Shier J 1988 A finite mesh technique for laser trimming of thin film resistors *IEEE J. Solid-St. Circ.* **23** 1005-9

[14] Gu B,Couch B,Oh J,Chase P 2003 Comparative study on laser trimming with different wavelengths *Proc. of the Int. Symp. on Microelectronics* pp672-5

[15] Birkett M and Penlington R 2013 Laser Trim Pattern Optimization for CuAlMo Thin-Film Resistors. *IEEE T. Compon. Hybr.* **3** 523-9

[16] Birkett M and Penlington R 2012 Laser Trimmming of CuAlMo Thin-Film Resistors:Effect of laser processing parameters *J. of Electronic Materials* **41** 2169-77

[17] Wroński M, Kamiński S, Miś E and Dziedzic A 2005 New trim configurations for laser trimmed thick-film resistors—theoretical analysis, numerical simulation and experimental verification *Microelectron Reliab.* **45** 1941-8

[18] Gofuku E, Ohnawa T, Kohara M and Nunoshita M 1993 Resistance adjustment with short pulse Nd:YAG laser for ruo2 based thick-film resistors based in polyamide film *IEEE T. Compon. Hybr.* **16** 592-7

[19] Kurihara Y,Yamada K, Kanai K and Endoh T 1990 Laser trimming of thick film resistors on aluminum nitride substrates. *IEEE T. Compon. Hybr.* **13** 596-602

[20] Kestenbaum A and Baer T 1980 Photoexcitation Effects During Laser Trimming of Thin-Film Resistors on Silicon *IEEE T. Compon. Hyb.* **3** 166-71

[21] Kestenbaum A and Baer T 1980 Trimming behavior and post-trim characteristics of Ta2N resistors on silicon *IEEE T. Compon. Hybr.* **CHMT-3** 637-46

[22] Dow R, Mauck M, Richardson T and Swenson E 1978 Reducing post trim drift of thin film resistors by optimising yag laser output characteristics *IEEE T. Compon. Hyb.* **1** 392-7

[23] Bulger G 1975 Stability analysis of laser trimmed thin film resistors *IEEE T. Parts Hyb. Pac.* **PHP-11** 172-7

[24] Fehlhaber P. 1971 Laser trimming of silicon-chromium thin film resistors. *Solid State Technol.* 1971 33-6

[25] Liou B 2011 Use of laser trimming for the fabrication of high precise n-p compensation-doped polycrystalline silicon thin film resistors. *J. Comput. Theor. Nanos.* **8** 2291-98

[26] Wang Z,Yan X,Sun Y and Zhao,X 2010 The influence of laser spot on resistance of chip resistors *Symp. of Photonics and Optoelectronics*

[27] Manolescu A and Manolescu A M 2007 Laser trimming modeling of thin film integrated resistors Int Semiconductor Conf. 473-6

[28] Deluca P. 2002 A review of thirty-five years of laser trimming with a look to the future. Proceedings of the IEEE. **90** 1614-19

[29] Ramirez-Angulo J and Geiger R 1988 New laser trimmed film resistor structures for very high stability requirements *IEEE T Electron Dev* **35** 516-8

[30] Ramirez-Angulo J, Geiger R and Sanchez-Sinencio E 1987 Characterisation, evaluation and comparison of laser-trimmed film resistors *IEEE J Solid-St Circ* **22** 1177-89

[31] Ramirez-Angulo J, Geiger R and Sanchez-Sinencio E 1986 A circuit analysis approach to the performance evaluation of laser-trimmed film resistors *Proc. of the Int. Symp. on Circuits and Systems (IEEE)* 1201-5

IOP Conf. Series: Materials Science and Engineering **104** (2015) 012003 doi:10.1088/1757-899X/104/1/012003

Thin film resistive materials: past, present and future

C Cherian Lukose, G Zoppi and M Birkett

Faculty or Engineering and Environment
Northumbria University, Newcatle upon Tyne, NE1 8ST, UK

E-mail: martin.birkett@northumbria.ac.uk

Abstract. This paper explores the key developments in thin film resistive materials for use in the fabrication of discrete precision resistors. Firstly an introduction to the preparation of thin films and their fundamental properties is given with respect to well established systems such as NiCr, TaN and CrSiO. The effect of doping these systems in both solid and gaseous forms to further refine their structural and electrical properties is then discussed before the performance of more recent materials systems such as CuAlMo and Mn_3AgCuN are reviewed. In addition to performance of the materials themselves, the effect of varying processing parameters such as deposition pressure and temperature and subsequent annealing environment, as well as laser trimming energy and geometry are also studied. It is shown how these parameters can be systematically controlled to produce films of the required properties for varying applications such as high precision, long term stability and high power pulse performance.

1. Introduction

Thin film resistive materials possessing moderate resistivity and low temperature coefficient of resistance (TCR) properties are of great importance to the microelectronics industry for the production of precision thin film resistors (TFR) [1]. When used in the manufacture of discrete resistors, thin film materials can offer enhance performance and reliability and reductions in size when compared to alternative thick film and wirewound technologies. TFR's are typically employed when high stability, high accuracy or low noise are required in applications such as electronic measuring devices, strain gauges, pressure sensors and accelerometers [2].

As well as possessing a suitable resistivity and TCR, TFR's must be sufficiently stable so that any changes in resistance during their operating life can be reliably expected not to exceed a pre-specified value. Moreover the TFR must be able to be manufactured to this specification at a realistic cost.

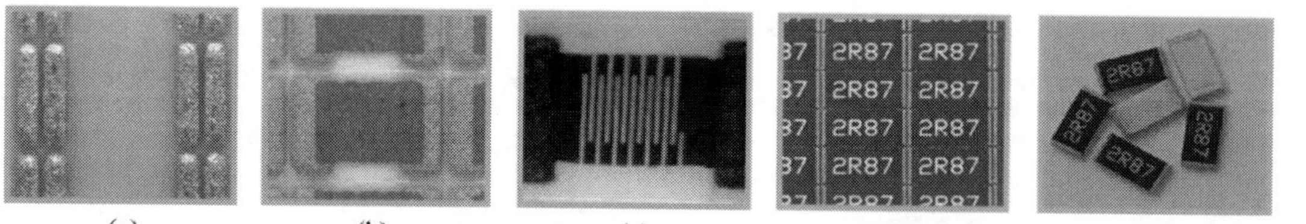

(a) (b) (c) (d) (e)

Figure 1. Typical SMD TFR process flow

Figure 1 shows a typical manufacturing process flow for a surface mount device (SMD) TFR. A ceramic substrate is patterned with high conductivity electrodes (Fig 1a) before being deposited with a sub-micron (μm) layer of resistive material (Fig 1b). The resistive film is then heat treated to obtain the required TCR and pre-value resistance [3] which is then subsequently adjusted to target value by removing sections of the film to increase its overall effective length, usually by a laser trimming machine [4] (Fig 1c). The substrate is then encapsulated in a protective coating (Fig 1d) before being snapped into individual components to allow application of external wraparound terminations to permit subsequent soldering during PCB manufacture (Fig 1e).

Perhaps the two most important stages of the TFR manufacturing process are the deposition and heat treatment of the thin resistive film. Both the chemical composition and structure of the film are functions of the deposition environment and essentially provide the base electrical characteristics of the TFR. These key properties can then be modified by subsequent heat treatment of the film to fine tune its performance to the required specification.

This paper discusses the key electrical properties and characteristics of a number of currently well established and potential future thin film resistive materials and the effects of varying deposition and heat treatment process parameters on their performance and long term reliability.

2. Key electrical properties

Some of the most important fundamental electrical properties of the resistive material are its sheet resistance, TCR and resistance stability.

2.1. Sheet resistance (Rs)

As the measured resistance, R, of the thin film material is directly proportional to its resistivity, ρ, and inversely proportional to its thickness, d, it is usual to use the sheet resistance of the film, R_s, which is defined as ρ/d, to describe thin film resistors [5]:

resistance, $$R = \frac{\rho l}{dw} \qquad (1)$$

however, $$\frac{l}{w} = \square \quad \text{(no. of squares in film pattern)} \qquad (2)$$

hence, $$R_s = \frac{R}{\square} = \frac{\rho}{d} \qquad (3)$$

2.2. Temperature co-efficient of resistance (TCR)

A parameter equally as important as sheet resistance is the temperature co-efficient of resistance (TCR) of the film. TCR describes the change in resistance of the film with change in its temperature $dR(T)/dT$. However, as $dR(T)/dT$ is nearly constant for most metal films it is common practice to use an average TCR parameter [6]:

$$TCR_{Av} = \frac{1}{R_{Tref}} \frac{\Delta R}{\Delta T} \times 10^6 \; ppm/^{\circ}C \qquad (4)$$

Where R_{Tref} is the initial reference resistance at 20°C and ΔR is the change in measured resistance with a change in temperature ΔT from 20 to 70°C.

2.3. Resistance stability, (ΔR/R)

Resistance stability or reliability is basically a measure of the change in resistance of the film against a certain specification over a period of time. There are a number of conditions under which the reliability of the component can be tested such as DC load, dry heat or humidity.

The severity of the effects of a test condition on the resistor depends greatly upon film thickness. For thicker films the resistance value is inherently lower and hence the dc load voltage that can be applied will also be lower. Moreover, humidity is also likely to have less effect on a more robust thicker film.

Perhaps the most convenient and universal test condition is that of dry heat (temperature) as the test can be performed by simply storing the film in an oven. Dry heat stability tests generally entail measuring the change in the resistance value of the film, ΔR, following storage at the upper category temperature, usually 155°C for axial leaded resistors and 125°C for surface mount resistors, for a period of 1000hrs [7]:

$$\frac{\Delta R}{R} = \frac{R_e - R_i}{R_i} \times 100\% \quad (5)$$

Where R_i is the initial resistance and R_e is the end resistance following storage.

3. Thin film resistive materials

Materials used in the manufacture of thin film resistors typically require resistivities in the range 100-2,000 μΩ-cm and a TCR of ±50ppm/°C. However, owing to their mechanical structure, bulk metals and alloys cannot have resistivities much in excess of the lower limit of this range and TCR is usually large and positive, see Table 1. Conversely, bulk semiconductors can readily satisfy these resistivity requirements, but this is usually at the cost of a very negative TCR.

Table 1. Electrical properties of some bulk metals and alloys [8].

Metal or alloy	Resistivity (μΩcm)	TCR (ppm/°C)
Aluminium	2.7	4200
Chromium	12.9	4500
Copper	1.7	4300
Molybdenum	5.7	4600
Nickel	7.3	6500
Tantalum	13.5	3800
Ni 80% / Cr20%	100	170

Fortunately when deposited in thin film form most metals produce resistivity values which are much greater than that when in bulk form and frequently without acquiring a large positive TCR. There are a number of mechanisms which can produce this phenomenon as illustrated in Table 2.

Table 2. Mechanisms causing metal films to have resistivities greater than the bulk [9].

Description	Mechanism for resistivity increase	effect on TCR
Ultra thin film	Conduction electron scattering	$\rightarrow 0$
Insulating phase	Inter grain barriers	$\rightarrow -\infty$
Porous film	Construction resistance	$\rightarrow -\infty$
Trapped gas	Impurity scattering	$\rightarrow 0$
Discontinuous	Particle separation	$\rightarrow 0$
Double layer	TCRs cancel	$\rightarrow 0$
New structure	Fewer carriers	$\rightarrow 0$

The thin film material systems available for resistor manufacture can therefore largely be divided into three main groups, depending upon the conduction mechanism present:

1. Metal alloys
2. Single metal systems
3. Cermets (metal-insulator)

3.1. Metal alloys

Metals are known to have a periodic crystal structure at low temperatures which provides a regular field for mobile electrons to flow [8]. Any distortions in this periodic nature of the lattice will distort the flow of electrons and thus increase its resistivity. This distortion can be caused by a number of defect types; lattice dislocations, contamination due to foreign atoms or atoms interstitially located within the lattice or non-stoichiometric ratio of constituents. Resistance can also be increased on a temporary basis due to lattice distortions caused by vibration increases with temperature; hence the reason TCR is generally positive for all metals. By alloying metals, the lattice distortions are dissolved in one another thus increasing the resistivity of the solution. In general the resistivity rises with increasing impurity concentration, reaching a maximum for an alloy concentration of approximately 50% impurity [10].

By far the most successful of the metal alloy systems used in the manufacture of bulk material resistors is nickel chromium, or Nichrome, with a composition of NiCr 80:20 wt.%. Over the years there have been a number studies undertaken on the properties of Nichrome thin film resistors [11-17]. Work has focused on the effects of varying substrate materials and process parameters on the electrical performance and structural properties of the fixed composition NiCr 80:20 binary alloys.

Among all the different substrates experimented, it was found that alumina substrates were the most suitable for the fabrication of thin film resistors due to their superior power handling capabilities [12]. By varying deposition parameters such as pressure, power and substrate temperature [17] films with sheet resistances of 10 to 1000 Ω/□, and TCR in the range +50 to +250 ppm/°C could be achieved. Subsequent annealing of these films led to typical long term stability figures of better than 0.5% [16]. Optimum annealing conditions were found to be in the range 200 to 600°C for 1.5 to 6 hours in air ambient [11-13]. Even to this day thin films of NiCr 80:20 still receive a substantial amount of interest in both resistor and strain gauge applications due to its high resistivity, low TCR and widespread commercial availability [14, 15]. To supply the increasing demand of high performance TFR`s in mid-1970`s, it was soon discovered that the TCR of the film could be reduced by increasing the chromium content in the film [18-24].

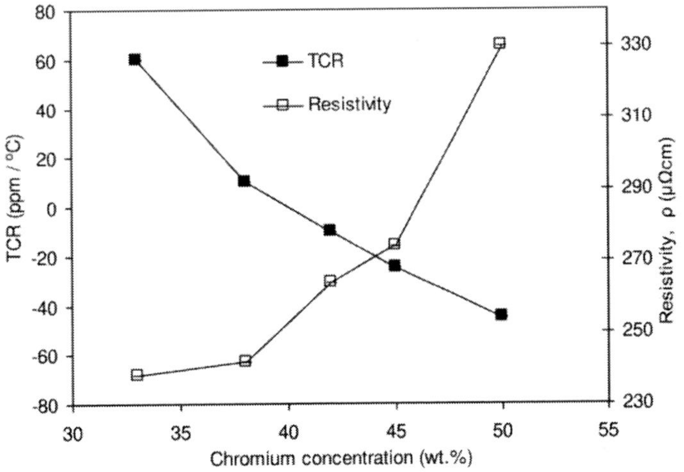

Figure 2. Variation of TCR and resistivity in NiCr films as a function of chromium concentration [23].

A plot of TCR and resistivity against chromium content for NiCr films deposited on oxidized silicon wafers is shown in figure 2. As can be seen the TCR steadily decreases while resistivity rises, with increase in chromium content in the film, passing the zero TCR line at approximately 40 wt.% having a value of around 250 μΩcm. These marked changes against the properties of 80:20 alloy are due mainly to two reasons; firstly the decrease in grain size of the film as the Cr content increased leading to a more amorphous structure having higher resistivity [22] and secondly the higher quantity of chromium oxide (Cr_2O_3) formation with negative TCR in the film [21]. The occurrence of amorphous and crystalline metastable phases are related to characteristics such as a large immiscibility gap between the two solid solutions (nickel fcc and chromium fcc solid solution), the existence of a relatively deep eutectic point and the presence of complex tetrahedral close-packed structures [19].

As chromium has a higher chemical affinity with oxygen than does nickel, then as its ratio in the film increases so does the amount of chromium oxide formed with residual oxygen in the deposition chamber [25]. Partial pressures in the range 2-6% reactive oxygen in the inert argon atmosphere were reported to give further increases in the content of chromium oxide in the film and hence a lower TCR [26]. There are a number of studies which report a near zero TCR for films with chromium content in the range 30 to 60 wt.%, both sputtered in argon and mixed argon/oxygen environments. It seems that in addition to film composition, accurate TCR control can be also be achieved through varying the deposition temperature and annealing treatment used. For films with negative as grown TCR, annealing in the temperature range 300 to 350°C appears to be essential if a near zero TCR is to be achieved [21, 26]. Heat treatment methods rely on the mechanisms such as grain growth and reduction of impurities to reduce resistivity and in turn increase TCR. In addition to the desirable change in the TCR properties of the film, increasing the chromium concentration of the film and sputtering in an oxygen partial pressure can also improve the long term resistance stability of the film. Some research shows that stability figures as low as 0.3% change following 1000 hrs at 155°C are not untypical for films sputtered in an oxygen partial pressure of $3x10^{-5}$ Torr [27]. Again the reason for this improved performance is attributed to the increase in chromium oxide formation at the surface of the film. As the chromium oxide grows it forms a protective passivation layer, similar to aluminium oxide, which protects the film from further attack [28].

Once it was realised that the addition of a third element, oxygen, into the film could give improvements in resistance stability, a number of investigations were carried on doping or modifying the NiCr alloy with a third element incorporated into the sputtering target itself. Probably the most widely documented of these elements is aluminium [29-32]. Additions of aluminium were proven to stabilise the film and reduce the TCR to values around zero [31]. Three regions have been found in the NiCrAl composition which result in highly stable resistive films with a very low TCR; one with less than 3at.% Al, one with 28-32 at.% Al and one with 45-60at.% Al. The role of aluminium is reported to be to stabilise the amorphous structure and state of ordering of the atoms in the deposited nickel chromium film. This can lead to stability figures of better than 0.08% following 1000 hrs storage in air at 155°C [30]. Moreover the addition of aluminium allows thin film resistors with lower sheet resistances in the range below 10 Ω/□ to be produced [33]. Of the other elements which have been proven to improve the performance of NiCr films, Silicon is perhaps the most widely documented [34-36]. Additions of around 5 to 10 wt. % Si can produce films with TCRs of ±20 ppm/°C and long term resistance stability of better than 0.05%. It is assumed that the improved thermal stability of the NiCrSi film is due to the diffusion limiting effect of silicon [35]. Not satisfied with the performance of ternary film compositions, there have also been several studies investigating the effect of adding two extra elements to the NiCr alloy to form quaternary composition films. Thin film resistors based on the Evanohm, Ni-Cr-Cu-Al alloy are the most popular and have been reported to possess very low TCRs of ±10ppm/°C [37, 38].

Apart from NiCr the most popular metal alloy system used in the production of thin film resistors is probably copper nickel. Cu-45 wt.% Ni alloy, known as "constantan" is a typical material of low resistivity around 50 to 60 μΩ-cm and with a low TCR value of less than +50 ppm/°C [39-41]. Films were deposited having resistivities below 50μΩ-cm making it suitable for low ohmic applications. However the lower limit for the sheet resistivity of the sputtered CuNi films was 1 Ω/□ [33]. In order to produce thicker films of CuNi with even lower resistivities, electroless deposition is required. Films of up 10μm in thickness are achievable producing CuNi films possessing resistivities of around 40 μΩ-cm. Following post deposition stabilisation the TCR of the films shifted to near zero. However the electroless deposition rate of the alloy was reported to be remarkably slow [40].

An alternative low resistivity material to CuNi is copper aluminium molybdenum (CuAlMo). This recently developed material has been proven to produce films with sheet resistances of < 1 Ω/□ and low TCR of ±15 ppm/°C. These properties when combined with its excellent long term stability make it ideal in precision current sense applications [42].

3.2. Single metal systems

As is evident for the nickel chromium films discussed above, the resistivity of the thin film is usually dominated by the background gases which are incorporated into the film during the deposition process, rather than the resistivity of the source material itself. With this realisation the focus of many researchers turned to the use of single metal systems as problems such as alloy composition control and fractionation could be automatically eradicated.

By far the most widely documented of the single metal systems is tantalum, reactively deposited in nitrogen to produce tantalum nitride (TaN) [43-55]. The as-deposited resistivity of pure tantalum is close to that of bulk tantalum (13 μΩ-cm). However when reactively sputtered in a nitrogen atmosphere the tantalum and nitrogen atoms combine to form a nitride which is trapped in the growing film. As shown in figure 3, this nitride produces an increase in resistivity and subsequent decrease in TCR [47].

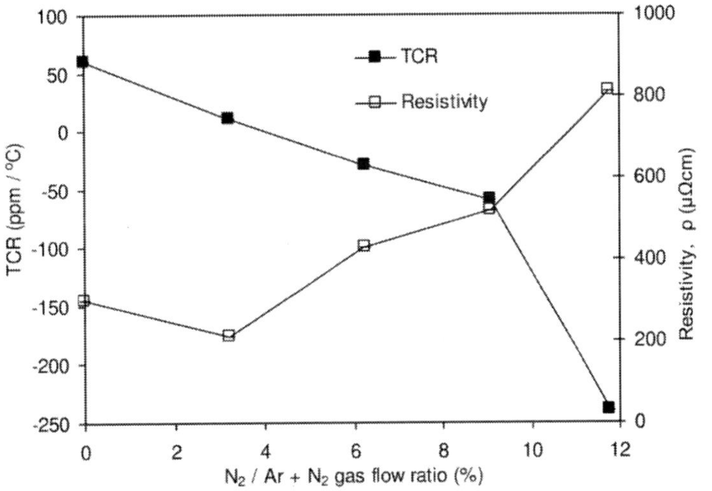

Figure 3. Variation of TCR and resistivity in TaN films as a function of nitrogen flow rate [47].

As can be seen TaN possesses similar electrical properties to that of NiCr and is generally used in the same mid-range area of resistor manufacture. However the intrinsic properties of linear TCR and physio-chemical inertness made TaN thin film material superior to the more popular NiCr for thin film resistors [43] and the mass production of TaN thin film resistors by magnetron sputtering with sheet resistivity of 50-100 Ω/□ and TCR of about -80 ppm/°C was announced in the early 1980's [44]. Recent investigations have suggested properties similar to NiCr, with films sputtered to thicknesses of between 300-1000 nm possessing resistivities of 100-450 μΩ-cm and giving sheet resistances of around 10 to 100 0Ω/□ and TCR values of around -50 to -100 ppm/°C. Again post deposition stabilisation at around 200 to 400°C in both air and N_2 atmospheres was reported to produce films with TCR's of -5 to +5 ppm/°C and a resistance stability of better than 0.1% [45-47]. As with NiCr, to further develop the properties of TaN, third element was introduced to produce a ternary system. There were two main issues with TaN; its negative as deposited TCR and the need to further improve its stability for precision resistor applications.

The first issue was tackled with the addition of copper [56, 57]. By varying the Cu concentration in a TaNCu sputtering target, as-deposited films of 150μΩ-cm with near zero TCR were produced. To further improve resistance stability, as with NiCr the incorporation of aluminium was used [58-61]. Reports showed that additions of 35-60 wt.% of Al to TaN could produce films of sheet resistance 50-600 Ω/□ with long term resistance stabilities of better than 0.05% [58]. Investigations were also conducted for binary systems of TaAl, with sheet resistance of around 150-300 μΩ-cm resulting. However the elimination of nitrogen from the film appeared to result in inferior stabilities of around 0.3% [59, 60].

3.3. Cermets

Electron conduction is greatly impeded by a potential barrier in the path of the flow and although the energy of the electron may be less than the barrier height there is a finite probability that the electron will pass the barrier. This form of conduction is called tunnelling and is an extremely useful mechanism for the production of high resistivity materials [5]. In the case of a cermet, a heterogeneous mixture of ceramic insulator and conductive metal material is formed producing the potential barrier, hence the name cermet. As the inclusion of ceramic, an insulator, has the effect of increasing resistance, cermet films usually find application at the higher sheet resistance values >1 kΩ/□.

The most successful cermet to date has been the chromium-silicon monoxide (CrSiO) system [62, 63]. Its main use in thin film resistor manufacture is in the 1k to 10 kΩ/□ range where standard alloy films such as NiCr and TaN are too thin. However, although CrSiO produces a thicker, more robust film at these higher ohmic ranges, its resistance stability and TCR are far inferior to that of the alloy and single metal systems [62].

3.4. Future candidates for Thin Film Resistors:

As discussed above, alloying different metals in suitable proportions (NiCr) or depositing metals in a reactive environment (TaN) is seen to bring resistivity and TCR in a suitable range for thin film resistor fabrication. But recent interest in a hybrid structure of Manganese alloy deposited in reactive nitrogen environment has revealed some unique properties [64]. This structure is called Manganese based Antiperovskite structure (AP) and is represented as Mn_3AX: where A represents a transition metal, present on the cubic corners and X represents an interstitial element like N or C, present at the body centred position [65], see figure 4.

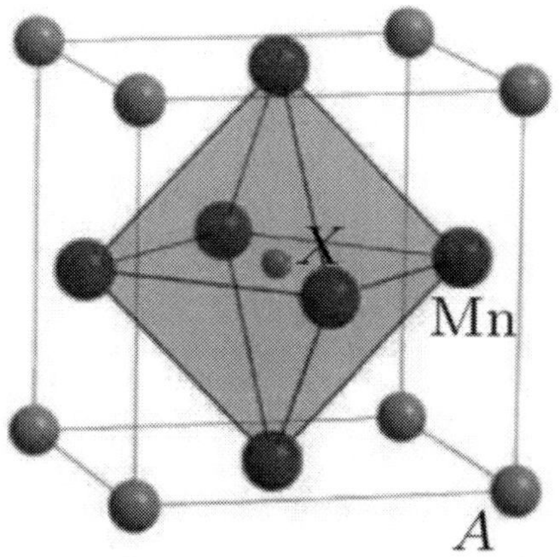

Figure 4. Mn based antiperovskite structure with a transition metal at the cubic corner and an interstitial atom at the body centre position [65].

IOP Conf. Series: Materials Science and Engineering **104** (2015) 012003 doi:10.1088/1757-899X/104/1/012003

Mn was always considered a suitable material for thin film resistor application because of its high resistivity and low TCR values [66]. Alloys based on Mn like Manganin and Zeranin have found widespread use in the production of shunt resistors [67], but because of a lack of proper research into the deposition of Mn based alloys, they have not been utilised in the manufacture of thin film resistors as extensively as NiCr or TaN.

Experiments have been conducted to study the AP structure with various transition metals like Ni [68, 69], Cu [70-72], Ag [73], Zn [74, 75], Ga [76], and depending upon the transition metals at the cell corners, this structure exhibits various interesting properties. Mn_3CuN is reported to show large magnetostriction [77], Mn_3ZnN shows negative thermal expansion [75], giant magnetoresistance in Mn_3GaC [76], superconductivity in Ni_3MgC [78], and lattice contraction in $Mn_3Zn(Ge)N$ [72]. Low TCR behaviour is so far shown to be exhibited by the Mn_3AN structure with Ni, Ag and Cu at the cubic corners [68], and electrical properties exhibited by these materials were better than the currently established thin film resistor materials with bulk resistivities in the range of 250 to 400 μΩ-cm [79], see figure 5a.

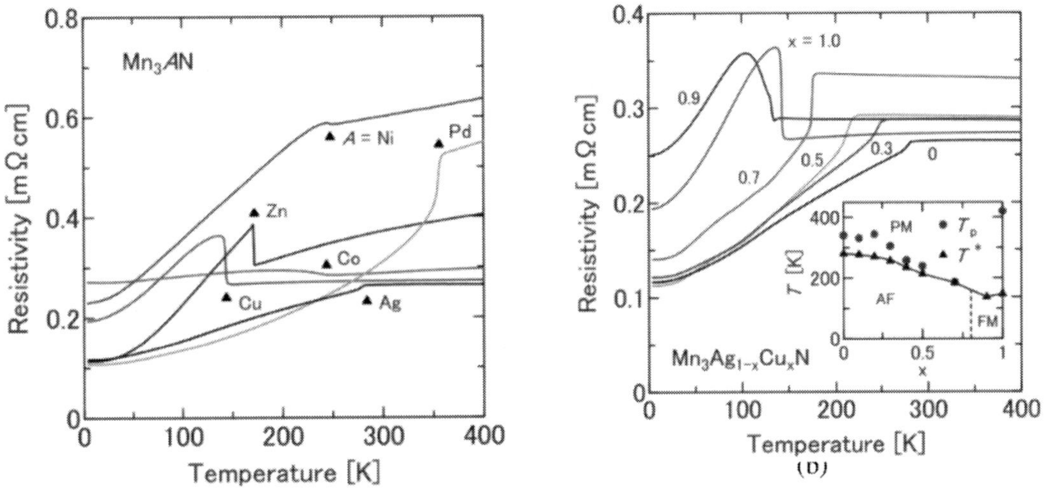

(a) (b)

Figure 5. (a) Temperature dependence of resistivity for different transition metals at the cubic corner of the AP structure, (b) Temperature dependence of resistivity for different concentrations of Cu in the AP structure [79].

Partial substitution of the transition metal by a third metal is seen to provide better control to fine tune these properties. For example, partial substitution of Cu by Ge in Mn_3CuN was observed to introduce negative thermal expansion and better control over the magnetic transition temperature of the structure [80]. Similarly, a better control over TCR was gained by partially substituting Ag by Cu in Mn_3AgN, resulting in an average TCR <1 ppm/°C between 294 to 304°K, see figure 5b [79].

However, most of these studies have conducted electrical analysis of this structure on bulk samples prepared by solid state sintering of individual metal constituents [64, 68, 73, 79, 81, and 82]. And the few studies done on thin film did not prioritise electrical characteristics; hence sheet resistance values are not readily available. But films are reportedly deposited in thickness ranging from 100 to 400 nm [69-72], so it could be expected for them to have sheet resistances in the range of 10 to 1000 Ω/□. Moreover, sintered bulk samples have been observed to have an extremely low drift rate of 9.1 μΩ/Ω/year (0.001%), which is one order of magnitude lower than the current industry standards [81]. It has also been shown that the direction of this resistance drift can be shifted from negative to positive by annealing the antiperovskite structure for 30 min in air at 500°C [82].

A sharp transition in physical and electrical properties is observed at magnetic transition temperature, above which magnetic orders re-arrange, and it supports the argument that re-arrangement of magnetic orders changes the band structure and is responsible for low TCR observed in these materials [79]. A low resistivity slope *($d\rho(T)/T$)* in combination with a large resistivity *(ρ_0)* is considered to be the reason for the low TCR of Mn_3CuN, whereas in Mn_3NiN, a subtle balance between carrier mobility and carrier density, (which are both temperature dependent) at the transition temperature is held accountable for its low TCR [65]. In Mn_3AgCuN, it was proposed that quasi particle states break down due to strong magnetic scattering, and the short range orders generated exhibit negative TCR which counteract the positive TCR from phonon scattering, resulting in a net near zero TCR [79]. But still a unified reason for low TCR behaviour of the manganese based AP nitride (Mn_3AN) structure is not yet present [65].

4. Conclusions

The previous review has shown that there are primarily four material systems currently used in the manufacture of thin film resistors and there are potential novel materials under research with promising properties for future development. A summary of the basic electrical properties of these films is presented in Table 3

Table 3 Electrical properties of sputtered thin film resistor material systems.

Film type	ρ (μΩcm)	Rs (Ω/□)	d (nm)	TCR (ppm/°C)	ΔΩ/Ω (%)	Stabilisation	
						temp (°C)	time (hrs)
NiCr NiCrO NiCrSi NiCrAl NiCrCuAl	100-500	10-1000	10-500	<5	0.1 0.01 0.08	200-400	2-5
CuNi CuAlMo	50 80	1-50 0.1-10	500-5000	<15	0.15 0.1	100-500	1-5
Ta TaN TaNAl	100-450	10-1000	300-1000	<10	0.3 0.1 0.05	200-400	2-5
CrSiO		1-10k		<25		200-600	2-5
Mn_3AgCuN	250-400	10-1000	100-400	<5	0.001	500	0.5

Alloys of nickel and chromium either in binary form or with small additions of doping elements such as aluminium or silicon are used in the mid-range of sheet resistance from approximately 10 to 1000 Ω/□ and possess excellent TCR and long term stability properties. Alloys of copper and nickel and copper, aluminium and molybdenum find use in the lower sheet resistance range of around 0.1 to 50 Ω/□ and have similar stability properties to that of NiCr.

Single metal systems of tantalum, reactively sputtered in nitrogen are also used in the same mid-range sheet resistance area as nickel chromium, but possess slightly worse TCR and stability properties. The main advantage of TaN is in its corrosion resistance under humid conditions, meaning it usually finds use in the high reliability military and aerospace applications.

Ceramic metallic systems of chromium silicon monoxide are used when higher sheet resistances in the kilo-ohm range are required. TCR and stability are poor in comparison to the alloy and single metal systems; however the inherent high resistivity properties of the cermet systems mean they continue to find widespread application.

Because of their suitable resistivity in the range 200 to 400 μΩ-cm and better TCR (<5 ppm/°C) than NiCr and TaN, hybrid structures like Mn based antiperovskite have the potential to be used to manufacture future thin film resistor of mid-range sheet resistance value. But researches concerning these materials are still in the initial stages and a clear understanding of the reasons behind their unique behaviour needs to be fully understood. It is very much necessary to conduct further studies on these structure and is hoped that this will help to overcome the challenges related to successful deposition of thin films of this structure.

5. References

[1] Kang B, Hur S, Kim D and Yoon S 2005 *Electrochem. Solid.* St **8** 92
[2] Fraga M and Pessoa R 2013 *Thin film resistors: Review and perspectives Resistors Theory of Operation, Behavior and Safety Regulations* ed R Dauo (New York: Nova) chapter 1 1-16
[3] Kennedy R 1999 Materials for thin film resistors Adv. Micro. 26 12
[4] Elshabini-Riad A and Barlow F 1998 *Thin Film Technology Handbook* (New York: McGraw-Hil) chapter 3 1-19
[5] Hall P 1997 *Thin Solid Films* **300** 256-264
[6] Jonas M and Peled A 1982 *Thin Solid Films* **90** 385-391
[7] Vishay Intertechnology 2013 *Drift Calculation for Thin Film Resistors-Technical note Document Number: 28809*
[8] Zhigal'skii G and Jones B 2003 *Electrocomponent Science Monographs – The Physical Properties of Thin Metal Films* Taylor and Francis Publications.1-34
[9] Maissel L 1968 Thin film resistance materials and characteristics *Solid State Technol.* **11** 27-32
[10] Rossiter P 1987 *Cambridge Solid State Science Series - The Electrical Resistivity of Metals and Alloys*, Cambridge University Press. **28** 137-271
[11] Kwon Y, Kim N, Choi G, Lee W, Seo Y and Park J 2005 *Microelectron. Eng.* **82** 314-320
[12] Degenhart H and Pratt L 1963 *Electronic Engineers International Convention*, **11** 59-68
[13] Petrovic S, Bundaleski N, Radovic M, Ristic Z, Gilgoric G, Perusko D, Mitric M, Pracek B, Zalar A and Rakocevic Z 2007 *Nuclear Instruments and Methods in Physics Research* B **256** 368 372
[14] Kazi I, Wild P, Moore T and Sayer M 2006 *Thin Solid Films* **515** 2602-2606
[15] Yan J and Zhou J 2007 *Mater. Sci. Tech. Ser.* **23** 195-202
[16] Nachrodt D, Paschen U, Have A and Vogt H 2008 IEEE *Electr. Device L.* **29** 212-214
[17] Vinayak S, Vyas H, Muraleedharan K and Vankar V 2006 *Thin Solid Films* **514** 52-57
[18] Birjega M, Constantin C, Florescu I and Sarbu C 1982 *Thin Solid Films* **92** 315-321
[19] Dhere N, Vaiude D and Losch W 1979 *Thin Solid Films* **59** 33-41
[20] Bhatt A, Lock C and Stevenson D 1984 DC sputtering of Ni-Cr thin film resistors *International Society for Hybrid Microelectronics* 370-376
[21] Au C, Jackson M and Anderson W 1987 *J. Electron. Mater.* **16** 301-306
[22] Lee B, Park G, You D and Lee D 2003 Composition control and electrical properties of Ni-Cr thin films prepared by co-sputtering method, *Conference on Electrical Insulation and Dielectric Phenomena (CEIDP), Annual Report* 72-74
[23] Phuong N, Kim D, Kang B, Kim C and Yoon S 2006 *J. Electrochem. Soc.* **153** G27-G29
[24] Nguyen M and Yoon S 2006 *J. Electrochem. Soc* **153** G606-G608
[25] Koltai M, Trifonov I and Czermann M 1983 Segregation phenomena in thin NiCr layers, *Vacuum* **33** 49-52
[26] Griessing J 1977 *Electrocomp. Sci. Tech.* **4** 133-137
[27] Hardy W and Murti D 1974 *Thin Solid Films* **20** 345-362
[28] Nocerino G and Singer K 1979 *J. Vac. Sci. Technol.* **16** 147-150
[29] Zelenska J, Chudoba V, Rehak J and Rohacek K 1991 *Thin Solid Films* **200** 239-246
[30] Schippel E *Thin Solid Films* **123** 57-62
[31] Schippel E 1987 *Thin Solid Films* **146** 113-138
[32] Singh A 1986 *Thin Solid Films* **138** 63-65
[33] Van Den Broek J, Donkers J, Van Der Rijt R and Janssen J 1998 *Philips J. Res.* **51** 429-447

[34] Andziulis A, Andziuliene B, Vaupsas J and Zadvydas M 2006 *Surf. Coat. Tech.* **200** 6212-6217
[35] Gawalek W 1983 *Thin Solid Films* **116** 205-210
[36] Lee B, Park G, Kim J and Lee D 2002 The effect of the process parameters on the electrical properties of Ni-Cr-Si alloy thin film resistors 2002 *Conference on Electrical Insulation and Dielectric Phenomena (CEIDP), Annual Report* 72-74
[37] Satrapinski A, Savin A, Novikov S and Hahtela O 2008 *Meas. Sci. Technol.* **19** 1-5
[38] Lee B and Lee D 2002 *J. Korean Phys. Soc.* **40** 339-343
[39] Hur S, Kim D, Kang B and Yoon S 2004 *J. Vac. Sci. Technol. B: Microelectronics and Nanometer Structures* **22** 2698-2701
[40] Ishikawa M, Enomoto H, Mikamoto N, Nakamura T, Nawafune H, Uegaki T, Mizumoto S, Matsuoka M and Iwakura C 2000 *T. I. Met. Finish.* **78** 86-88
[41] Hur S, Kim D, Kang B and Yoon S 2005 *J. Electrochem. Soc.* **152** 472-476
[42] Birkett M, Penlington R, Wan C and Zoppi G 2013 *Thin Solid Films* **540** 235-241
[43] Au C, Anderson W, Schmitz D, Flassayer J and Collins F 1990 *J. Mater. Res.* **5** 1224-1232
[44] Yoshida S 1982 Mass production of HCP tantalum nitride film resistors sputtered by planar magnetron, *Proceedings of the 32nd Electronic Components Conference San Diego, CA, XII, published by IEEE*, 530-535
[45] Radhakrishnan K, Ng. Ing and Gopalakrishnan R 1999 *Mater. Sci. Eng.* **57** 224-227
[46] Lu Y, Weng R, Hwang W and Yang Y 2001 *Mater. Chem. Phys.* **72** 278-280
[47] Cuong N, Kim D, Kang B, Kim C, Yu K and Yoon S 2006 *J. . Electrochem. Soc.* **153** G164-G167
[48] Min K, Chun K and Kim K 1996 *J. Vac. Sci. Technol.* **14** 3263-3269
[49] Golan G, Axelevitch A, Margolin R and Rabinovitch E 2001 Microelectr. J. **32** 61-67
[50] Cuong N, Phuong N, Kim D, Kim B, Kim C and Yoon S 2006 *J. Vac. Sci. Technol.* **24** 682-685
[51] Scandurra A, Indelli G, Pignataro B, DiMarco S, Stefano M, Ravesi S and Pignataro S 2007 *Surf. Interface Anal.* **40** 758-762
[52] Das A, Grabbe C and Hufnagel R 2008 *Complexities in the deposition of thin film resistors, Capacitor and Resistor Technology Symposium (CARTS)* 339-352
[53] Kang S, Yoon S, Suh S and Yoon D 2008 *Thin Solid Films* **516** 3568-3571
[54] Hong S and Ravi R 2005 *Microelectron. Eng.* **83** 206-212
[55] Na S, Park I, Park S, Jeong G and Suh S 2008 *Thin Solid Films* **516** 5465-5469
[56] Wang C, Hsieh J and Li C 2004 Electrical and piezoresistive properties of TaN-Cu nanocomposite thin films *Thin Solid Films* **469-470** 455-459
[57] Wang C, Hsieh J, Li C, Fu Y and Chen T 2005 *Surface & Coatings Technology* **193** 173-177
[58] Reddy P, Bhagavat G and Jawalekar R 1980 *Thin Solid Films* **70** 27-35
[59] Wuu D, Chan C, Horng R, Lin W, Chiu S and Wu Y 1999 Structural and electrical properties of Ta-Al thin films by magnetron sputtering, *App. Surf. Sci.* **144-145** 315-318
[60] Yang B and Jia Y 1991 Tantalum/aluminium alloy resistive film containing aluminium 50at%, *Vacuum*, **42**(16) 1073
[61] Yukiko O, Tajiri S, Aozono T, Okamoto A, Ogawa S and Mima H 2007 *J. . Vac. Soc. Jp.* **50** 173-174
[62] Jankowski A 1998 *Thin Solid Films* **332** 272-276
[63] Dong X, Wu J, Mao D and Mao L 2000 *Shanghai Jiaotong Daxue Xuebao/Journal of Shanghai Jiaotong University* **34** 1610-1614
[64] Wen Y, Wang C, Nie M and Sun Y 2015 *J. magn. Magn. Mater.* **391**, 22-25
[65] Tong P, Wang B-S and Sun Y-P 2013 Mn-based antiperovskite functional materials: Review of research. *Chinese Phys. B* **22** 1-13 067501
[66] Ammar A H 1996 Electrical transport properties of manganese thin films. *ELSEVIER Physica B* **225** 132-136
[67] Braudaway D W 1999 Precision Resistors: A Review of Material Characteristics, Resistor Design, and Construction Practices. *IEEE T. Instrum. Meas.* **48**(5). doi: 0018–9456/99
[68] Sun Y, Wang C, Chu L, Wen Y, Nie M and Liu F 2010 *Scripta Materialia*, **62** 686-689
[69] Na Y, Wang C, Chu L, Ding L, Yan J, Xue Y and Chen X 2011 *Elsevier Materials Letters*, **65** 3447-3449
[70] Aoyama M, Takenaka K and Ikuta H 2013 *Journal of Alloys and Compounds* **577** S314-S317
[71] Na Y, Wang C, Sun Y, Chu L, Nie M, Ji N and Wang J 2011 *Mater. Res. Bull.* **46** 1022-1027
[72] Sun Y, Wang C, Na Y, Chu L, Wen Y and Nie M 2010 *Mater. Res. Bull.* **45** 1230-1233.
[73] Oe T, Urano C, Kaneko N, Hadano M and Takenaka K 2013 Standard-resistor compounds with adjustable operating temperature *Appl. Phys. Lett.* **103** 173518-1 doi: 10.1063/1.4826611

[74] Hamada T and Takenaka K 2012 Phase instability of magnetic ground state in antiperovskite Mn3ZnN: Giant magnetovolume effects related to magnetic structure. *J. of Appl. Phys.* **111** 07A904. doi: 10.1063/1.3670052

[75] Qu B Y and Pan B C 2010 Nature of the negative thermal expansion in antiperovskite compound Mn[sub 3]ZnN *J. of Appl. Phys.* **108** 113920. doi: 10.1063/1.3517824

[76] Kamishima K, Goto T, Nakagawa H, Miura N, Ohashi M, Mori N, Sasaki T and Kanomata T 2000 *Phys. Rev.* B **63** 024426

[77] Asano K, Koyama K and Takenaka K 2008 Appl. *Phys. Lett.* **92** 161909

[78] He T, Huang Q, Ramirez A.P, Wang Y, Regan K.A, Rogado N, Hayward M.A, Haas M.K, Slusky J.S,. Inumara K, Zandbergen H.W, Ong N.P and Cava R 2001 *Nature* **411** 54

[79] Takenaka K, Ozawa A, Shibayama T, Kaneko N, Oe T and Urano C 2011 Extremely low temperature coefficient of resistance in antiperovskite Mn[sub 3]Ag[sub 1−x]Cu[sub x]N. *Appl. Phys. Lett.* **98** 022103

[80] Na Y, Wang C, Chu L, Ding L, Ji N, Wang J-p and Chen X 2011 *Mater. Lett.* 65 2401-2403

[81] Oe T, Urano C, Hadano M, Ozawa A, Takenaka K and Kaneko N-h 2013 Optimization ofMn3Ag1−xCuxN Antiperovskite Compound Fabrication for Resistance Standard. *IEEE T. Instrum. Meas.* **62** 1450-1453.

[82] Oe T, Urano C, Kaneko N-h, Eisaki H, Yoshida Y, Yamamoto A and Takenaka K 2014 Antiperovskite Manganese Nitride Standard Resistor *IEEE T. Instrum. Meas.* **5**

IOP Conf. Series: Materials Science and Engineering **104** (2015) 012004 doi:10.1088/1757-899X/104/1/012004

Designing and Testing Energy Harvesters Suitable for Renewable Power Sources

B. Synkiewicz, P. Guzdek, J. Piekarski and K. Zaraska

Institute of Electron Technology, branch Cracow, Cracow, Poland

Abstract. Energy harvesters convert waste power (heat, light and vibration) directly to electric power . Fast progress in their technology, design and areas of application (e.g. "Internet of Things") has been observed recently. Their effectiveness is steadily growing which makes their application to powering sensor networks with wireless data transfer reasonable. The main advantage is the independence from wired power sources, which is especially important for monitoring state of environmental parameters. In this paper we describe the design and realization of a gas sensor monitoring CO level (powered by TEG) and two, designed an constructed in ITE, autonomous power supply modules powered by modern photovoltaic cells.

1. Introduction

The number of various electronic devices in our environment is growing fast. It creates problems concerning energy supply and increases electromagnetic radiation pollution, especially in dangerous form of microwaves. A possible solution to this problem is to use energy which has usually been treated as a waste. It is available in great amounts, since most popular devices have relatively low energetic effectiveness – most of them show only 15% efficiency. A good example is general lighting, only a few years ago, thanks to the widespread usage of LED's, popular lamps have achieved efficiency greater than 10%. Energy harvesters are a new group of electronic devices designed to acquire and store this "waste" energy, which can be used for control and measurement equipment, wearable devices, personal communicators and many others [1-12].

Devices described in this paper acquire small amounts of power (from miliwatt range), but they collect it permanently. This allows to store energy suitable for high power devices which work for a short time but with long intervals. Designed and described harvesters are suitable for powering wide range of sensors working in measurement systems with wireless data transfer including emergency systems. As a source of power we use modern photovoltaic cells (especially Graetzel cells –DSSC, organic – OPV and amorphous silicon cells) which can work in artificial lighting conditions. As a base for calculations we have used EN 12464-1:2012 recommendations, which require 500 lux for office space lighting and 300 lux for others spaces in buildings as a minimum. In such conditions even relatively small cells are suitable for powering sensors usable in building automation such as ambient temperature sensors, light level sensors, smoke detectors, presence detectors, access control points, etc. Small power thermogenerators (TEG) are usable in many applications too. They are especially efficient for heating systems controllers, because they eliminate a need of wired grid supply systems. They are suitable for example for gas sensors and smart power meters.

2. Review of designs

Our main goal was designing and testing devices manufactured from standard electronic parts, which have application in white goods area.

2.1. Harvester based on thermogenerator (teg)

An example of such application are devices measuring level of CO in output gases of gas boilers and ovens. They can be used in heating systems and powered by standard thermogenerator. Such device eliminates a need of standard grid supply and the oven can work autonomously. Similar device can be used in solar collectors for powering the system monitoring the amount of the generated heat. In this case hot water is the power source.

A block diagram of the device is shown in Figure 1:

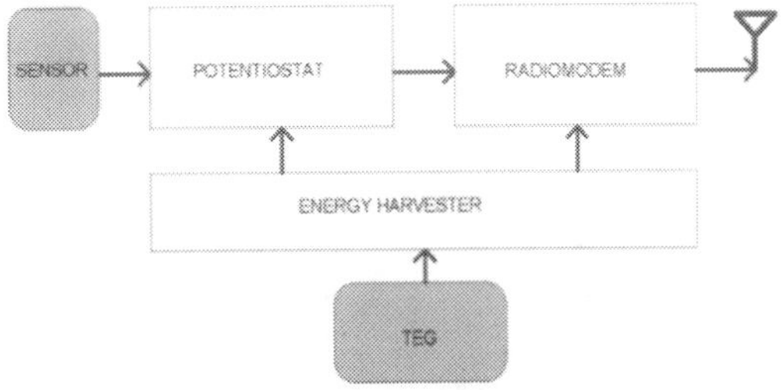

Figure 1. Block diagram of the CO detector

Level of CO is measured by electrochemical sensor manufactured by „Alphasense" company. Diagram of potentiostatic measurement module for the sensor is shown in Figure 2. It is based on micropower operational amplifiers ADA4505-2 from "Analog Devices" [13].

Detailed description of the module is available in application note CN0234 [14]. It states that the module in the "wake" mode needs 110µA, and after detecting CO presence the current grows up to 460 µA, (supply voltage level is 5V) generating an alarm signal. We can estimate power needs of a module at ca. 2.5mW. Above listed data were confirmed during measurements of the working devices.

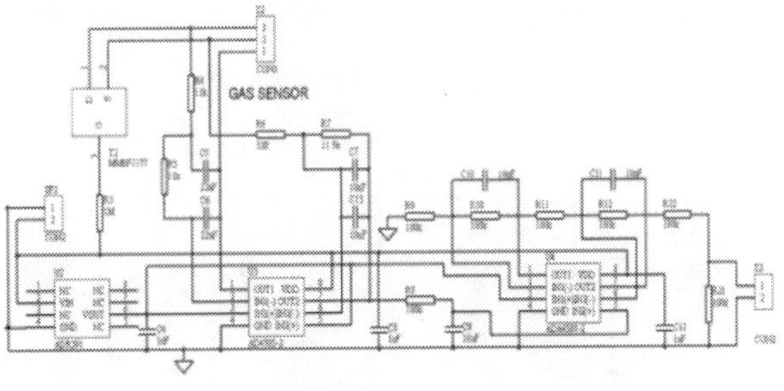

Figure 2. Potentiostat based on ADA4505-2

For its proper work the device requires a storage of small amount of power for wireless alarm. It can be estimated using the data of the radiomodem used in the device (in this case Telit LE50-433) [16]. During transmission the device requires 50mA for 0.1s. Start from the „sleep" mode requires 10mA for 5s. It implies that it needs minimum 0.1mWh of energy stored before start of the alarm procedure. Such amount of power can be stored by a 0.22F/5.5V supercapacitor.

LTC3107 [23] from „Linear" can work as a harvester, it has a very low input voltage (starts from Vin=20mV). It is a result of using a switching input via TR1 input transformer. The diagram of harvester is shown in Figure 3.

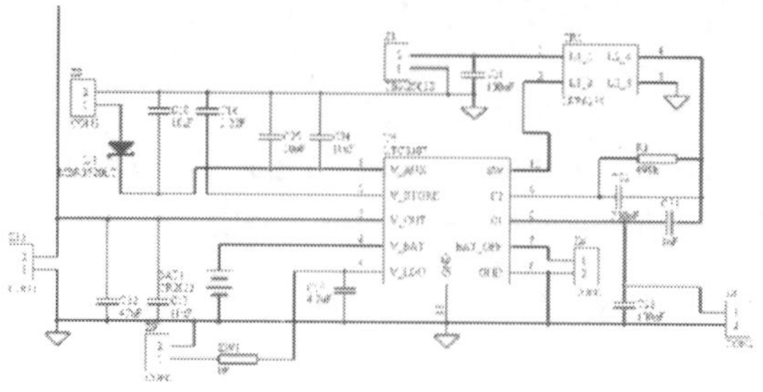

Figure 3. Harvester based on LTC3107.

Telit radiomodem works in a telemetric mode. This opens a possibility ~~to~~ of automatic activation of the radiomodem when the alarm signal is on. Before activation the radiomodem is in "sleep" mode and requires under 10μW of power. Alarm signal generates interrupt which obliges ~~a~~ the radiomodem to send a signal which specifies a source of the alarm [17].
Figure 4 shows a diagram of the radiomodem, and Figure 5 presents the complete device.

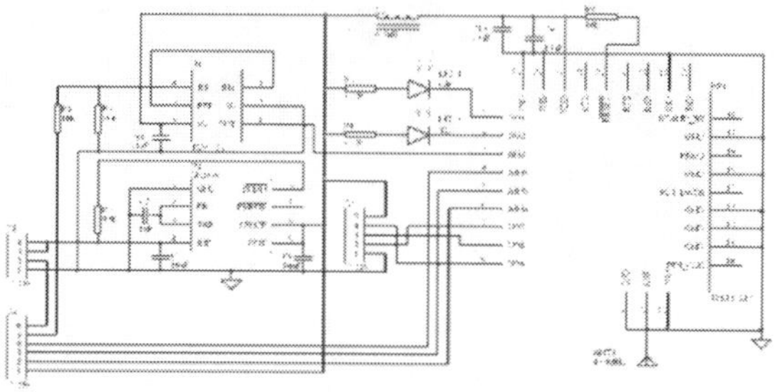

Figure 4. Radiomodem Telit LE50 433MHz

Figure 5. Complete device (enclosure without a cap)

During tests we measured thermogenerators manufactured by "Marlow" – TG-12-4 [18], "Customthermoelectric" 1261-7L31-04CL [19] and "Wellen Technology" TEG 10-46 [20]. Tests confirmed that they supplied the required power (ca. 2.5 mW) when the temperature difference between „cold" and „hot" side was greater than 25-30°C. This corresponds to the hot side temperature within 45-50°C range. Such working conditions are acceptable for practically all ovens or boilers currently present on the market. It is possible to use similar device for power generated by solar hot water collectors.

Diagram of the power generated by TEG's is shown in Figure 6.

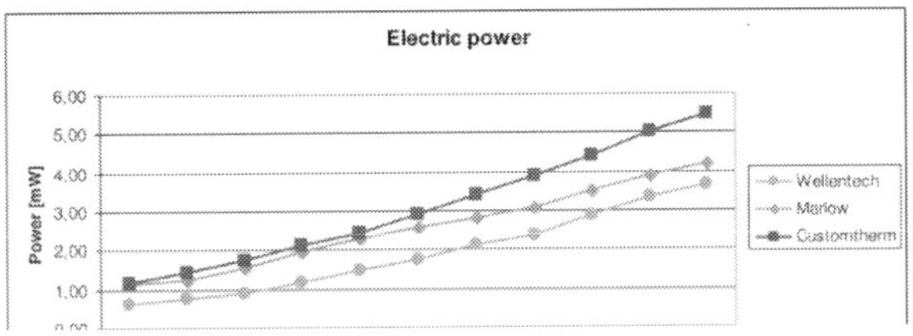

Figure 6. Power generated by TEG's.

Supercapacitors are optimal for energy storage. They have a big reliability - up to 500 000 charge/discharge cycles and a wide range of working temperatures -40/+60°C. At present we can use small-dimensions capacitors from Cooper-Bussmann with capacity 5F (Vmax= 5.5V). Amount of energy stored in such capacitors allow a device to work even if gaps in heating are long.

2.2. Devices powered by photovoltaic cells.

During tests we designed and manufactured power supplies based on IC's from „Linear Technology" (LT3105) „Texas Instruments" (bq25505) and „ST Microelectronics" (SPV1040 i SPV1050). They was DC/DC converters working with maximum power point tracking algorithm (MPPT).

Figure 7 shows example of such converter.

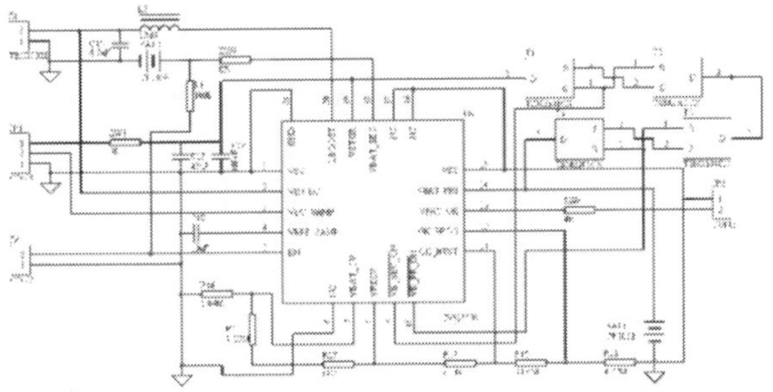

Figure 7. Power supply with MPPT based on bq25505.

It is based on chip bq25505 [15], it contains Li-Ion battery charger and power source selector (multiplexer battery / primary battery). In this design we used coin cell CP1654 from „Varta Microbatteries". Power source selector is based on P-MOS transistors (T1, T3, T4, T5). The resistor ladders R13, R14, R15 and R16, R17, R18 set control points to charging/discharging block and power source selector. Maximum regulated power is on level ca. 500mW. Maximum power point tracking is in accord with „Perturb&Observe" algorithm. MPPT block in chips SPV1040 and SPV1050 works similar.

The maximum power point in LTC3105 [24] is programed using a resistor, which sets value of the maximum power point voltage. This IC contains a regulator which aproximate the optimal amount of current draw from input photovoltaic cell.

As power sources we tested photovoltaics cells suitable for working with artificial lighting manufactured in below listed technologies:

- thinfilm (CIGS) – SP4.2-37 cell from Powerfilmsolar - USA
- amorphous silicon („Amorton") – AM-1816 cell from Panasonic - Japan
- DSSC (Graetzel cell) – from Gcell – UK
- OPV (organic photovoltaic cell) - from „InfinitePV" - Danemark

We measured cells lighted by fluorescent light with intensity from range 200-1100lux (in accord with EN 12464-1:2012). Normalized output current (μA/cm^2) is presented on figure 8.

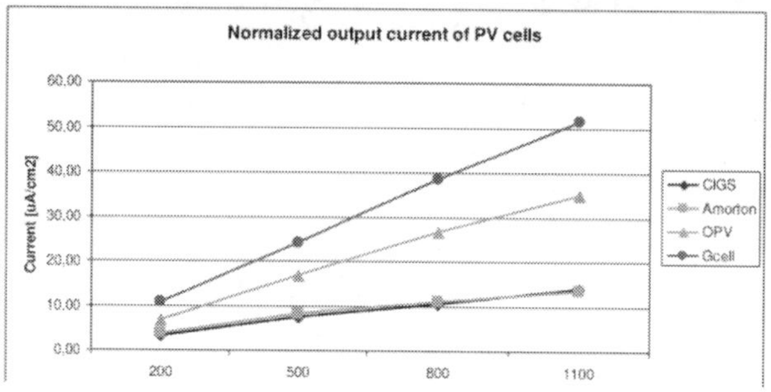

Figure 8. Normalized output current of PV cells.

Based on such tests we can select DSSC cells as optimal for working with artificial lighting. It is recommended to test perovskite and tandem OPV cells when their commercial versions will be available.
When a small dimensions cell (ie. credit card size) is required, a good alternative is to replace a MPPT controller by simple circuit named "Ideal diode". It cuts off connection between cell and battery (or supercapacitor) when cell is in the dark, preventing battery discharge by the darkened cell. In normal conditions battery is directly charged by cell, and in such conditions we spare energy needed for working internal MPPT regulator of the controller. Figure 9 presented a diagram of such simple circuit.

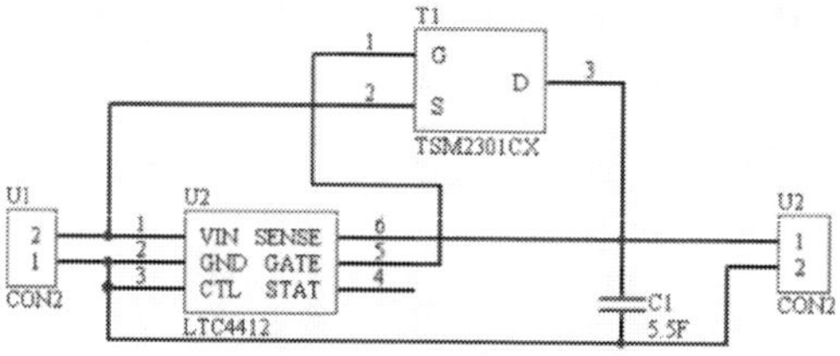

Figure 9. „Ideal diode" circuit.

Above mentioned circuit is based on LTC4412 [21] and works well in most small power applications. We recommended application of the multivibrator for long periods – ie. TPL5100 [22] – it allows periodical activation of load for a short time. It allows practically permanent charging battery of supercapacitor by attached PV cell.

3. Conclusion

When we consider as an example acredit card size GCell, exposed to 500 lux light during 12 hours, we find that harvesting a 13.5mAh charge is possible. The radiomodem for standard session requires 0.02 mAh. Rest of the charge we can use for powering a controller and input/output devices. When we use a modern CMOS devices, the amount of energy harvested is suitable for sophisticated actions. Such example shows that even small power PV cells can be reliable and efficient power source for many usable devices. It is expecially valid for standard office building where lighting is practically permanent. The same opinion is valid for thermogenerators in winter season. We suppose that harvesters will grow in number and particular application areas. They will beimportant especially for „Internet of Things" devices, because they are alternative to expensive wired supply grids or to permanent battery replacement. Very probable is further growth in usage of non silicon photovoltaic devices, because their manufacturing cost will be low and their „manufacturing energy return time" is short. Manufacturing technology is based on printing layers of polimers and cheap. Currently six european research teams are working in such area. We can estimate that as a result of such efforts, this kind of energy sources will be cheap and reliable. Very probable are efforts to design „printable" harvesters which will be manufactured in the same technology as the energy sources (photovoltaic cells).

4. References

[1] Starner T Paradiso J.A 2004 Human Generated Power for Mobile Electronics in Piguet, C. (ed) *Low Power Electronics* CRC Press **45**-1– 45-35

[2] Harry Zervos 2013 Thermoelectric Energy Harvesting 2013-2023: Devices *Applications, Opportunities PRNewswire* (New York)

[3] Chung-Yang Sue, Nan-Chyuan Tai 2012 Human powered MEMS-based energy harvest devices *Applied Energy* **93** 390–403.

[4] Shengulette J 2012 Pace is picking up for Energy Harvester's Walking Charger *Rochester Democrat and Chronicle, Gannett Co., Inc., McLean* (USA) .

[5] Paradiso J A, Starner T 2005 Energy Scavenging for Mobile and Wireless Electronics *IEEE Pervasive Computing* **4** 18-27.

[6] Saha C R, O'Donnell T, Wang N, McCloskey P 2008 Electromagnetic generator for harvesting energy from human motion *Sensors and Actuators A* **147** 248–253.

[7] Ahmed Munaz, Byung-Chul Lee, Gwiy-Sang Chung 2013 A study of an electromagnetic energy harvester using multi-polemagnet *Sensors and Actuators A* **201** 134– 140.

[8] Korla S, Leon R A, Tansel I N, Yenilmez A, Yapici A, DemetGul M 2011 Design and testing of an efficient and compact piezoelectric energy harvester *Microelectronics Journal* **42** 265–270

[9] Tkaczyk E 1997Compact and Lightweight Energy Conversion Using Electrostrictive Polymers *Proc. Prospector IX: Human-Powered Systems Technologies Space Power Inst., Auburn Univ* 313-329.

[10] Kymissis J, Kendall C, Paradiso J, Gershenfeld N 1998 Parasitic Power Harvesting in Shoes *Proc. 2nd Int'l Symp. Wearable Computers, IEEE CS Press* 132–139.

[11] Shaahank Priya, Daniel J Inman 2009 Energy harvesting Technologies *Springer* (New York).

[12] Krupenkin T, Ashley Taylor J 2011 Reverse electrowetting as a new approach to high-power energy harvesting *Nature Communications* **2** 448.

[13] Datasheet of the ADA4505-2 http://www.analog.com , access february 2015

[14] Application note CN0234 from Analog Devices, http://www.analog.com , access february 2015

[15] Datasheet of the bq25505 http://www.ti.com, access february 2015

[16] Telit_xE50-433/868_RF_Module_User_Guide_r11 http://www.telit.com access february 2015

[17] Telit_Star_Network_Protocol_Stack_User_Guide_r9 http://www.telit.com access february 2015

[18] Datasheet of the TG12-4 http://www.marlow.com , access october 2014

[19] 1261G-7L31-04CL_spec_sht http://www.customthermoelectric.com access october 2014

[20] Datasheet of the TEG 10-46 http://www.wellentech.com access october 2014

[21] Datasheet of the LTC4412 http://www.linear.com, access february 2015

[22] Datasheet of the TPL5100 http://www.ti.com, access may 2015

[23] Datasheet of the LTC3107 http://www.linear.com, access february 2015

[24] Datasheet of the LTC3105 http://www.linear.com, access february 2015

IOP Conf. Series: Materials Science and Engineering **104** (2015) 012005 doi:10.1088/1757-899X/104/1/012005

Estimation and harvesting of human heat power for wearable electronic devices

P Dziurdzia[1], I Brzozowski[1], P Bratek[1], W Gelmuda[1], A Kos[1]

[1] AGH University of Science and Technology
Al. Mickiewicza 30, 30-059 Krakow, Poland

E-mail: piotr.dziurdzia@agh.edu.pl

Abstract. The paper deals with the issue of self-powered wearable electronic devices that are capable of harvesting free available energy dissipated by the user in the form of human heat. The free energy source is intended to be used as a secondary power source supporting primary battery in a sensor bracelet. The main scope of the article is a presentation of the concept for a measuring setup used to quantitative estimation of heat power sources in different locations over the human body area. The crucial role in the measurements of the human heat plays a thermoelectric module working in the open circuit mode. The results obtained during practical tests are confronted with the requirements of the dedicated thermoelectric generator. A prototype design of a human warmth energy harvester with an ultra-low power DC-DC converter based on the LTC3108 circuit is analysed.

1. Introduction

Among different ubiquitous and free energy sources, such as: light, electromagnetic background, heat and vibrations, only the latter ones can be generated by human beings and next converted into useful electrical energy to supplying smart electronic devices, wireless sensor nodes, as well as wearable devices [1], [2]. Small and irregular pieces of human energy have to undergo harvesting processes which mostly require careful modeling and customized design of ultra-low power management circuits in order to increase the overall energy conversion efficiency [3], [4].

The feasibility study and specific needs of self-powered smart wearable devices should be confronted with available energy, therefore the performance of the human power sources have to be theoretically investigated and acknowledged in practice under experiments.

In our recent research work we have been conducting investigations toward self-powering sensor bracelet for monitoring important life parameters like pulse, respiration, and blood oxygenation. The bracelet is going to be equipped with an energy harvesting module converting available human warmth into electricity to support and prolong the life time of the primary battery. Figure 1 shows considerations on suitable locations for attaching the sensor bracelet; they need to take into account not only the comfort of the user, but also the available human heat power and easy access to measuring points of the life parameters.

The paper presents theoretical considerations on a method for estimation of heat power released by a human being during day-to-day activities, and then the results of practical measurements at different body locations. It is followed by a discussion on modeling of interfacing mechanical assembly and electronic circuit capable of converting ultra-low-power signals form energy harvester. The rating of

IOP Conf. Series: Materials Science and Engineering **104** (2015) 012005 doi:10.1088/1757-899X/104/1/012005

the heat power is a useful indicator for wearable electronic devices that are to be powered with the energy harvested from human warmth, such as biomedical autonomous systems, healthcare sensors, etc.

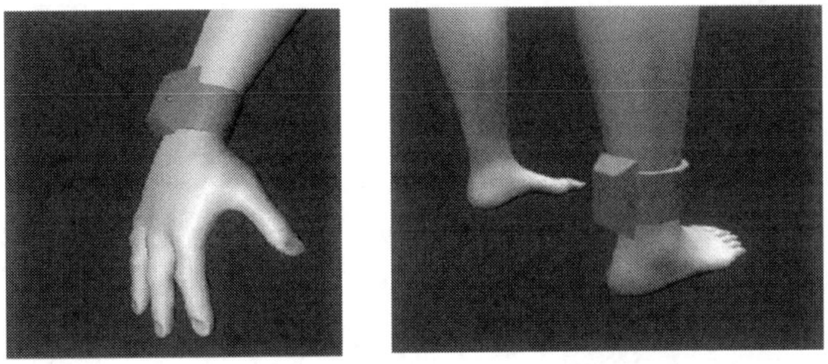

Figure 1. Different locations considered for attaching the sensor bracelet.

Harvesting of dissipated human heat energy has been investigated by many research teams for the recent years. A few successful demonstrators of autonomous energy sources made of tiny thermoelectric generators (TEG) were presented. They were next implemented to supplying body area networks and wearable sensors [5], [6].

Specifics of the human power sources as well as the design process of wearable harvesters entail laborious measuring and collecting of data, and also a careful process of electrothermal modeling and simulations. It is mainly due to the fact that human warmth harvesting process deals with an ultra-low power source which is additionally hampered by relatively low conversion efficiency. According to authors' calculations and simulations, the output electrical energy $P_{electrical}$ from TEG is equal only to about a few percent of the input human heat power Q_{human} taken from an unit of body area, which is shown in figure 2 [7].

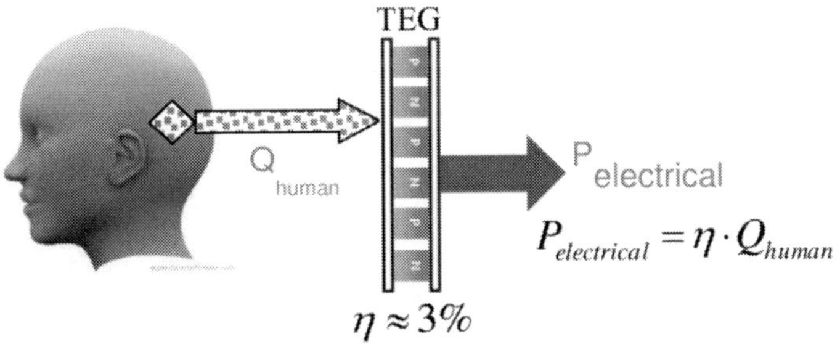

Figure 2. Human heat harvesting process.

2. Estimation of human heat power

Quantitative analysis of heat transfer in human tissues has been a subject of research work performed for last few decades; some highly detailed models have been obtained, suitable for implementation to numerical simulations [8]. Whereas the heat power dissipated by the humans differs between individuals and the location on the body area, the output electrical energy $P_{electrical}$ depends not only on the conversion efficiency η of the TEG. It is influenced heavily by the quality of the attachment of the TEG to the body too (which also includes thermal resistance of the skin), and the thermal resistance which is imposed on the human heat flow on the way from the TEG to the ambient (which can be expressed by thermal resistance of the heat sink).

2.1. Electrothermal modeling

Apart from that thermoelectric modules are used as TEGs in target heat energy harvesting devices, they can also be applied to experimental estimation of human heat power when they are operating in open circuit mode. Heat transfer in human warmth harvesters attached to the body can be modeled as one-dimensional flow of human heat power Q_{human} through a series of thermal resistances: expressing attachment to the skin R_{th_attach}, thermoelectric generator R_{th_TEG}, and the heat sink $R_{th\_heatsink}$. The heat starts flowing at temperature T_{human}=36.6 °C, and moves to the ambient at temperature T_a, which is depicted on the equivalent electrothermal model presented in figure 3.

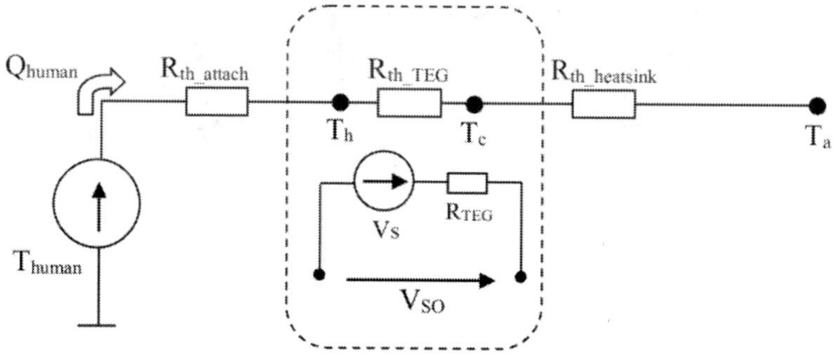

Figure 3. Equivalent electrothermal model for heat transfer and energy conversion analysis.

The heat power Q_{human} flowing through the TEG causes a temperature gradient ΔT_{TEG} across thermoelectric module's plates, and due to the Seebeck effect it generates a voltage V_S according to the formula (1).

$$V_S = \alpha \Delta T_{TEG} = \alpha(T_h - T_c) \tag{1}$$

where: α [V/°C] is the total Seebeck coefficient for a given thermoelectric module, T_h and T_c [°C] are the temperatures at the hot and cold plates of the TEG, respectively.

The Seebeck voltage is equal to the output voltage of the TEG working in open circuit mode (2). Moreover, the temperature gradient can be expressed as a temperature drop over the thermal resistance R_{th_TEG} , caused by the flowing heat power Q_{human} (3). Combining (1), (2) and (3), one can get the formula (4) for heat power, expressed by the measured open circuit Seebeck voltage.

$$V_{SO} = V_S \big|_{I_{TEG}=0} \tag{2}$$

$$\Delta T_{TEG} = Q_{human} \cdot R_{th\_TEG} \tag{3}$$

$$Q_{human} = \frac{\Delta T_{TEG}}{R_{th\_TEG}} = \frac{V_{SO}}{\alpha \cdot R_{th\_TEG}} \tag{4}$$

Although the Seebeck coefficient, as well as TEG thermal resistance, are strongly temperature dependent, we can assume that for wearable applications they are constant, thus the human heat power is proportional to the Seebeck voltage.

2.2. A testing setup and measurements

The testing setup for estimation of human heat power is presented in figure 4. The measurements were performed by means of Data Logger, with 16-bits A/C converter, and at 1ms time interval. It was collecting the Seebeck voltages from the thermoelectric generator under test and attached directly to the body surface. The TEG was composed of a commercially available thermoelectric module consisting of 127 thermopiles connected electrically in series and thermally in parallel, and the heat sink having thermal resistance of 2 K/W. The surface area of the TEG directly attached to the skin was a square 4cm x 4cm.

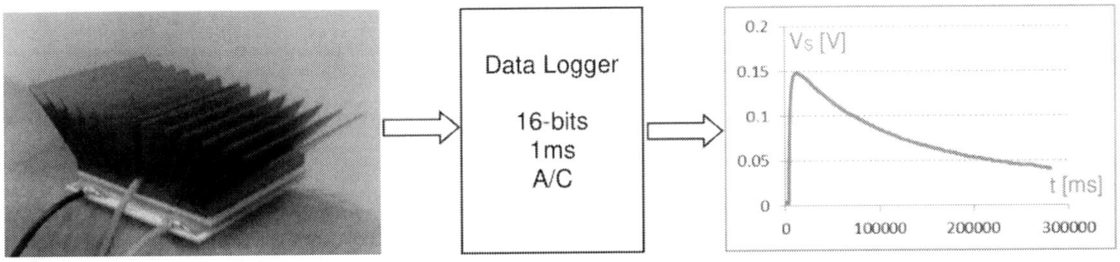

Figure 4. Testing setup.

In table 1, a series of V_{SO} measurements at steady states, and calculated ΔT_{TEG}, Q_{human}, and R_{th_attach} in different locations of the body area is shown. For estimation of constant coefficients α and R_{th_TEG} we used previously evaluated electrothermal model of the thermoelectric module and the temperature characteristics for Seebeck coefficient α and the thermal conductance λ (which is in fact the reciprocal of the R_{th_TEG}) [9]. From figure 5 it is evident that for considered very small temperature range around T=35°C: α=51.75 mV/K, and R_{th_TEG}=2.28 K/W

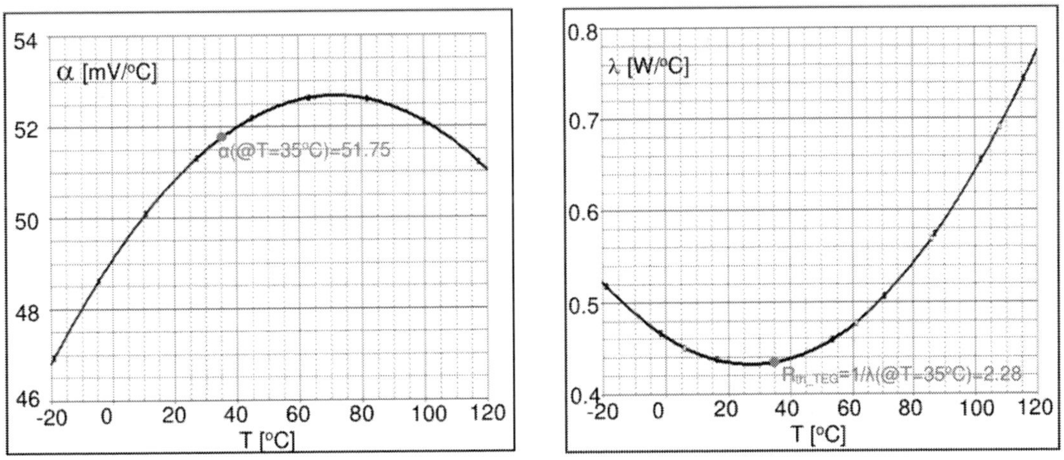

Figure 5. Temperature characteristics of the Seebeck coefficient α, and the thermal conductance λ.

Table 1. Measured open circuit voltage and calculated: temperature gradient, heat power, and attachment thermal resistance at different locations of the human body (surface area of the thermoelectric module is 40 mm x 40 mm).

Attachment	V_{SO} at steady state [mV]	ΔT_{TEG} [°C]	Heat power Q_{human} [mW]	R_{th_attach} [K/W]
wrist up side	10.0	0.19	80	115.0
wrist down side (close to artery)	13.1	0.25	110	86.8
forehead	14.7	0.28	112	76.9
leg	8.0	0.15	70	144.9
arm	11.2	0.22	90	102.3

The measurements took place at ambient temperature T_a=26.5 °C; for estimation of the R_{th_attach}, we assumed T_{human}=36.6 °C. The results shown in table 1 met our expectations to some extent. The highest levels of heat power were generated in a close proximity to the artery and the head. On the other hand the highest attachment thermal resistance is observable under the muscles and adipose tissue. If the average human body surface is 1.8 m^2, and assuming an average heat power dissipating on a 16 cm^2 TEG surface in the range of 100 mW, it would mean that the human being is dissipating by his skin more 100 W of the heat power at resting position.

3. Harvesting of human warmth and converting into electricity

Even if the thermoelectric generator can provide reasonable power harvested from the human warmth, the main difficulty (regarding its utilization) manifests in converting an ultra-low Seebeck voltage to a level that is necessary to supplying electronic circuits and sensors.

One of the most suitable electronic integrated circuits, with the lowest input voltage, is offered by Linear Technology. DC-DC LTC3108 converter can be applied directly to a very low voltage sources; the only limitation is that the minimum required input voltage under operation is 20 mV. It means that the open circuit Seebeck voltage should be at least 40 mV (under operation, the internal resistance of the TEG is matched with input resistance of the converter).

A feasibility study shown below reveals some hints and limitations when human heat harvesting is considered. During simulation experiments we used an accurate electrothermal model of the thermoelectric module implemented into electronic circuit simulator LTSpice, and we also assumed that the TEG is attached to a wrist close to the artery.

Figure 6 shows the simulation setup. All the external conditions (ambient and human body temperatures) were set to the same values as when the measurements were performed. For the first case we applied a variable thermal resistance instead of estimated earlier R_{th_attach}=86.8 K/W; its value was swept by 'control' signal. Our intention was to determine a relation of the Seebeck voltage against thermal resistance of the attachment, which was presented in figure 7.

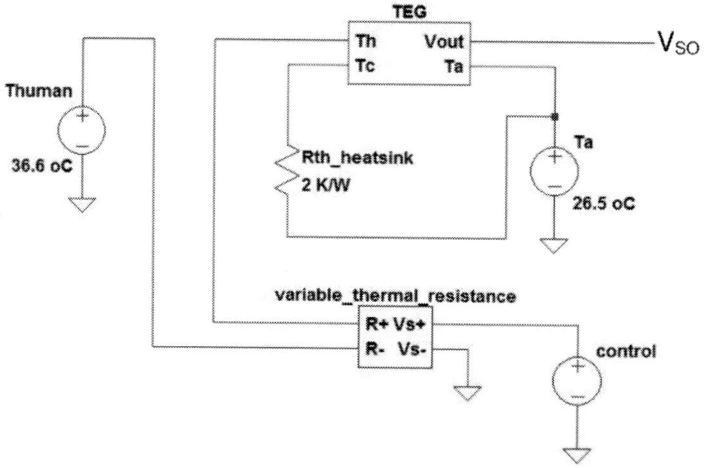

Figure 6. Scheme of the simulation setup for determination of the Seebeck voltage relation against the thermal resistance of the attachment.

As it could be expected, the quality of the attachment of the TEG to the body strongly influences on the output Seebeck voltage. The required minimum voltage needed to start the converter could be attained if the thermal resistance R_{th_attach} had been decreased to 25 K/W. Another option of increasing the output voltage is the heat sink. But as it is shown in figure 8, decreasing the $R_{th\_heatsink}$, even to the extreme low level, only slightly improves the V_{SO}.

IOP Conf. Series: Materials Science and Engineering **104** (2015) 012005 doi:10.1088/1757-899X/104/1/012005

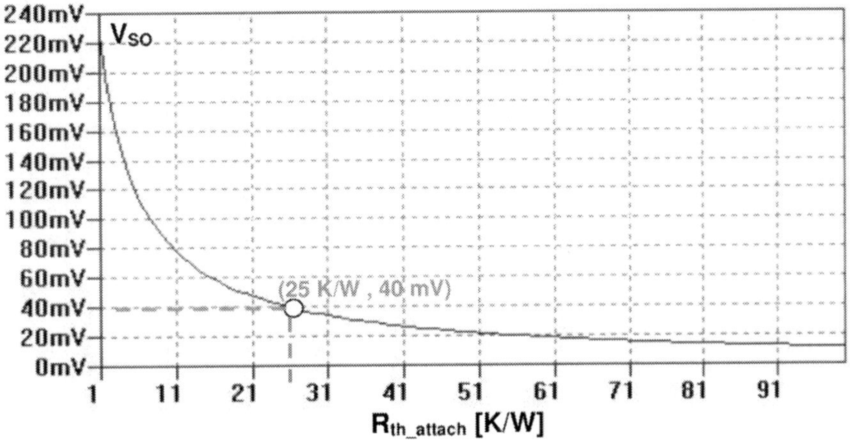

Figure 7. Open circuit Seebeck voltage against thermal resistance R_{th_attach}.

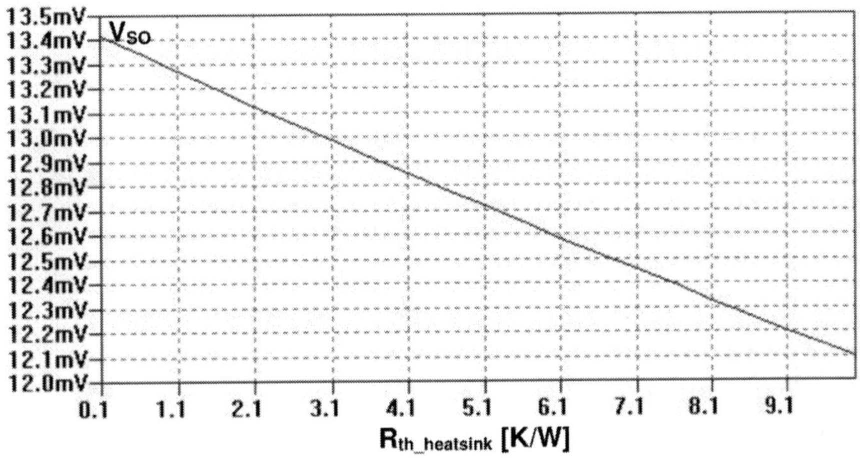

Figure 8. Open circuit Seebeck voltage against thermal resistance $R_{th\_heatsink}$.

At the end we performed experiments on a setup shown in figure 9, which was combined of the thermal part – the model of the TEG, and the electrical one – the boost-up voltage converter.

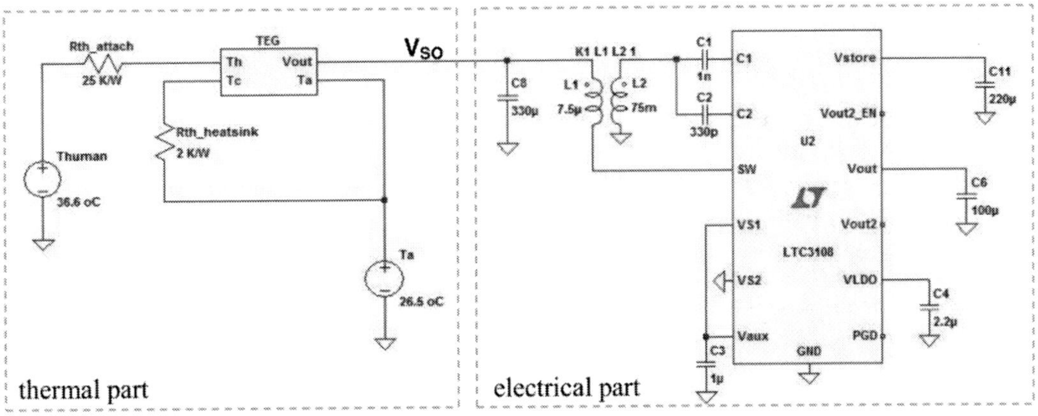

Figure 9. Setup for combined simulations with boost-up voltage converter.

IOP Conf. Series: Materials Science and Engineering **104** (2015) 012005 doi:10.1088/1757-899X/104/1/012005

Our goal was to prove that by lowering the attachment thermal resistance to 25 K/W, the LTC3108 converter could be triggered by the very low output Seebeck voltage. From the figure 10 it is evident that for presumed case, the converter circuit is able to deliver a regulated constant voltage at its output VLDO=2.2 V.

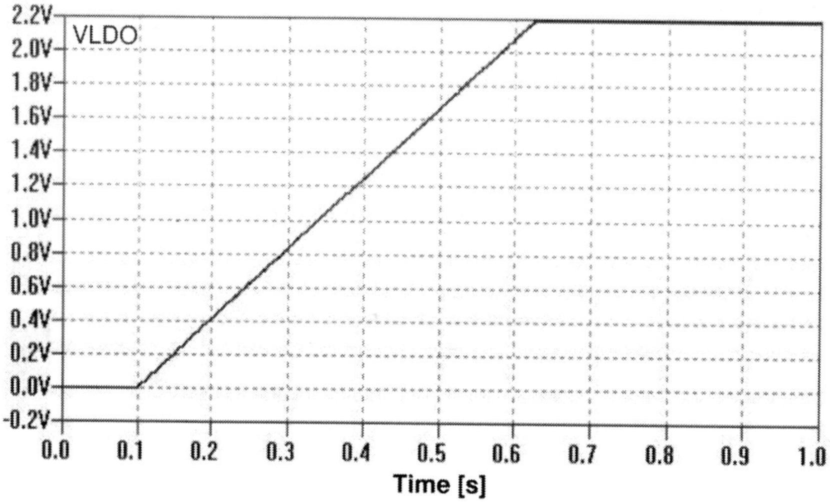

Figure 10. Regulated output voltage from the converter supplied by human heat.

4. Conclusion

In the paper a whole process of investigations toward a prototype human warmth harvester was described. The measured values of heat power sources and attachment thermal resistances in different points of the body area were confronted with the requirements of the commercially available ultra-low voltage converter, namely with its required minimum start-up input voltage. It should be noted that we considered only steady states (TEG working in equilibrium), therefore the stable output Seebeck voltage was much lower than the maximum levels observed during transient states (for example, right after attaching the TEG to the body). Due to the relatively large thermal time constants, reaching the equilibrium by the whole harvester assembly takes a few hundreds of seconds in the case of the tested TEG.

During the experiments on a real circuit with LTC3108 we observed that it was able to deliver a regulated VLDO=2.2 V for a while, which next disappeared in the thermal equilibrium. The output Seebeck voltage could sustain operation of the converter only in the thermal transient state, right after attaching the TEG to the wrist. It was mainly due to a thermal capacitance of the heat sink, whose temperature was getting higher as the heat was flowing from the body through the TEG to the ambient, thus at the same time lowering the temperature gradient at the plates of the thermoelectric module, and in consequence decreasing the Seebeck voltage below the minimum start-up level required by the LTC3108.

The presented combined electro-thermal models indicate that for a given single thermoelectric module and external conditions, namely the human and ambient temperatures, the only possible way to increase the output Seebeck voltage is improving (lowering) the attachment thermal resistance, which appeared higher in the experiments than we expected. We suspect that due to a very thin TEG plate that was attached directly to the skin, the human heat could leak to the upper plate, thus creating some path allowing the heat flux to bypass the TEG. Another natural solution to achieve required higher output Seebeck voltage in the thermal equilibrium is connecting more thermoelectric modules in series.

5. References

[1] Joseph A. D. *2005* Energy Harvesting Projects, *Published by the IEEE CS and IEEE ComSoc,* 1536-1268/05

[2] Beeby S, White N 2010 Energy Harvesting for Autonomous Systems, *(Artech House 2010, ISBN-13: 978-1-59693-718-5)*

[3] Priya S, Inman D J 2009 *Energy Harvesting Technologies (Springer 2009, ISBN 978-0-387-*76463-4)

[4] Joon Kim K, Cottone F, Goyal S, Punch J 2010 *Energy Scavenging for Energy Efficiency in Networks and Applications (Bell Labs Technical Journal 15(2), 7–30 (2010)* pp. 7-29

[5] Leonov V, Fiorini P, Sedky S, Torfs T, Van Hoof C. 2005 ThermoelectricMEMS Generators as a Power Supply for a Body Area Network*: Proc. of the 13th International Conference on Solid-State Sensors, Actuators and Microsystems (Seul, Korea, June 5-9, 2005)* 291-294

[6] Leonov V. 2013 Thermoelectric Energy Harvesting of Human Body Heat for Wearable Sensors *(IEEE Sensors Journal, vol. 13, no. 6, June 2013)* 2284-2291

[7] Dziurdzia P, Mirocha A 2009 From Constant to Temperature Dependent Parameters Based Electrothermal Models of TEG, *Proc. of the 16-th International Conference Mixed Design of Integrated Circuits and Systems,* Łódź, Poland, 25-27 June 2009 555-559

[8] Prishvin M, Manukyan L, Jeladze V, Petoev I, Tabanadze V, Kakulia D, Zaridze R 2008 Numerical Simulation of Heat Transfer in Human Tissue According to Improved Vascular Structure Model*: Proc. of the 13th Workshop on Direct and Inverse Problems of Electromagnetic and Acoustic Wave Theory DIPED 2008, Tbilisi* 143-148

[9] Dziurdzia P, Kos A 2001 Tool for Fast Modelling Active Heat Sink, Proc. of the XVIIth Annual IEEE Semiconductor *Thermal Measurement and Management Symposium SEMITHERM, San Jose, USA, 20-22 March 2001* . 174-179

[10] Linear Technology *Journal of Analog Innovation, Vol. 10 no 3 October 2010*

Acknowledgments

The work was supported by the National Centre for Research and Development (NCBiR), project grant No. DOB-BIO6/12/24/2014.

Optimisation of industrial production of low-force sensors – adhesive bonding of force-centring ball

T Maeder, C Jacq, M Blot, and P Ryser

Laboratoire de Production Microtechnique (LPM), École Polytechnique Fédérale de Lausanne (EPFL), CH-1015 Lausanne, Switzerland

thomas.maeder@epfl.ch

Abstract. This work addresses the issue of attaching the force-centring part (a round ball) to the load cell of a force sensor, a piezoresistive thick-film Wheatstone bridge deposited onto a ceramic cantilever. As the current soldering process requires expensive metallisation steps for both the ball and the cantilever, and subjects the solder pads used for mounting the cantilever to an additional reflow cycle, an alternative adhesive bonding process was developed, allowing both simpler production and the use of other ball materials such as ceramic and glass. The self-centring action of solder capillary forces was ensured by structuring the adhesive so as to form a mechanical cuvette allowing centring of the ball by gravity. The selected adhesive materials exhibited good printability and bonding, as well as surviving the subsequent soldering and cleaning process steps.

1. Introduction

1.1. Simple cantilever-type thick-film force sensor

Mechanical load (pressure and force) sensors based on the piezoresistive properties of thick-film (TF) resistors have found wide application due to their robustness and simple, straightforward manufacture [1-5]. We have developed [5-7] a basic low-range force sensor (figure 1) with an alumina cantilever that also acts as the load cell. For lower forces, LTCC (low-temperature cofired ceramic) technology, having a lower elastic modulus, being available in thin sheets and amenable to 3D structuration, has allowed fabrication of a more responsive variant of the sensor by replacing the alumina cantilever with an LTCC one (figure 2) [8-11].

1.2. Motivation of present work

Production of these sensors is quite straightforward, with the overall process flow listed in table 1. Most processes fall within the bounds of standard TF / LTCC technology and other usual procedures in electronics, such as surface-mount technology (SMT) using solder attachment. However, the solder pads of the cantilever, which are used for mechanically bonding and electrically interconnecting it to the base, currently experience three solder reflow operations: 1) pre-tinning those pads, 2) soldering the force-centring ball, and 3) soldering the cantilever to the base. This exposes the pads to considerable leaching by and reaction with solder, an issue that has been considerably exacerbated by the introduction of lead-free soldering based on tin-rich solders [12,13].

Therefore, the object of the present work, which has been carried out in the frame of a student project [14] and complemented by additional experiments is to examine the possibility of replacing the ball soldering process (table 2) by adhesive bonding (table 3), which promises several advantages:

- Gluing eliminates one solder reflow step, decreasing the thermal / metallurgical budget of the bottom cantilever solder pads to 2 vs. 3 reflow operations.
- No special metallisation of the ball or cantilever is necessary (the thick-film conductive tracks on the cantilever do not allow soldering by themselves).
- A wider range of ball materials may in principle be used, including low-cost glass balls, which also have the advantage of having a low thermal conductivity (lower influence of the temperature of the contact surface on the signal) and being electrically insulating (lower influence of noise sources coupled through the contact surface)

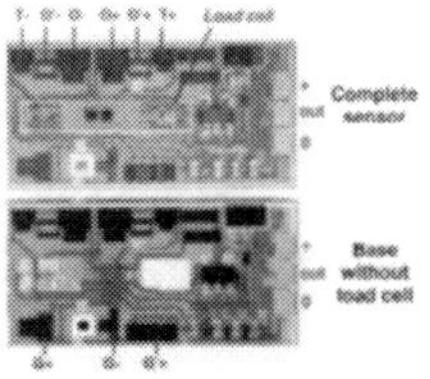

Figure 1. Basic low-force sensor with alumina cantilever (the letters denote positions of laser trimming cuts); from [7].

Figure 2. Force sensor, variant with LTCC cantilever (in blue); from [10].

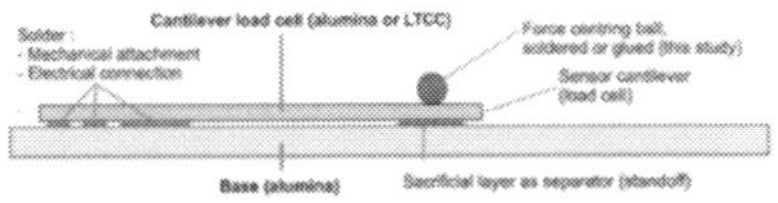

Figure 3. Assembly principle of cantilever onto base, with solder pads (left side) as a permanent electrical and mechanical connection [7] and a sacrificial layer (right, tip side) as a temporary separator.

Table 1. Overall force sensor fabrication process flow.

#	Main process phase	Description
1	Deposition of circuit on base	Standard TF deposition onto alumina substrate
2	Deposition of piezoresistive load-cell bridge on cantilever	TF deposition onto 0.25-0.635 mm thick alumina substrate, or LTCC process with co-fired TF layers
3	Coarse trimming of load cell	Laser offset trimming & stabilisation by annealing
4	Pre-tinning of load cell pads	Solder screen printing & reflow
5	Assembly of force-centring ball onto load cell	Soldering (current) or adhesive bonding – ***object of present work***
6	Individualisation of cantilevers	Breakage of pre-scored cantilever substrate
7	Mounting of components onto base, including cantilever	SMT process, with lead-free SAC (Sn-Ag-Cu) solder & sacrificial-layer cantilever separator (figure 3)
8	Active sensor trimming	Laser trimming with force applied by custom jig
9	Cleaning	Ultrasound bath – solder flux cleaning & sacrificial-layer removal
10	Individualisation of sensors	Breakage of pre-scored sensor base substrate
11	Final steps	Optional lead attachment and inspection

Table 2. Ball attachment onto cantilever - process steps for soldering.

#	Process step	Description
1	Additional thick-film metallisation layer on top side of cantilever	Standard TF process[a b]
2	Metallisation of steel ball	Etching, followed by electroless nickel / immersion gold (ENIG) metallisation[c]
3	Deposition of "wet" solder	Screen / stencil printing
4	Mounting of ball	Mechanical "sieve"[d]
5	Solder reflow[e]	Reflow oven

[a] A conductor layer on the cantilever top side is either not present (LTCC) or too thin to allow soldering (alumina), requiring deposition of a specific solderable metallisation.
[b] The metallisation of the attachment pads on the bottom side of the cantilever must also be better (i.e. more expensive) than in case of adhesive bonding, because it has to survive three vs. two reflow steps, but this only requires a thicker layer, not an additional one.
[c] Necessary to allow solderability of the chromium steel bearing ball (EN 1.3505 / 100Cr6).
[d] Balls are simply "sieved" over a plate with holes corresponding to their correct position on each cantilever. A standard component pick-and-place machine has also been used.
[e] Second solder reflow operation to the cantilever solder pads, after tinning (see table 1).

Table 3. Ball attachment onto cantilever - process steps for adhesive bonding.

#	Process step	Description
1	Deposition + cure of 'cuvette' formulation	Screen printing + thermal cure
2	Deposition of wet 'glue' formulation	Screen printing same screen as (1)
3	Mounting of ball	Mechanical "sieve" (see table 2, note d)
4	Curing of adhesive	Box oven[bc]

2. Requirements and selected approach

2.1. Requirements

To be validated as a replacement for soldering, the adhesive bonding process must meet the following requirements:

1) Correct positioning of the balls (ensured by capillary forces for the soldering process)
2) Sufficient lateral shear strength, at least ~4 N (a conservative requirement, as the maximum vertical nominal force range measured by the sensor is 2 N)
3) Simple, straightforward process steps, as few as possible, with low cost of materials
4) Compatibility with subsequent steps: acceptable strength conserved after cantilever mounting (solder reflow at ~260°C) and cleaning (exposure to solvents)
5) Acceptable compression strength of ball (if the current steel ball is replaced)

2.2. Adhesive deposition process and basic materials category

We chose screen printing as the adhesive application method, for several reasons: good integration within the TF production process (mostly screen printing), parallel deposition (large amount of cantilevers, 180 per substrate), and deposition capability for thick films in one print (thickness needed to achieve a good contact area with the round ball, see also figure 4). This choice, however, restricts the acceptable range of adhesives to those materials that have a sufficiently long screen / pot life: adhesives curing with ambient moisture or two-part ones with fast ambient temperature cure should be avoided. Additionally, the requirements outlined in section 2.1 further constrain the available choices: the adhesive must have a high mechanical strength, be thermally stable to survive subsequent soldering of the cantilever, and resist immersion in cleaning solvent mixes.

Given these considerations, we chose our adhesive materials amongst high-temperature curing epoxies, which exhibit the right combination of strength, stability and long pot life.

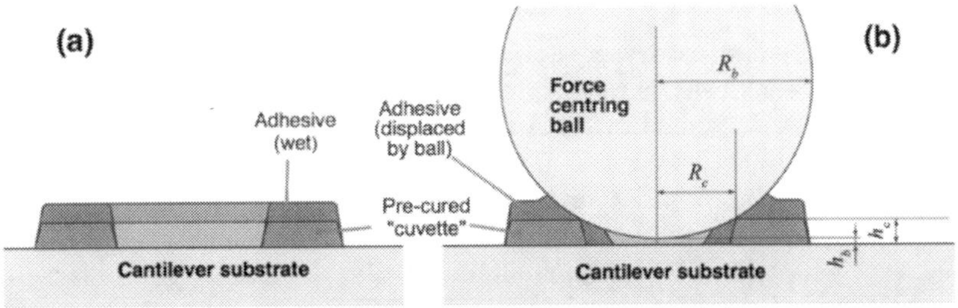

Figure 4. Positioning principle: a) deposition of "cuvette" + "adhesive" layers; b) principle of centring by cuvette & bonding by adhesive (displaced by ball).

2.3. *Positioning of the balls*

The chosen heat-cure epoxies do have one drawback: positioning of the ball is difficult, even with a combination of precise placement onto the cantilever and a thixotropic composition, as the adhesive will first become quite liquid upon heating, before the cure sets in (see also discussion). Therefore, we conceived a double-layer method, described in table 3 & figure 4, consisting of first depositing and curing a resin rim forming a "cuvette", then depositing the actual adhesive layer for the ball using the same mask. The pre-hardened cuvette solves two issues: 1) it acts as a mechanical centring device for the ball, and 2) it provides additional material needed for bonding over a larger area, i.e. a higher bond strength.

From figure 4, the condition for proper guiding of the ball (radius = R_b) by the cuvette (inner radius of layer = R_c; height = thickness = h_c) is for the ball to rest on the cuvette rim, i.e. the thus defined height of the bottom of the ball with respect to the cantilever surface, h_b, should not be below zero:

$$0 \le h_b = h_c + \sqrt{R_b^2 - R_c^2} - R_b \quad (1)$$

If R_c is small vs. R_b, (1) may be approximated by the following expression, which illustrates the roughly quadratic dependence of minimal cuvette thickness h_c on its contact radius R_c:

$$0 \le h_b \cong h_c - \frac{R_c^2}{2R_b} \quad (2)$$

3. Experimental

3.1. *Paste formulation*

The starting resins used were two-component Epo-Tek epoxies (Epoxy Technology, Billerica, USA): 354T, H70S, H70E and H70E2. To adjust rheology, various additives were used: -Al_2O_3 powder (Alfa Aesar #45482, ~0.8-1 μm particle size, Karlsruhe, Germany), ethylcellulose EC-300-48 (300 mPa·s for 5% in 80:20 toluene : ethanol, 48% ethoxyl, #200654, Sigma-Aldrich, Buchs, Switzerland), amyl acetate, and 1-octadecanol (C18E0). To prepare the pastes, both components of the epoxy resin and additive(s) are first hand mixed, then run through an Exakt 50 three-roll mill (Exakt, Norderstedt, Germany).

3.2. Printing geometry

The same design was always used for the cuvette and adhesive, as this avoids a change of screen between both steps. The R_c parameter (figure 4) was varied from 150 µm to 275 µm in 25 µm increments. The basic design for the first studies is a circle (see figures 4 and 5). In a second test series, the effect of the design is examined, comparing this circle to a triangle and a square; in these latter cases, R_c corresponds to the radius of the inscribed circle.

3.3. Sample fabrication

Round, 1 mm diameter balls of five materials (steel – 100Cr6, alumina, yttria-stabilised zirconia, soda-lime glass and borosilicate glass) were bonded onto blank alumina substrates, using the procedure detailed in table 3. Screen printing (Aurel 900, Aurel Automation, Modigliana, Italy) was carried out with several stainless steel wire mesh screens (table 4) in order to optimise the compromise between thickness of deposited paste and feature definition. Throughout this work, screens are designated by two numbers, with the first being the meshing density of the stainless steel wire ('mesh'), in lines per inch, and the second the emulsion film thickness ('µm'). Please refer to reference work on screen printing [15, 16]) for more details.

Table 4. Printing screens.

Designation	Wire pitch [µm]	Emulsion thickness [µm]
325 mesh / 40 µm	78	40
200 mesh / 60 µm	127	60
135 mesh / 80 µm	188	80

3.4. Characterisation

Cuvette thickness was measured with an optical profilometer (Surface Profile Measurement Station, Breitmeier, Ettlingen, Germany). Centring of the balls was qualitatively evaluated optically with a stereo microscope. Although this method seems rather crude, ball misalignment is easy to spot in practice, as the lateral freedom of the ball increases rapidly when cuvette thickness drops below the minimum value needed to achieve correct guidance.

Mechanical strength of the adhesive bond was measured by shearing with a Royce Instruments (Napa, USA) 552 tester, with a 50 kgf (~490 N) shear head, which was also used to test the compressive strength of glass balls. Bond shear strength was also tested after thermal cycles and exposure to solvents, simulating the subsequent soldering and cleaning steps in sensor fabrication.

4. Preliminary tests – single paste formulation for both 'cuvette' & 'glue' layers

In this first test series, the effect of cuvette radius and thickness as well as paste formulation is ascertained, with the same formulation used throughout, i.e. identical 'cuvette' and 'glue' inks.

4.1. Preliminary tests of cuvette formation

A first optimisation of paste formulation and cuvette parameters was carried out for a screen-printed resin cuvette (circular shape), printed with a 325 mesh / 40 µm screen. The fine screen used here was found to be necessary for correct deposition of the smaller circles (tests with the coarse screens tended to yield full disks for small R_c values). 'Cuvette' paste formulations were formulated by loading Epo-Tek 354T with various amount (% of total mass) Al_2O_3 powder. The resulting print quality is shown in figure 5, for 7 resin layers, each cured 10 min at 150°C and for alumina loading levels of 0% (neat resin), 30% and 50%. While the neat resin is thixotropic during screen printing, it becomes quite liquid and flows excessively upon heating before setting, which prompted us to load it with alumina. A loading of 50% of total mass was found to yield the best shape retention, albeit with some flow; going further to 60% loading caused gaps in the printed pattern to frequently occur (the corresponding image in figure 5 is for single print with a coarser screen). The printed thickness with the 325 mesh / 40 μm screen varied somewhat, at typically 20 μm for one layer, and with 15-20 μm for each additional one (~120 μm for 7 layers) – see also table 5.

The result of the evaluation of ball centring is given in table 5, for 0% and 50% alumina loading, as a function of the number of deposited cuvette layers (thickness). Accounting for process variations, the results for 50% alumina roughly agree with the prediction of (2). The approximately one more layer needed for the neat resin (0%) illustrates the effect of the resin flowing (figure 5) during heat cure in shifting the effective R_c to higher values.

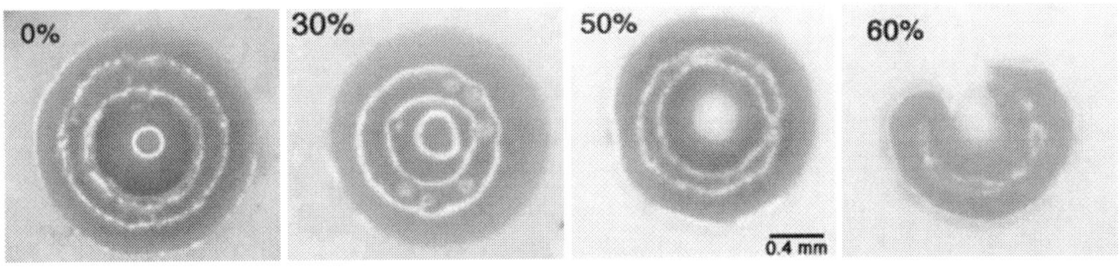

Figure 5. Cuvette formation of Epo-Tek 354T loaded with various amounts (by overall mass) of Al_2O_3 powder, after curing for 10 min at 150°C. 0-50%: 7 layers with 325 mesh / 40 μm screen; 60%: single print with 135 mesh / 80 μm.

Table 5. Centring of the balls for Epo-Tek 354T with 0% & 50% alumina loading.

Layers	Thickness		Correct centring (Y = yes; N = no), for 0% / 50% mass Al_2O_3					
(cuvette)	[μm]	R_c =	150 μm	175 μm	200 μm	225 μm	250 μm	275 μm
1	~20		Y / Y	Y / Y	Y / Y	Y / Y	N / N	N / N
2	~40		Y / Y	Y / Y	Y / Y	Y / Y	N / Y	N / N
3	~55		Y / Y	Y / Y	Y / Y	Y / Y	Y / Y	N / Y
4 & more	•70		Y / Y	Y / Y	Y / Y	Y / Y	Y / Y	Y / Y

4.2. Effect of radius and thickness on shear strength – neat 354T & 50% alumina

The cuvettes were then overprinted with one 'glue' layer having the same composition as the cuvette, for two series: neat 354T (0%) or 50% Al_2O_3, without changing the screen (325 mesh / 40 μm). Steel balls were placed and the assemblies were cured 10 min at 150°C.

The shear strength values for 50% Al_2O_3 are all low, between less than 1 N to max. ~4 N: while this formulation is adequate to form the cuvette shape, the high filler level and low flow reduces bonding with the ball, i.e. it does not work well as a 'glue' layer.

Neat 354T is the opposite, i.e. behaves poorly as a 'cuvette' layer, but achieves good bonding with the ball. This allows seeing the effect of the cuvette parameters (figure 6) on strength. Although only conditions ensuring centring (table 5) were chosen, low cuvette thicknesses yielded low strength values. Moreover, this dependence is much more pronounced for large R_c values. Obviously, neat 354T cuvettes flow excessively, and thus strongly depart from the idealised shape shown in figure 4, yielding bonding only on the bottom surface, which is unfavourable for achieving high shear strength. This issue is compounded by the small amount of resin added by the wet 'glue' layer in this test, due to the fine screen used (325 mesh / 40 μm).

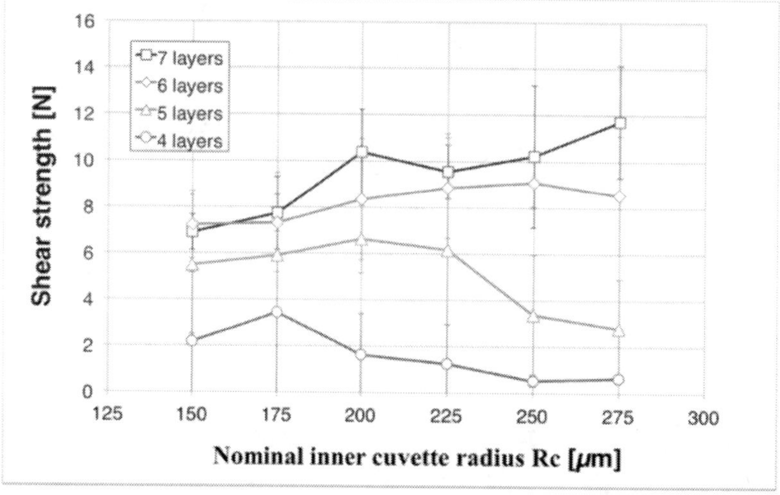

Figure 6. Shear strength of steel ball vs. nominal inner cuvette radius and number of cuvette layers (neat Epo-Tek 354T 'cuvette' and 'glue' layers, 325 mesh / 40 µm screen). Values & error bars: averages ± standard deviation; lines shown to guide the eye.

4.3. Conclusions of first tests

From this first test series, we may arrive at the following conclusions:

- It is more favourable to use different formulations for both 'cuvette' and 'glue' layers, as they require partly contradictory properties (i.e. low vs. higher flow upon cure).
- A high degree of embedding of the ball is conducive to high shear strength; this is favoured by using a high R_c value, depositing a thick cuvette, and presumably also by having a sufficiently thick final "glue" layer to accommodate any imperfections.

5. Tests with separate 'cuvette' and 'glue' compositions

Based on the results of the first series of tests, we first retain the 354T resin loaded with 50% Al_2O_3 for forming the cuvette, and separately optimise the adhesive for the 'glue' layer. After tests with the different screens (table 4), the coarsest one, 135 mesh / 80 µm was chosen, allowing to rapidly build up cuvette thickness and also giving a thick 'glue' layer. Such a screen restricts R_c to • 200 µm, which is not an issue as we aim for high R_c values. Curing was as before, i.e. 10 min at 150°C, for both 'cuvette' and 'glue' layers.

5.1. Positioning issue with a large amount of glue – evacuation of excess adhesive

A first test was carried out with steel balls, 2 'cuvette' layers, and neat 354T as 'glue'. With a higher thickness of 'glue' layer from the coarser screen, another issue came up: if the ball was placed off-centre, some positioning error remained after curing, although the layers were thick enough to satisfy the centring criterion (see 4.1). This was interpreted as a result of constricted flow and the high amount of glue (figure 7): as the ball progressively settles into the cuvette, the gap with the cuvette wall becomes smaller, and further settling is impeded, eventually retaining some of the positioning error, an issue expected to be even more acute with lighter balls, e.g. glass.

As we want to retain the favourable coarse screen and large amount of adhesive (fewer process steps and high shear strength due to extensive encapsulation of the ball), we addressed this issue by promoting evacuation of the 'glue' layer as the ball settles, in two ways: 1) using a lower-viscosity 'glue' composition, and 2) using a square or triangular geometry (figure 8), conserving the positioning function but allowing easier outflow of adhesive at the corners.

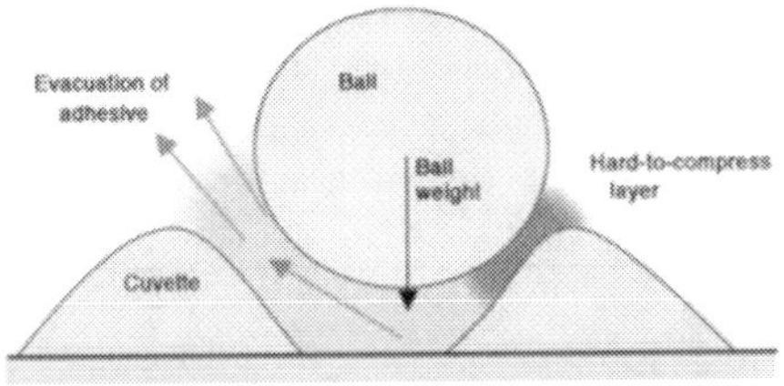

Figure 7. Illustration of impeded adhesive evacuation during settling of the ball.

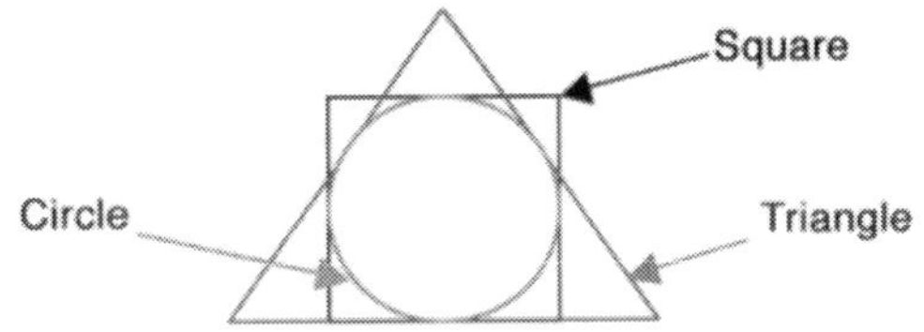

Figure 8. Triangle and square shapes, tested in addition to basic circle.

5.2. Neat H70S glue - effect of cuvette height and ball material for circular cuvettes

We first examine the approach of keeping a circular cuvette (two 354T-alumina layers, also with one for steel balls), but with a lower-viscosity H70S adhesive, and examining the effect of ball material. For steel balls, a single 'cuvette' layer was also tried.

The resulting shear strength values, shown in figure 9, all lie considerably above the specified 4 N minimum. The steel balls behave somewhat as in the first tests: increasing R_c yields higher strength for deep cuvettes (C2) and lower strength for shallower ones (C1), although the results remain acceptable here thanks to the thick deposits resulting from printing with the 135 mesh / 80 µm screen, and the better-defined cuvette shape resulting from using separate 'cuvette' and 'glue' formulations.

With two cuvette layers, the steel balls tend to yield the best strength values. It is tempting to ascribe this to their higher density, causing them to better sink into the wet 'glue' layer, achieving better embedding into the cuvette. While this may be a factor, there is no discernible effect of density on strength for the other materials (density: steel > zirconia > alumina > both glasses), and surface tension effects presumably also play a role. Interestingly, rupture tends to occur not at the ball – glue interface, but between the 'glue' and 'cuvette' layers, possibly meaning that the alumina loading of the latter may be somewhat too high. As expected, the low glue viscosity yielded good ball centring.

In spite of the apparently good results shown here, the tested configuration is not practicable, as H70S is insufficiently viscous at room temperature, and seeps through the coarse screen, requiring frequent cleaning: a somewhat more viscous resin is necessary, which may however reintroduce the issue of glue evacuation described in 5.1.

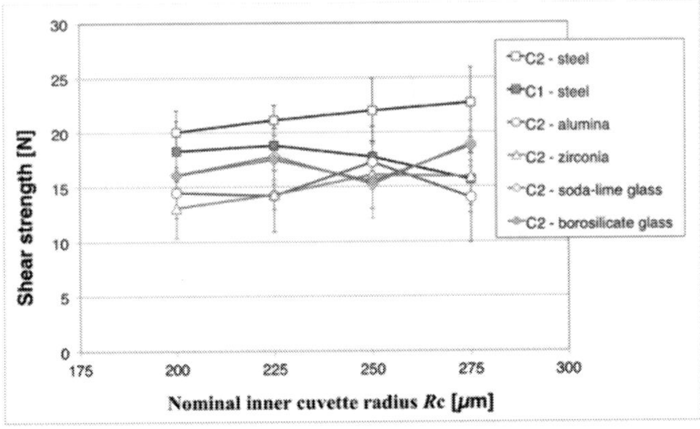

Figure 9. Shear strength for circular geometry (larger designs) vs. ball material, inner cuvette radius R_c and (for steel) number of cuvette layers (C1 = 1, C2 = 2), for Epo-Tek 354T & 50% Al_2O_3 cuvette and H70S glue, both printed with 135 mesh / 80 µm screen. Values & error bars: averages ± standard deviation; lines shown to guide the eye.

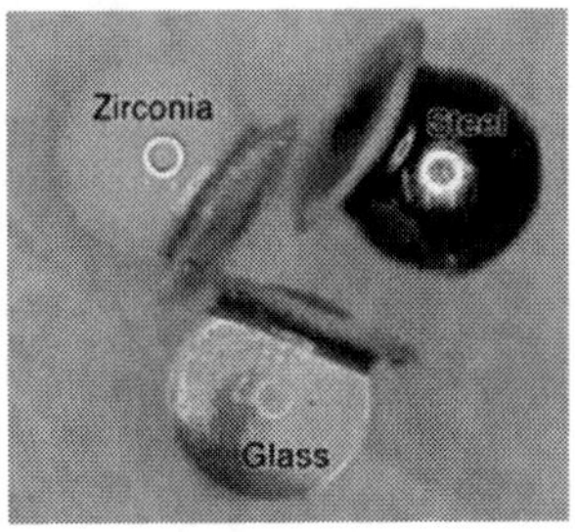

Figure 10. Shorn-off balls, showing rupture mostly between 'cuvette' and 'glue' layer.

5.3. Optimisation of 'glue' layer with 135 mesh / 80 μm screen and square geometry

To ensure good evacuation of adhesive (as we seek a somewhat more viscous one than H70S), we decided to carry out the next tests with triangular and square cuvettes (and glue) printed layers (figure 8), choosing R_c = 250 μm, one of the larger values. While first tests [14] showed favourable centring and embedding behaviour for both triangles and squares, the square form was preferred in terms of visual centring and screen printing (as the screen mesh consists of two sets of wires at a 90° angle to each other, all lines are equivalent for a square, but not for a triangle).

We compared here six 'glue' compositions, with 2 'cuvette' layers of 354T & 50% Al_2O_3, as before, and using soda-lime glass (both smooth and matte finish) and steel balls: the original H70S (1), H70E (2), H70E2 (3), and H70S mixed with 1% (4), 3% (5) and 6% (6) C18E0. C18E0 was mixed in the molten state (~60-70°C) into the resin component of H70S, and the beaker was cooled with water while mixing; this results in a fine suspension of C18E0, enhancing the viscosity at room temperature (i.e. C18E0 acts as a filler) but enhancing flow during heating, as C18E0 then melts and rather acts as a solvent, in the same vein as our previous work with C18E0 and other such waxy materials [17, 18].

Shear strength was measured both in the cured state (10 min at 150°C), and after further heat treatment, 5 and 15 min at 260°C, which considerably exceeds the actual time of further soldering operations (<1 min at ~260°C). Tests were also made to simulate cleaning operations: 30 min ultrasound bath, immersed in solvents (universal thinner, solder flux cleaner and isopropanol).

The results (figure 11) show high strength for all tested compositions, except H70S loaded at 3% and 6% with C18E0 together with the glass balls, which rupture by separation with the ball: C18E0 presumably migrates out of the resin and forms a low-strength layer there. This issue was not seen with 1%, which nevertheless sufficiently increased the viscosity of H70S to make it printable. Heat treatment was not deleterious; on the contrary, strength levels rather tended to increase by further curing of the resins. Immersion in solvents [14] (results not shown here) had no visible effect.

Based on these results, we chose the commercial adhesive with the best printability, H70E2.

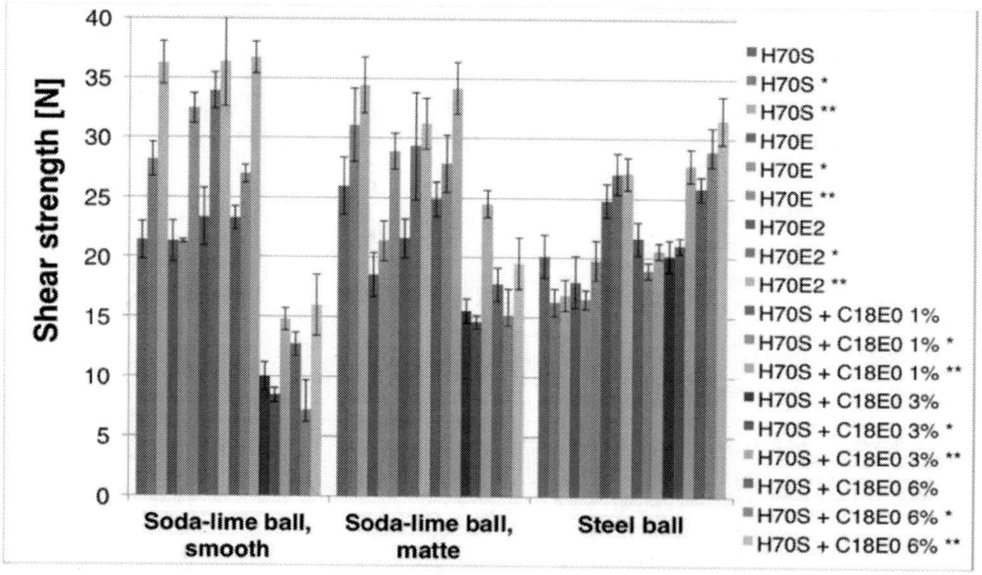

Figure 11. Shear strength (average ± standard deviation) for square geometry (R_c = 250 μm) vs. ball material, 'glue' formulation and heat treatment: initially cured 10 min at 150°C, then treated 5 min (*) or 15 min (**) at 260°C. Cuvette = 2× 354T & 50% Al_2O_3; printing with 135 mesh / 80 μm screen.

5.4. Further optimisation of 'cuvette' layer with 135 mesh / 80 μm screen and square geometry

As the previously-retained 'cuvette' composition (354T & 50% Al_2O_3) tended to form somewhat irregular shapes with the coarse 135 mesh / 80 μm screen, a finer evaluation of filler level was made: 40%, 45%, 50% and 55% Al_2O_3. As seen in figure 12, such formulations result in bleeding (low filler level), or in insufficient smoothing out of the screen irregularities (high level), with 50% turning out to be the best compromise, as before; therefore, a more advanced formulation was developed, with 41% Al_2O_3 in conjunction with 1.7% EC-300-48 and 3% amyl acetate added into the resin component of 354T (incorporated as a solution in acetone, which evaporates during mixing). EC-300-48 considerably increases the room-temperature viscosity, which is partly counteracted by the amyl acetate solvent. Upon heating, amyl acetate evaporates, leaving EC-300-48 only and thus restricting bleeding; this clearly results in the best print quality, as shown in the last image of figure 12. This optimised 'cuvette' formulation did not significantly impact the bond shear strength, as the results of a confirmation test show in figure 13.

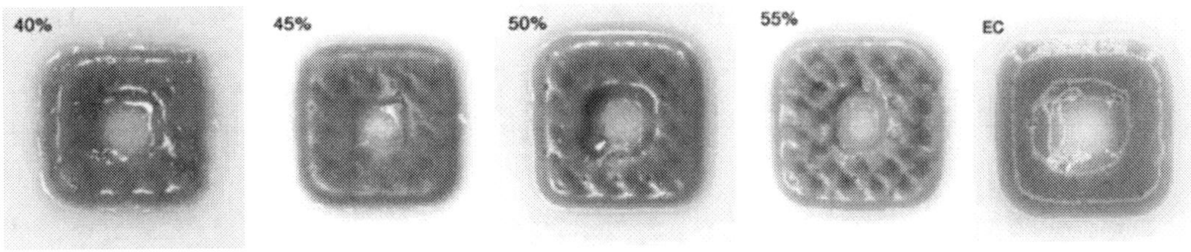

Figure 12. Cuvette formation (one layer printed with 135 mesh / 80 μm screen) of Epo-Tek 354T, loaded with various amounts (percentage of overall mass) of Al_2O_3 powder, after curing for 10 min at 150°C; "EC" = with 41% Al_2O_3 and 1.7% EC-300-48.

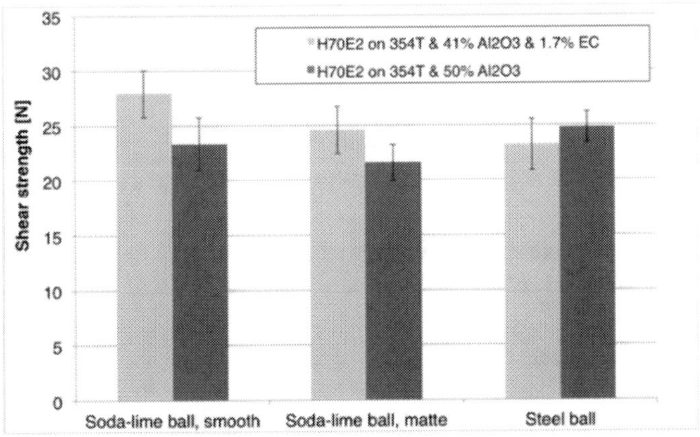

Figure 13. Shear strength (same conditions as figure 11) for as-cured H70E2 'glue' layer, comparing original 'cuvette' formulation (354T with 50% Al_2O_3 filler) with optimised one (354T with 41% Al_2O_3 and 1.7% EC-300-48, with additional 3% amyl acetate that evaporates during curing).

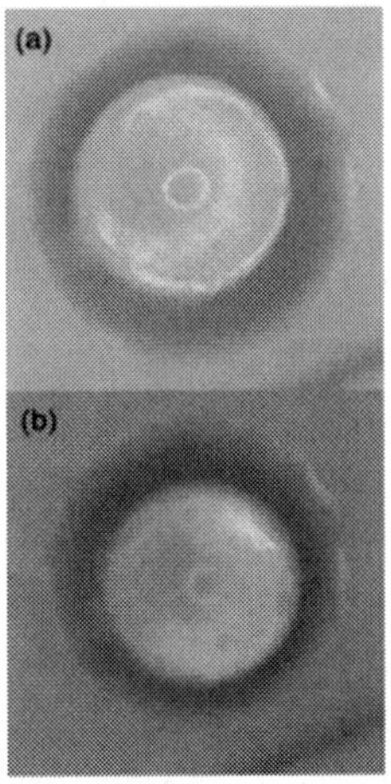

Figure 14. Glass balls after compression test at 30 N (a) and 50 N (b), with no visible damage (white circles in centre due to reflections).

6. Compression of glass balls

Finally, a compression test was carried out on the soda-lime glass balls against a relatively hard tool steel surface. The glass balls were adhesively bonded onto an alumina substrate, which was held by the tester vice with several backing substrates to enhance its strength.

In a single cycle, no visible damage to the ball was observed up to ~50 N, i.e. 25× the maximal force range of the sensor (2 N), as attested by figure 14. At this point, the pile of substrates backing the ball broke, ending the test. Performing 5 cycles with 15 N [14] also did not visibly damage the ball. Therefore, the low-cost glass balls are deemed acceptable for our application.

7. Conclusions and outlook

A simple and straightforward adhesive bonding process has been developed to replace solder attachment, and shown to yield acceptable shear strength levels. Compared to soldering of the ball, adhesive bonding entails significantly simpler processing. Additionally, the bottom cantilever mounting pads experience two vs. three reflow steps, i.e. only those necessary for 1) pre-tinning and 2) solder attachment, reducing solder oxidation and metallisation leaching issues, translating into lower fabrications costs and higher process reliability.

The process consists in two printing steps of two different epoxy-based compounds, in order to duplicate the twin functionality of the solder: centring of the ball (by a first 'cuvette' formulation) and actual attachment (with a 'glue' one). The resulting penalty is relatively small, however, as they are performed with the same screen, requiring only summary cleaning and flushing of the screen when switching from one process to the other. Other alternatives (e.g. precise pick-and-place of the ball with a UV glue) could also be used.

Acknowledgments

The authors are indebted to M. Garcin for help in carrying out the experiments.

References

[1] Morten B and Prudenziati M 1994 Piezoresistive thick-film sensors *Handbook of Sensors and Actuators vol. 1: Thick film sensors* ed M Prudenziati (Amsterdam: Elsevier) pp 189-208
[2] White NM and Turner 1997 Thick-film sensors: past, present and future *Measurement Science and Technology* **8** 1-20
[3] White NM 2007 Advances in thick-film sensors *Proc. 14th Int. Conf. on Solid-State Sensors, Actuators and Microsystems - Transducers / Eurosensors'07* (Lyon, France) (Piscataway: IEEE) pp 107-11
[4] Partsch U, Lenz C, Ziesche S, Lohrberg C, Neubert H and Maeder-T 2012 LTCC-based sensors for mechanical quantities *Informacije MIDEM - Journal of Microelectronics, Electronic Components and Materials* **42** 260-71
[5] Maeder T 2011 Ceramic modules for sensors, fluidics & packages in harsh environments *MEMS Technology Review* **8** 10-3
[6] Maeder T, Fahrny V, Stauss S, Corradini G and Ryser P 2005 Design and characterisation of low-cost thick-film piezoresistive force sensors for the 100 mN to 100 N range *Proc. XXIX Int. Conf. of IMAPS Poland* (Koszalin) pp 429-34
[7] Maeder T, Jacq C and Ryser P 2014 Thick-film load-sensing bridges – effect of temperature and mechanical boundary conditions *Procedia Engineering* **87** 180-3
[8] Jurków D, Maeder T, Dąbrowski A, Santo Zarnik M, Belavič D, Bartsch H and Müller J 2015 Overview on low temperature co-fired ceramic sensors *Sensors and Actuators A* **233** 125-46
[9] Birol H, Maeder T, Nadzeyka I, Boers M and Ryser P 2007 Fabrication of a millinewton force sensor using low temperature co-fired ceramic (LTCC) technology *Sensors and Actuators A* **134** 334-8
[10] Jacq C, Maeder T and Ryser P 2011 Signal stability of LTCC cantilever force sensors *Procedia*

Engineering **25** 551-4

[11] Jacq C, Maeder T, Haemmerle E, Craquelin N and Ryser P 2011 Ultra-low pressure sensor for neonatal resuscitator *Sensors and Actuators A* **172** 135-9

[12] Nousiainen O, Kangasvieri T, Rönkä K, Rautioaho R and Vähäkangas J 2007 Interfacial reactions between Sn-based solders and AgPt thick film metallizations on LTCC *Soldering & Surface Mount Technology* **19** 15-25

[13] Achmatowicz S and Zwierkowska E 2006 Lead free thick film circuits *Materiały Elektroniczne* **34** 5-47

[14] Blot M 2014 Collage de billes sur le capteur de force MilliNewton [Adhesive bonding of balls on the MilliNewton force sensor] *Student project in microtechnology, EPFL, Laboratoire de Production Microtechnique, Lausanne, Switzerland*

[15] Sergent JE 2007 Screen printing, *Ceramic interconnect technology handbook*, ed FD Barlow A Elshabini (Boca Raton, USA: CRC Press), Chapter 5, pp 199-233

[16] Miller LF 1969 Paste transfer in the screening process, *Solid State Technology* **12** 46-52

[17] Maeder T, Jacq C, Ammon L and Ryser P 2014 Tuneable PTC effect in polymer-wax-carbon composite resistors", Microelectronics International 31 (3), 143-148, 2014.

[18] Maeder T, Jiang B, Vecchio F, Jacq C, Ryser P and Muralt P 2012 Lamination of LTCC at low pressure and moderate temperature using screen-printed adhesives *Proc., 8th Int. Conf. on Ceramic Interconnect and Ceramic Microsystems Technologies (CICMT)* (Erfurt, Germany) pp 348-352

IOP Conf. Series: Materials Science and Engineering **104** (2015) 012007 doi:10.1088/1757-899X/104/1/012007

Challenges in packaging of IR detectors – technology of elastic electrical connections

Marcin Myśliwiec, Arkadiusz Lewandowski, Wojciech Wiatr, Jerzy Weremczuk, Zbigniew Szczepański, Ryszard Kisiel

Warsaw University of Technology, Institute of Microelectronics and Optoelectronics, Nowowiejska 15/19, 00-665 Warszawa, Poland

E-mail: kisiel@imio.pw.edu.pl

Abstract. This paper describes the study on flexible connections application in IR detector package. Such a connection is used between two modules in single package. One module is IR detector operating at -73 °C, other is electronic controlling circuit operating at room temperature. Flexible electrical connection must provide good electrical parameters in both DC and RF regime as well as poor thermal conductivity. 35 μm epoxy-phenolic foil with 5 μm Ni film and special RF compatible Ground-Signal-Ground geometry was successfully applied as connection. Low eutectic temperature InSn solder was used for soldering flexible connection to both IR detector and electronic circuit. Two types of soldering processes were investigated: bump and lap type. Flexible connection was tested for DC operation with long term stability test and RF parameters were determined. Authors found that both joining techniques are suitable for application in IR photodetector devices. However lap type joint is easier to apply in technological process, therefore it will be applied in final assembly.

1. Introduction

Middle and long wave IR photodetectors (MWIR, LWIR), such as those developed and produced by Polish company VIGO, have found important applications in IR spectrometry, quantum cascade laser based gas analyzers, laser radiation alerts and many other IR systems [1].

In such a device the control circuit operates at room temperature, while the IR detector has to operate at a temperature as low as -73 °C. The electrical connections between IR detector and control circuit are currently realized using 25 μm Au wires. The usage of Au wires causes excessive heat transfer to the IR detector and increases its operating temperature.

Figure 1. Concept of IR detector package with Ni microstrip line connection.

IOP Conf. Series: Materials Science and Engineering **104** (2015) 012007 doi:10.1088/1757-899X/104/1/012007

We propose a new solution of the electrical connection, which consists of: a Ni coplanar waveguide (5 µm thick) line on epoxy-phenolic substrate (Figure 1) and two solder joints: one between IR detector and coplanar waveguide and second between coplanar waveguide and electronic circuit. Ni line has worse thermal conductivity than Au (which reduce the heat transfer) and sufficiently good electrical properties. We designed the line in Ground-Signal-Ground (G-S-G) geometry and we calculated its dimensions for microwave operation. The construction of IR detector determine the maximum soldering temperature to 120 °C. The second reason for limiting soldering temperature is epoxy-phenolic foil on which Ni path are attached. Vitrification temperature of this foil is 130 °C, that limits soldering temperature.
To prove above concept Ni coplanar line in G-S-G geometry (250 µm – 100 µm – 250 µm) were prepared. In addition we made special test sample holders for Ni foil coplanar line from FR-4 laminate with Ni/Au paths and corresponding geometry. We made solder joints in two different ways. One way was called lap type. In this connection surface of FR-4 laminate and Ni coplanar line was firstly covered by flux and then wetted by InSn solder. The other was called bump type. In that connection after covering both surfaces with flux, 100 µm bumps made form InSn solder were applied between Ni coplanar line and the FR-4 test holder. After soldering process (all contacts are soldered in single step), we performed electrical measurements of DC and RF up to 3 GHz. The aim of this measurements was to evaluate which joining technique is more suitable for connection of IR detector and electronic circuit.

2. Experimental details

In Figure 2a we show geometry of design of coplanar line designed to operate at GHz frequences regime. For single step soldering of all Ni coplanar line connections we designed and manufactured special termode (Figure 2b). This tool is equipped with heater and thermocouple. To achieve optimal soldering conditions we performed calibration of termode. We determined that for eutectic temperature of InSn solder (118 °C) heater should be set to 140 °C. For reducing soldering time, temperature at heater should be slightly increased.

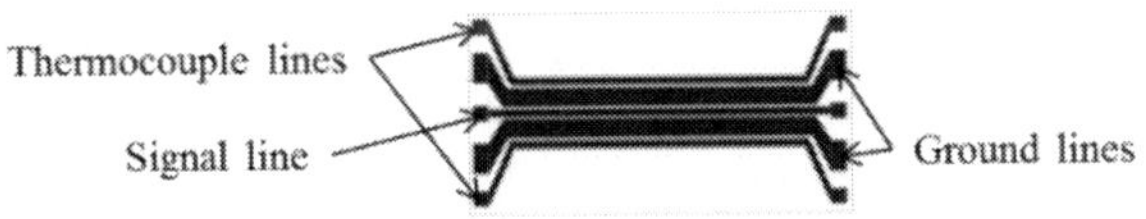

(a)

(b)

Figure 2. (a) Ni coplanar line design pattern, (b) thermode for multiple contact formation.

We prepared two types of solder joints: lap type and bump type. We processed contacts for lap type solder joint in following steps:

1. application of acetic acid based flux on Ni coplanar line contacts,
2. wetting, with miniature soldering iron with precise temperature control, Ni contacts of coplanar lines. Temperature of soldering tip was 170 °C,
3. identical steps for FR-4 laminate contact pads,
4. clean of both surfaces from flux residue.

Than no-clean low solid flux was put between soldered contacts. Soldering process parameters were as follows: temperature of FR-4 test substrate: 50 °C, temperature of termode: 145 °C, applied pressure 1.2 N, soldering time 15 seconds.
Preparation steps For bump type solder joint:

1. application of colophony based flux on Ni coplanar line,
2. placement of 100 µm InSn bumps on Ni coplanar line contact pads,
3. preheating at 150 °C for bump joint formation.

Again no-clean low solid flux was applied between bumps and FR-4 contact pads. Soldering process parameters were as follows: substrate temperature 100 °C, termode temperature 140 °C, applied pressure 0.5 N and soldering time 15 seconds.

3. DC measurements

We measured the DC electrical parameters of both joints types in temperature range from room temperature down to -35 °C. Additionally we tested stability of those parameters after storage at low temperature (-35 °C). Results of resistance dependence on temperature are presented in Figure 3. Changes of resistance versus temperature for both connection types are the same. Resistance of lap type joint is slightly higher. This might be result of differences in contact area of both types of solder joints. In bump joint area of connection is connected with volume 100 μm InSn solder bump. In lap type joint area of connection is result of contact pads wetting distance. This distance was limited to Ni coplanar line pads. Therefore area of contact should be smaller in lap type joint.

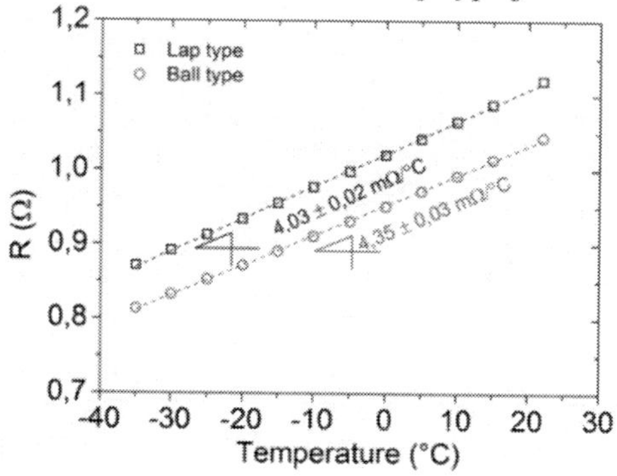

Figure 3. Results of DC characterization. Changes in resistance in function of temperature.

4. RF measurements

In order to characterize the interconnects in an environment similar to the one in which they will operate in the IR detector, we mounted them onto 1.5 mm thick FR-4 frames with the same CPW pattern as in the detector. We further cut a rectangular hole in the FR-4 frame underneath the interconnects in order to separate the CPW strips from the ground plane. The FR-4 frame was also put on a an additional 5 mm thick plexiglass frame, also with a rectangular hole, in order to better emulate the operation of the interconnects in the free space.

We characterized RF properties of the above structures in the frequency range 0.01-3 GHz with a vector network analyzer (VNA). The VNA was calibrated with the multiline through-reflect-line TRL calibration [2] which employed a custom designed (see [3]) set of six transmission lines, an additional thru connection, and an offset short as a reflect standards. The calibration standards were manufactured on a 1.5 mm thick FR-4 substrate placed on a 5 mm thick plexiglass spacer with the same CPW pattern as in the IR detector.

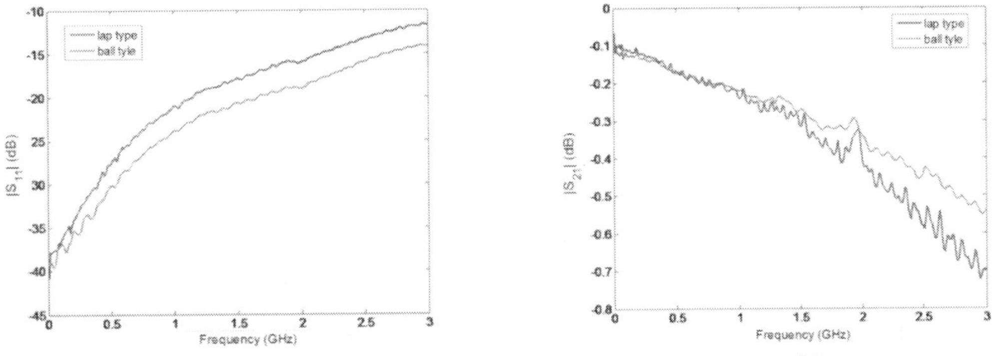

(a) (b)

Figure 4. Scattering parameters of the flexible interconnects: magnitude of (a) S_{11} and (b) S_{21}.

Measurement results are shown in Figure 4, were we plot the magnitude of S_{11} and S_{21}. We see that for both types of assembly technologies (lap type and bump type), the interconnects are well matched up to 1 GHz (return loss larger than 20 dB), and have insertion loss smaller than 0.1 dB. At frequencies larger than 3 GHz, the interconnect assembled with the bump technology presents slightly better matching and also smaller insertion loss. This can be explained with the fact that the current flow in the bump interconnect takes place on the surface of a metallic ball, which has larger conductivity than the lap type joint.

5. Conclusions

Both joining technologies fulfill demands of operation at DC and RF regime. Differences in RF and DC operation are really small and can be neglected. From technological point of view it is important to have assembly technology as simple as possible to reduce cost of single device. In our case more simple but only slightly worse technology is lap type solder joint. Taking all pros and cons we decide to apply lap in final device of IR photodetector.

6. References

[1] VIGO Systems S.A. webpage www.vigo.com.pl
[2] Roger Marks, 1991, A multiline method of network analyzer calibration *IEEE Trans. Microwave Theory Techn.* **vol. 39** pp. 1205–1215.
[3] Aarkadiusz Lewandowski, Wojciech Wiatr, Leszek J Opalski, Rafał Biedrzycki, 2015 Accuracy and bandwidth optimization of the over-determined offset-short reflectometer calibration *IEEE Trans. Microwave Theory Techn.* **vol. 63** pp. 1076-1089.

Acknowledgement

The work was supported by The National Centre for Research and Development, NCBR, Poland, project PBS2/B3/20/2013, INTIR.

Low frequency noise spectroscopy of high operating temperature HgCdTe infrared detectors

Ł Ciura[1], A Kolek[1], W Gawron[2], D Stanaszek[3] and D Stępień[2]

[1] Department of Electronics Fundamentals, Rzeszow University of Technology, 35-959 Rzeszow, Poland
[2] Institute of Applied Physics, Military University of Technology, 00-908 Warsaw, Poland
[3] VIGO System SA, 129/133 Poznańska St., 05-850 Ożarów Mazowiecki, Poland

E-mail: lciura@prz.edu.pl

Abstract. Low frequency noise spectroscopy (LFNS) is a tool, complementary to other methods, such as deep level transient spectroscopy, to characterize traps in semiconductor devices. LFNS method was described, illustrated and applied to HgCdTe high operating temperature infrared detectors. The fluctuation model which connects generation-recombination process (Lorentzians) with trap level in doped semiconductor was briefly introduced. Several trap levels were obtained for detectors with different band gaps (E_g in the range 219 meV to 320 meV at 300 K). One of the trap levels $E_t \approx 140$ meV, is common for almost all examined specimens. This level, which lies in the middle of the band gap, is the main source of generation–recombination noise observed in examined HgCdTe HOT detectors.

1. Introduction

Low frequency noise in various radiation detectors is a key issue since it introduces detectivity limitation of a biased detector. Low frequency noise could have either $1/f$ or Lorentzian spectrum connected with electrically active defects acted as generation-recombination (g-r) centres (traps). Low frequency noise spectroscopy (LFNS) can be used to identify the traps that contribute to the low frequency limitation of device performance. Moreover, extracted trap parameters can be compared to the values obtained by deep level transient spectroscopy (DLTS), impedance spectroscopy (IS) or optical methods, such as photoluminescence.

LFNS has been successfully applied for studying trap levels in Au/Ni-doped nanowires [1], [2] Ge JFETs [3]. Several publications compare results between DLTS and LFNS [4] [5], [6], [7]. It was shown that extracted trap levels are consistent in both methods. LFNS has some advantages over DLTS, since the latter requires high excitation signal (reverse voltage bias), proper samples with depletion region and with small leakage currents. It is also possible with LFNS to roughly estimate the trap density, when dopant concentration and sample volume are known [8]. This problem vanishes in LFNS because current fluctuations are the source of the information about traps in contrast to DLTS, where capacitance transients are analysed. Nevertheless, there are also some disadvantages because only the traps located near Fermi level can produce g-r noise. It is also necessary to take into account that not all g-r noise comes from the bulk. The g-r noise could be surface [9] or contact layers [10] related. LFNS requires physical model to interpretation of the location and density of the traps. This leads to the conclusion that LFNS should be considered as a complementary tool to other methods of semiconductor trap characterization [11].

In this work, LFNS is described and applied to high operating temperature (HOT) HgCdTe midwavelength infrared detectors (MWIR) with various band gaps of the absorber. Mercury cadmium telluride is a still basic material for infrared detectors [12]. The trap levels were extracted according to Kleinpenning *et al.* [13] model which is briefly introduced.

2. Samples and Method

Detectors were grown by Metal Organic Chemical Vapour Deposition on GaAs substrate with CdTe buffer in VIGO System S.A. laboratories. The basic architecture of specimens is presented in figure 1 [14]. The capital letters denote wide band gap material, while "+" sign means high doped region. Generally, the devices are $P^{+}pN^{+}$ diodes with lightly p-type narrow gap absorber and heavily doped/wide band gap hole P^{+} and electron N^{+} contacts. P-type layers are As doped while N-type layers are I doped. Additional n^{+} cap layer provides low resistance between metallization and hole contact. The surface is CdTe passivated. Detailed fabrication process was previously reported [15]. Several detectors with slightly different thickness and composition of particular layer were examined. The key difference is the absorber composition which equals $x = 0.321$ (specimens #3, # 8), 0.304 (specimen #2), 0.291 (specimen #4), and 0.247 (specimens #1 and #6). It corresponds with the band gaps (estimated at 300 K) $E_g = 320$ meV, 296 meV, 280 meV, and 219 meV, respectively.

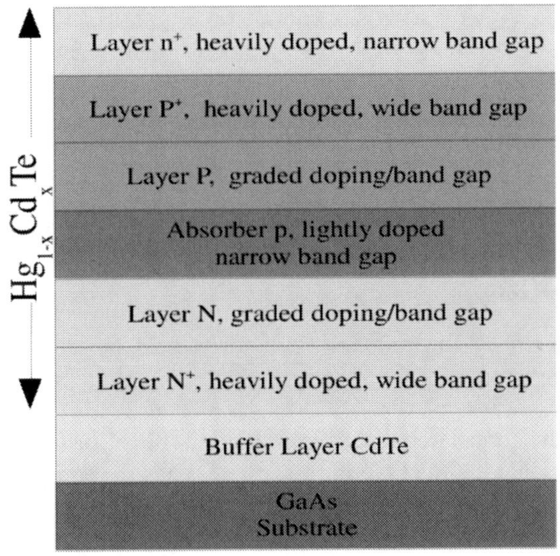

Figure 1. The basic architecture of HOT IR detectors described in the paper.

LFNS relies on examination of the temperature dependence of a corner frequency f_c which characterizes a Lorentzian spectrum $S_I = A[f_c(1+f/f_c)]^{-2}$, contributed by g-r noise, where constant A is connected with noise amplitude. The Lorentzian inclusions appear in the power spectral density of the detector's noise, S_I, for biased detector. The total measured noise usually includes g-r noise, $1/f$ noise and white noise (shot, thermal noise or system noise). Then, the product $f \cdot S_I$ could be written in the form

$$f \cdot S_I = fA[f_c(1+f/f_c)]^{-2} + B + f \times C \quad (1)$$

where B is the intensity of $1/f$ noise at 1 Hz and C is the white noise therm. Equation (1) is depicted in figure 2 (a), where the frequency dependence of the particular contributions is shown. It is convenient to analyse the product $f{\cdot}S_i$, because the local maxima of the Lorentzians appear at frequency $f = f_c$ on the flat noise background constituted by $1/f$ noise. The corner frequency follows a thermally activated manner

$$f_c = f_0 \exp(-E_A / kT) \tag{2}$$

where E_A is an activation energy of the g-r process and f_0 is a constant. The activation energy can be easily extracted from a set of plots, $f{\cdot}S_i(T)$, because the local maximum changes the position with temperature. This is illustrated in figure 2 (b), where measured products, $f{\cdot}S_i(T)$, are shown at different temperatures for biased specimen MCT #8. It is clearly to see that the local maximum, which initially appears at $f = 10$ Hz ($T = 145{,}8$ K), changes the position with temperature.

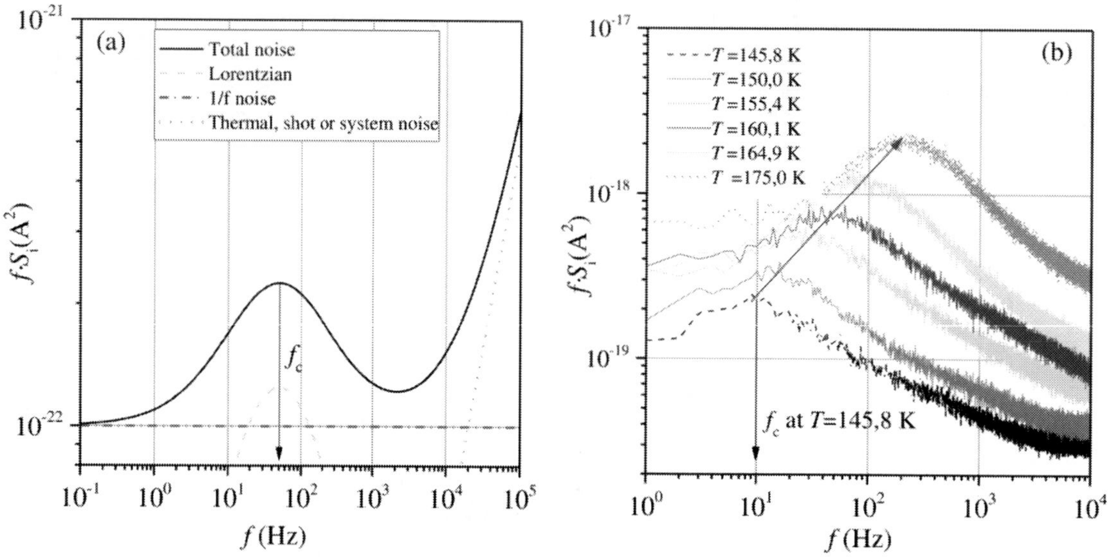

Figure 2. (a) The sketch of the product $f{\cdot}S_i$ with typical noise contributions. (b) The measured products $f{\cdot}S_i$ versus temperature for the specimen #8 at -50 mV.

The Arrhenius plot of the corner frequency from figure 2 (b) is illustrated in figure 3. Fitting procedure with equation (2) shows that the fluctuation (g-r) process has the activation energy $E_A = 235$ meV.

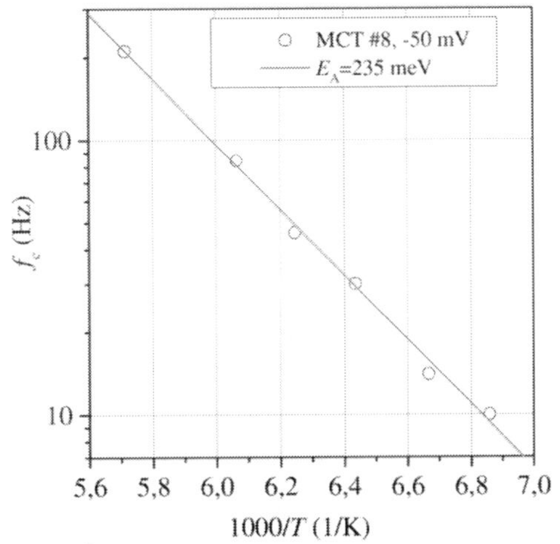

Figure 3. The Arrhenius plot of the corner frequency from figure 2 (b). The solid line indicates the activation energy $E_A = 235$ meV.

3. Results and Discussion

The connection of activation energies with a trap level requires a model of g-r process, which occurs in semiconductor material. A few authors developed such a model which helps to interpret the LFNS results in the context of trap parameters such as a trap level E_t, a capture cross-section σ or a trap the density N_t. Kleinpenning *et al.* [13] considered g-r noise in n-type GaAs layers. For electron trap level, where density of free electron n is larger than the density of trapped electron n_t and the Fermi level E_f lies below E_t by a few kT ($|E_t - E_f| >> kT$ – energy is measured up from valence band), the corner frequency can be expressed as [13]:

$$f_c = \frac{1}{2\pi} v_{th} \sigma N_c \exp(-|E_t|/kT) \tag{3}$$

where v_{th} is the thermal velocity, N_c is the effective density of states in conduction band and σ is the capture cross section. Scholz *et al.* developed similar expression [5] which was successfully adopted for GaN MOSFETs studies [7]. The temperature dependence of $v_{th}\sigma N_c$ should be considered. Typically, T-dependence of $v_{th}N_c$ is T^2 [11]. The temperature dependence of a capture cross section is a separate problem, because it could also follow a thermally activated manner $\sigma=\sigma_0\exp(-E_\sigma/kT)$. This kind of behaviour was observed for DX centres in AlGaAs layers [6]. Using LFNS, it is not possible to obtain the activation energies E_σ and E_t separately, contrarily to DLTS, where E_σ could be extracted. The assumption about position of Fermi level with the respect of the trap level could be satisfied due to the existence of band bending in the deletion layer even in high doped material. It is postulated by Kleinpenning *et al.* [11] who observed g-r noise in high doped n-GaAs where Fermi level lies well above trap level.

LFNS was performed for several HOT IR HgCdTe detectors in the temperature range 65 K - 300 K. The values of activation energies and trap levels, obtained for fluctuations processes in particular specimens, are gathered in Table 1. It is assumed that in lightly p-doped absorber the hole trap plays a major role. Consequently, equation (3) for the hole trap, can be rewritten in the form $f_c = 1/2\pi v_{th}\sigma N_v \exp(-|E_t|/kT)$, where N_v is effective density of states in valence band. The trap level was obtained using above equation and assuming that T – dependence of thermal velocity and effective density of states in valence band is T^2.

Table 1. The activation energies and trap energies of fluctuation processes obtained by LFNS for HOT HgCdTe detectors.

Specimen	Activation energy [meV]	Trap level [meV]
MCT#1	30, 110	15, 90
MCT#6	155, 180	140, 165
MCT#2	80, 155	60, 135
MCT#4	140	130
MCT#3	155	140
MCT#8	40, 110, 155, 235	35, 90, 140, 210

One can notice that the trap level $E_t \approx 140$ meV is detected for almost all detectors specimens including devices with the narrow (specimen #6) and the wide band gap absorber (specimens #2, #3, #4, #8). This level is responsible for g-r noise and is common in our HgCdTe specimens. Similar trap level was reported by Hu *et al.* [16] with the use of of admittance spectroscopy.

4. Conclusions

Low frequency noise spectroscopy is a powerful method to characterize traps in semiconductor devices, complementary to optical and other electrical methods such as deep level transient spectroscopy or admittance spectroscopy. LFNS is sensitive and easy to apply but interpretation of the results could be a difficult task because the Lorentzian inclusions, appearing in the spectrum, could have various origins related to, e.g., surface and contacts. LFNS requires a model of generation-recombination processes occurring in semiconductor to extract the position of trap levels. However, LFNS results support DLTS results in many studies. LFNS performed on HOT HgCdTe detectors with various absorber band gap reveals a common trap level $E_t \approx 140$ meV (probably hole trap) which lies in the middle of the band gap.

5. References

[1] Sharma D, Motayed A, Krylyuk S, Li Q and Davydov A V 2013 Detection of deep-levels in doped silicon nanowires using low-frequency noise spectroscopy *IEEE Trans. Electron Devices* **60** 4206–12.
[2] Motayed A, Krylyuk S and Davydov A V 2011 Characterization of deep-levels in silicon nanowires by low-frequency noise spectroscopy *Appl. Phys. Lett.* **99** 5–8.
[3] Grassi V, Colombo C F and Camin D V 2001 Low frequency noise versus temperature spectroscopy of recently designed Ge JFETs *IEEE Trans. Electron Devices* **48** 2899–905.
[4] Loreck L, Dambkes H, Heime K, Ploog K and Weimann G 1984 Deep-level analysis in (AlGa)As–GaAs 2-D electron gas devices by means of low-frequency noise measurements *IEEE Electron Device Lett.* **5** 9–11.
[5] Scholz F, Hwang J M and Schroder D K 1988 Low frequency noise and DLTS as semiconductor device characterization tools *Solid. State. Electron.* **31** 205–17.
[6] Kirtley J R, Theis T N, Mooney P M and Wright S L 1988 Noise spectroscopy of deep level (DX) centers in GaAs-AlxGa 1-xAs heterostructures *J. Appl. Phys.* **63** 1541–8.
[7] Nguyen X S, Lin K, Fitzgerald E a., Chua S J, Zhang Z, McSkimming B, Arehart A R, Speck J S, Ringel S A, Fitzgerald E a. and Chua S J 2015 Correlation of a generation-recombination center with a deep level trap in GaN *Appl. Phys. Lett.* **106** 102101.

[8] Kunets V P, Morgan T Al, Mazur Y I, Dorogan V G, Lytvyn P M, Ware M E, Guzun D, Shultz J L and Salamo G J 2008 Deep traps in GaAs/InGaAs quantum wells and quantum dots, studied by noise spectroscopy *J. Appl. Phys.* **104.**

[9] Ivanov P A, Levinshtein M E, Palmour J W and Rumyantsev S L 2000 Noise spectroscopy of local surface levels in semiconductors *Semicond. Sci. Technol.* **15** 164–8.

[10] Vandamme L K J 1994 Noise as a diagnostic tool for quality and reliability of electronic devices *IEEE Trans. Electron Devices* **41.**

[11] Jones B K 1994 Low-frequency noise spectroscopy *IEEE Trans. Electron Devices* **41** 2188–97.

[12] Martyniuk P, Antoszewski J, Martyniuk M, Faraone L and Rogalski A 2014 New concepts in infrared photodetector designs *Appl. Phys. Rev.* **1** 041102.

[13] Kleinpenning T G M, Jarrix S and Lecoy G 1995 Generation-recombination noise in submicron semiconductor layers: Influence of the edges *J. Appl. Phys.* **78** 2883.

[14] Stanaszek D, Piotrowski J, Piotrowski a, Gawron W, Orman Z, Paliwoda R, Brudnowski M, Pawluczyk J, Pedzińska M, a V S S and Mazowiecki O 2009 Mid and long infrared detection modules for picosecond range measurements *System* vol 7482, ed G W Kamerman, O K Steinvall, K L Lewis, R C Hollins, T J Merlet, G J Bishop and J D Gonglewski (International Society for Optics and Photonics) 1–11.

[15] Piotrowski J, Pawluczyk J, Piotrowski A, Gawron W, Romanis M and Kłos K 2010 Uncooled MWIR and LWIR photodetectors in Poland *Opto-Electronics Rev.* **18**

[16] Hu X, Fang J, Wang Q, Zhao J, Lu H, Gong H, Zhang S and Lu F 1998 A deep level induced by gamma irradiation in $Hg_{1-x}Cd_xTe$ *Appl. Phys. Lett.* **73** 91.

Acknowledgement

The work was supported by the Polish research supporting agencies National Science Centre (NCN), under Project No. DEC-2014/13/N/ST7/03074 (Ł.C.), and The Polish Ministry of Science and Higher Education under Key Project No. POIG.01.03.01-14-016/08.

IOP Conf. Series: Materials Science and Engineering **104** (2015) 012009 doi:10.1088/1757-899X/104/1/012009

Integration of electronic system with electro-thermally cooled IR detector: thermal analysis

E Raj[1], Z Lisik[1], L Ruta[1], B Guzowski[1], P Kalinowski[2], Z Orman[2]

[1] Lodz University of Technology, Department of Semiconductor and Optoelectronic Devices, Wolczanska 211/215, 90-924 Lodz, Poland
[2] VIGO System S.A., Poznanska 129/133, 05-850 Ozarow Mazowiecki, Poland

E-mail: ewa.raj@p.lodz.pl

Abstract. The paper presents thermal investigations of an idea of encapsulation in common, miniature package both preamplifier circuit and infrared photodetector cooled with the aid of four-stage thermoelectric module. Conducted numerical simulations show that the presence of electronics at printed circuit board with either ceramic or laminate substrate negligibly affects operation of thermoelectric cooler and hence, the operating conditions of the detector. Higher thermal loads are reported when placing wideband electronic system on the additional thermoelectric module. Even though, such solution reduces temperature difference between infrared sensor and the other part of electronics, additional cooler increases the amount of heat that must be dissipated from the package to the ambient.

1. Introduction

The history of infrared (IR) photon detectors reaches almost 100 years [1, 2], nowadays. In this class the semiconductor IR detectors should be pointed out as the one with rapid development. They offer selective wavelength dependence of the response per unit incident radiation power, high signal-to-noise performance and a very fast response. Even though, recently, more and more attention is given to near room temperature applications [2], higher performance parameters are available only when the photodetectors are cooled down to 200 K or even below. Temperature range of 240 K to 190 K can be assured by miniature thermoelectric coolers (TEC) [3]. Application of TEC allows creating system operating in room temperature from the point of view of end user [4], although IR detector is being cooled.

Further improvement in performance parameters of the detection modules demanded by modern optoelectronic applications can be obtained by integration of infrared sensors with wideband electronic systems. On the one hand, placing the preamplifier in the close proximity of the detecting structure allows increasing operational frequency and reducing the noises [1, 3]. On the other hand, in the case of electro-thermally cooled photodetectors, heat dissipating electronic system creates an additional heat load for the detector and its thermo-electric module (TEM). Havier thermal conditions result from convection mechanisms developing easier within bigger space of the package dedicated for the integrated solution (temperature at highest stage of TEM can be of 15 K higher due to convection and radiation [5]) and from the wire connection between the detector and the electronic board mounted in the locations with 100 K temperature difference [6]. Especially, the latter factor can be a serious problem due to the fact that heat amount transported through electric short connections can easily exceed several mW and double or triple the heat losses of the IR detector. Hence, higher temperature at the highest stage of the TEM is obtained and it affects detector operation.

The paper presents thermal investigations of miniature package with mounted thermoelectric cooler for photodetector and preamplifying system at printed circuit board (PCB) located next to it. Different assembly configuration and different materials solution regarding PCB substrate have been examined together with additional TEM option. Moreover an influence of ambient temperature on the whole system operation is analysed and discussed.

2. Models of multi-stage thermoelectric coolers

To analyse thermal coupling of IR photodetector and preamplifying electronic circuit integrated in common miniature package, thermal model of multi-stage thermoelectric cooler (TEC) has been elaborated. Within the model, the thermo-electric behaviour has been replaced with thermal compact approach, already presented in [7] and by the authors in [5] for single and multi-stage solutions, respectively. The general idea here is to substitute the sets of semiconductor pellets with uniform domains with equivalent thermal conductivity and heat sources corresponding to heat absorption and dissipation due to Peltier and Joule heating effects. It simplifies using the model with computational fluid dynamics (CFD) code. These heat sources are defined as:

$$Q_{absorb.i} = 2 \cdot N_i \cdot T_{ji\_C} \cdot \alpha \cdot I \tag{1}$$

$$Q_{diss.i} = 2 \cdot N_i \cdot T_{ji\_H} \cdot \alpha \cdot I \tag{2}$$

$$Q_{J.i} = 2 \cdot N_i \cdot I^2 \cdot R_{eq} \tag{3}$$

where: $Q_{absorb.i}$ and $Q_{diss.i}$ denotes heat rates absorbed and dissipated at the junctions of TEM ith-stage, T_{ji_C} and T_{ji_H} – temperature at the cold and hot junction of TEM ith-stage, N_i – number of thermocouples at ith-stage, I – Peltier current, R_{eq} – equivalent resistance of semiconductor pellet and parts of copper connections.

As exemplary structures, two TECs of RMT Ltd. Company have been chosen [8]: 2MX04-046-0510t and 4MD04-116-10. First of them is a two-stage stage solution selected for electronic circuit; while the second one is a four-stage TEC dedicated for photodetector mounting. Performance parameters of both thermoelectric modules are gathered in table 1. As an example, figure 1 shows modelled geometry of the two-stage TEC 2MX04 046-0510t, which contains 15 and 31 thermocouples at the first and the second stage, respectively. For both stages the width and depth of the equivalent thermal domain are the same; even though, the number of pellets differs.

Table 1. Performance parameters of chosen thermoelectric coolers at 300 K in vacuum [8].

Margin	**2MX04-046-0510t**	**4MD04-116-10**
ΔT_{max} [K]	91	127
Q_{max} [W]	1.59	0.30
I_{max} [A]	1.1	0.5
U_{max} [V]	3.9	8.4
N [-]	15/31	4/11/31/70

where N is a number of thermocouples at each stage.

Figure 1. Model outline of 2MX04-046-0510t thermoelectric cooler.

The elaborated model has been verified by comparing the results of simulations with the data supported by TEC producer, which are available in TECcad software [9] in the form of temperature values at particular junctions of the cooler. Figure 2 presents curves obtained for Peltier current of 0.7 A and three different values of temperature at the hot side. The data from the manufacturer are plotted with the aid of markers that for better visibility are connected with the aid of dotted lines. In CFD 3D simulations, the line, along which the data have been analysed, is the TEC's axis of symmetry. Good agreement between the numerical and experimental data taken from TECcad software confirms correctness of the elaborated model.

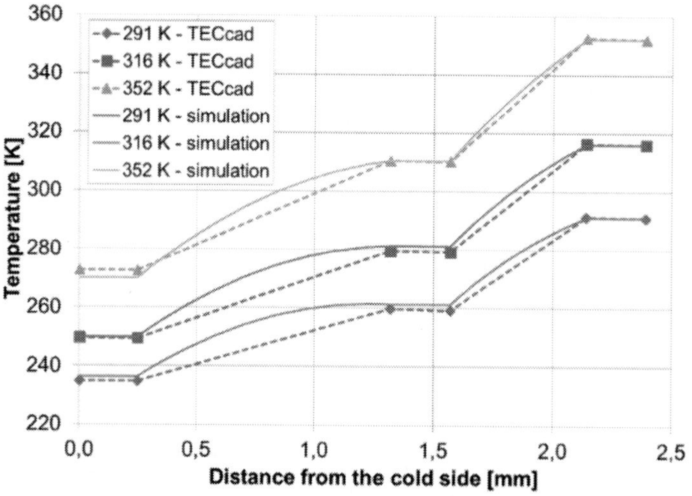

Figure 2. Comparison of temperature distribution within axis of 2MX04-046-0510t thermoelectric cooler with data from TECcad software [9].

3. Model of miniature package with integrated electronic circuit

Next, the elaborated model of four-stage TE cooler dedicated for IR detector has been integrated with electronic circuit within miniature FlatPack. A preamplifying system in a form of silicon die of the dimensions 0.7 x 1.2 x 0.25 mm is placed at printed circuit board (PCB) and located next to the photodetector TEC. Different solutions regarding PCB (assembly configurations) has been investigated:

- DCB (Direct Copper Bonding) – two side copper clad AlN ceramics where thicknesses of AlN and copper layers equal 0.63 mm and 0.3 mm respectively;
- Elevated DCB – the DCB board as above mounted at the copper dais, so the total thickness equals to 6.9 mm – the photodetector's TEC height;
- I-Tera – two side copper clad I-Tera laminate where thicknesses of isolator and copper layers equal 0.2 mm and 0.012 mm respectively;
- Additional TEM – additional two-stage TE cooler with AlN ceramics covered with copper conducting mosaic.

Geometry of an exemplary modelled structure is shown in figure 3.

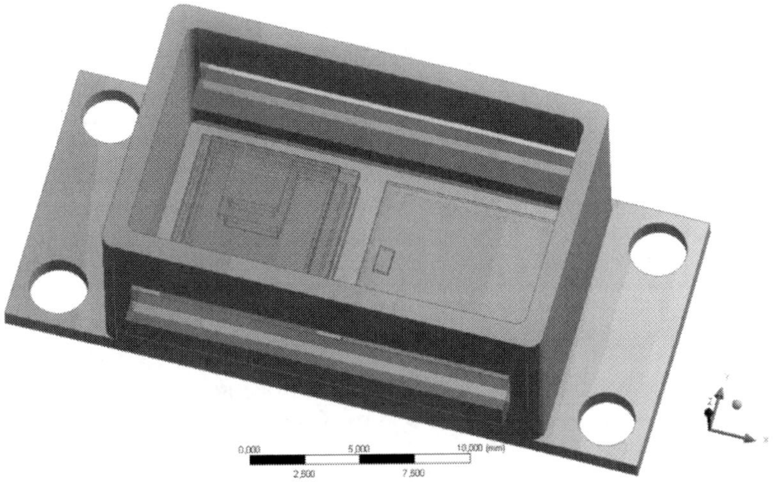

Figure 3. Miniature package with thermoelectric cooler dedicated for photodetector and PCB with preamplifier.

Simulations have been conducted with the aid of ANSYS CFX software for the following assumptions and boundary conditions:

- ambient temperature varies in the range 10 to 70°C;
- heat sink of thermal resistance 2 K/W is mounted at the bottom surface of the package;
- all other surfaces exchange heat with surrounding due to the convection with heat transfer coefficient of 10 W/m^2K;
- the package is filled with xenon gas;
- parameters of solid state materials used in modelled structure are gathered in table 2;
- gravity acceleration is directed opposite to the x-axis (the worst situation [5]);
- model of photodetector's TEC is elaborated for heat load at the highest stage of 10 mW and Peltier current of 0.45 A [5, 6];
- model of additional TEC is elaborated for heat load of 300 mW and Peltier current of 0.7 A;
- heat losses in preamplifier equals to 300 mW.

Exemplary temperature distributions at the internal surfaces (TEC and PCB), and velocity fields within the cross-sections of the modelled structures for three assembly configurations compared to situation without electronic circuit are presented in figure 4. The minimum temperature at the highest stage of the cooler is also marked in the figure. It can be noticed that the presence of the electronic circuit on any type of PCB almost does not affect the temperature distribution at the highest stage of TEC. Nevertheless, elevated DCB configuration influences velocity field within the package and slightly suppresses convection. The poorest conditions are achieved while using additional TEM for preamplifier

IOP Conf. Series: Materials Science and Engineering **104** (2015) 012009 doi:10.1088/1757-899X/104/1/012009

circuit due to higher heat load at the bottom of the package that affects operation of the detector's four-stage TEC. It should be emphasised that in this situation heat sink with lower thermal resistance is required to keep the temperature at the package at the same value as it was in configurations without additional TEM.

Table 2. Parameters of solid state materials used in modelled structure.

Material	Thermal conductivity [W/mK]	Specific heat [J/kgK]	Density [kg/m^3]	Emissivity [-]
tungsten/copper	190	163	16330	0.05
copper	401	380	8920	0.05
bismuth telluride	1,4	188	7850	0.36
silicon	150	710	2330	0.60
AlN ceramics	170	920	3300	0.90
I-Tera laminate	0.32	950	1850	0.90

The influence of the ambient temperature on thermal conditions within the area of detector location and maximum temperature of preamplifying silicon structure is presented in figure 5 and 6, respectively. We can see that both temperature at the highest stage of TE cooler, as well as temperature of electronic element increase with increase of ambient temperature. Curves in figure 5 confirms that from the point of view of thermal load on detector's TEC, all investigated configurations are comparable, despite of additional TEM case. Temperature of electronic element differs significantly only in the case of mounting preamplifier on the additional cooler. In the case of PCB with much bigger thermal resistance (9.77 K/W for laminate based solution and 0.08 K/W for ceramic based solution) like I-Tera laminate solution in comparison to DCB board, temperature of semiconductor element is only 2.5°C higher.

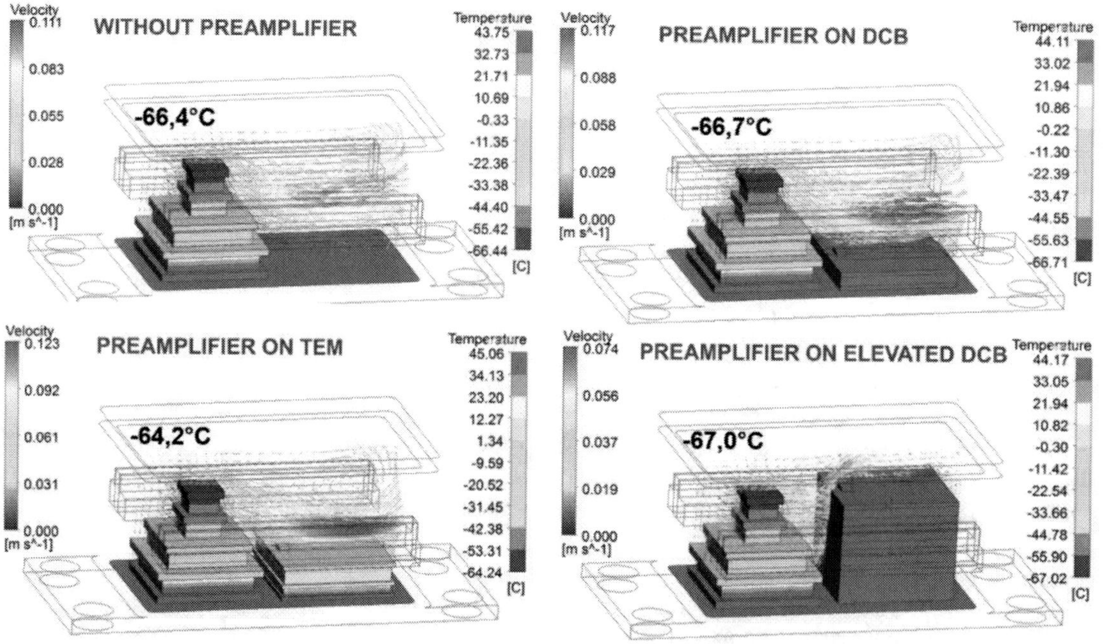

Figure 4. Influence of different assembly configuration on temperature and velocity fields in the modelled package.

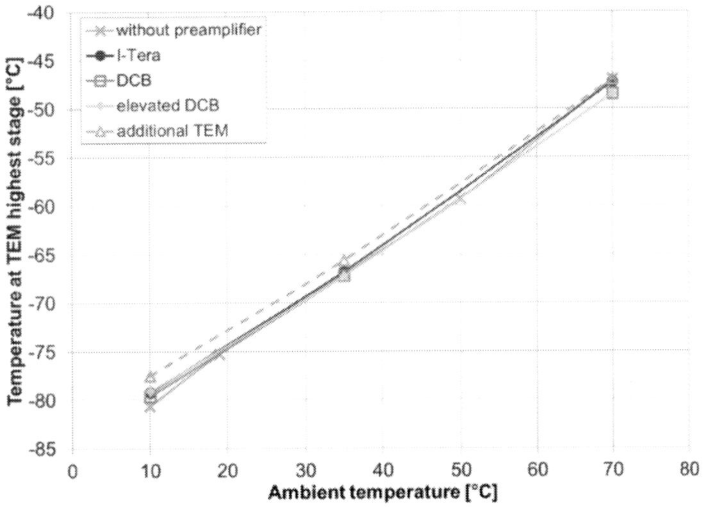

Figure 5. Temperature at TEM highest stage versus ambient temperature for different assembly configurations.

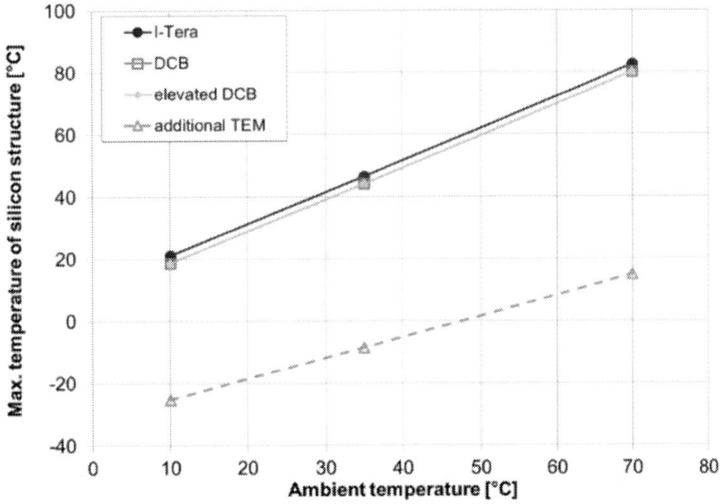

Figure 6. Temperature of silicon structure versus ambient temperature for different assembly configurations.

4. Conclusions

Encapsulation in common miniature package of both preamplifier circuit and IR photodetector is a beneficial solution due to high performance parameters that can be obtained. Unfortunately, thermal coupling between these two electronic parts exists. Heat can be transported through gaseous medium filling the package, through the constructional elements of the package and through the solid connection between the detector and the wideband electronic system. In the paper, the latter factor has been assumed to be constant, but two first heat transport paths have been investigated. Results of conducted simulations have proven that the electronic system placed next to the thermoelectrically cooled detector has only slight influence on the TEC operation. Comparable thermal operating conditions for IR semiconductor structure can be assured when using PCB with either ceramics or laminate substrate for electronic system. Furthermore, the reduction of free convection within the package can be assured by elevated PCB solution. Assembling electronic circuit on additional TEC, even though reduces the temperature difference between the detector and the remaining electronic part, generates additional heat loads and hence influences the operation of the detector cooler. Further investigation regarding thermal coupling through solid connections is demanded.

5. References

[1] Rogalski A 2002 *Infrared Physics & Technology* **43** 187-210
[2] Piotrowski J and Piotrowski A 2006 *Opto-Electronics Review* **14** 37-45
[3] 4-stage Cooling for Photodetector applications - new solution on standard TO-8 headers 2009 *TEC Microsystems* www.tec-microsystems.com [access 05/2015]
[4] Redus R H, Huber A C, Pantazis J A 2001 *Nuclear Instruments and Methods in Physics Research A* **458** 214-219
[5] Raj E, Lisik Z, Ruta L, Kalinowski P and Orman Z 2015 Numerical Analysis of Thermoelectrically Cooled IR Detector *Proc. Int. Conf. Microtechnology and Thermal Problems in Electronics* (23-25.06.2015 Lodz, Poland)
[6] Lisik Z, Raj E, Ruta L and Powierza J 2015 Flexible connection for high frequency signal transfer in integrated circuits *Proc. KKE* (8-12.06.2015 Darlowo, Poland) 741-746 [in Polish]
[7] Chen M and Snyder G J 2013 *Int. Journal Heat and Mass Transfer* **60** 689-699
[8] Thermoelectric cooling solutions, *catalogue of RMT Ltd.* www.rmtltd.ru [access 05.2015]
[9] *TECCad Lite* software delivered by RMT Ltd., www.rmtltd.ru [access 05/2015]

Acknowledgments
The work was supported by the National Centre for Research and Development according to the agreement no PBS2/B3/20/2013 within the Second Applied Research Programme.

IOP Conf. Series: Materials Science and Engineering **104** (2015) 012010 doi:10.1088/1757-899X/104/1/012010

Experimental approach to validation of an analytical and numerical thermal analysis of a travelling wave tube

W. Wiejak[1], A.Wymysłowski[2]

[1] PIT-RADWAR S.A., ul. Krakowska 64, 50-425 Wrocław, Poland
[2] Wroclaw University of Technology, ul. Janiszewskiego 11/17, 50-372 Wrocław, Poland

E-mail: artur.wymyslowski@pwr.edu.pl

Abstract. Travelling Wave Tube (TWT) is an electronic vacuum microwave device, which is used as a high power microwave amplifier, mainly in telecommunication purposes, e.g. radar systems. TWT's is an alternative solution in comparison to semiconductor devices in case of high power and high frequency applications. Thermal behaviour of TWT is one of the key aspects influencing its reliability and working parameters. The main goal of the research was to perform analytical, experimental and numerical analysis of a temperature distribution of a low band TWT in case of a typical working condition. Because the theoretical analysis seems to be very complex thus it was decided to compare the experimental results with the numerical simulations as well as with the simplified analytical formulas. As a first step of the presented research, the analytical analysis and numerical modelling of the helix TWT was carried out. The objective of the thermal analysis was to assess the temperature distribution in different parts of the helix TWT assembly during the extreme standard and working conditions. As a second stage of the research the numerical results were validated by the experimental measurements, which were carried out using a specially designed TWT test samples and corresponding experimental measurement tools.

1. Introduction

Travelling Wave Tube (TWT) is an electronic vacuum microwave device, which is used as a high power microwave amplifier, mainly in telecommunication purposes, as a driver or output tube in advanced radar systems and electronic test equipment. TWT seem to be an alternative and more reliable solution in comparison to semiconductor devices in case of high power and high frequency applications. The typical operating frequency range of TWT would be from 1 to 5GHz and each tube delivers a peak output power in frequency range without adjustment. There are a number of different construction of TWT devices of which the so-called helix TWT are the most popular. The main components of a helix TWT are: electron gun, helix slow wave structure (SWS), input and output coupler and collector. Additionally, a ceramic-metal construction provides exceptional mechanical strength. Electron beam is focused by periodic permanent magnet (PPM) structure.

Thermal behaviour of TWT is one of the key aspects influencing its reliability and working parameters. In fact the standard TWT is treated as a high power device and the supplied power is dissipated mainly through conduction and convection phenomena while the heat dissipation through the radiation can be neglected. In practical applications, depending on a type of TWT, they can be cooled either by a forced liquid or air circulation. Additionally, in some special TWT types, they would require mixed air-liquid cooling or specially designed heat conduction to the device base [1].

The main goal of the research was to perform experimental validation of the analytical and numerical analysis of a temperature distribution of a low band TWT in case of a typical working conditions. Because the theoretical analysis seems to be very complex thus it was decided to compare the experimental results with the numerical simulations as well as with the simplified analytical formulas. One of the advantages of numerical simulations over the analytical formulas is the ability to include real geometry and boundary conditions without a need of any physical phenomena simplification or neglecting any geometrical details. On the other hand the numerical analysis in comparison to experimental one, allows for application of an advanced numerical optimization algorithms, which makes it a very practical engineering tool during the prototyping stage. As a first step of the presented research, the analytical analysis and numerical modelling of the helix TWT was carried out. As a final stage of the research the achieved results were validated by the experimental measurements, which were carried out using a specially designed TWT test samples and corresponding experimental measurement setup [2].

2. Travelling Wave Tube

Traveling Wave Tubes (TWT) belong to high power microwave devices, which work in a density modulated mode, rather than the current modulated mode. This means that they work on the basis of clumps of flying electrons, rather than using a continuous stream of electrons. As the key driving source are electrons, thus part of TWT devices is based on the so-called electron optics and interaction of electrons with an electromagnetic field [3].

TWT devices containing helix delay structures seem to be the most popular. The main components of a helix TWT device are: electron gun, helix slow wave structure (SWS), input and output coupler and collector. Additionally, a ceramic-metal construction provides exceptional mechanical strength. Electron beam is focused by periodic permanent magnet structure. An example structure of such a device and a corresponding microwave amplifier are presented in the figure 1.

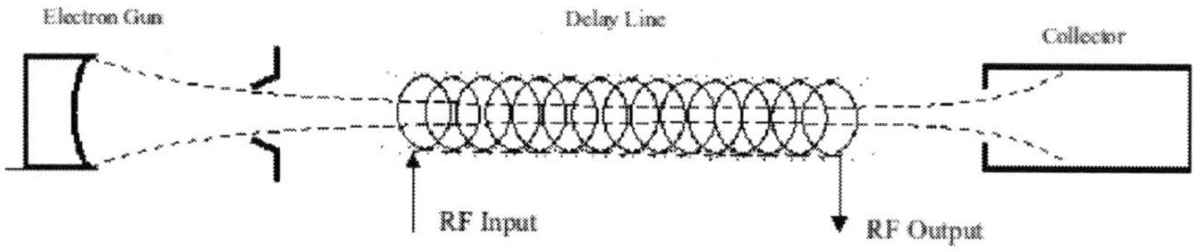

a.

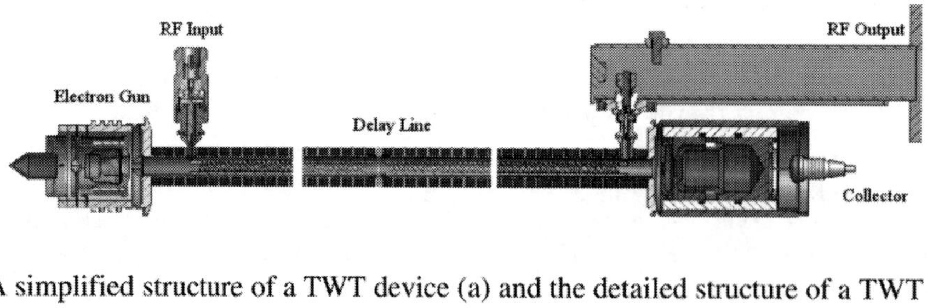

b.

Figure 1. A simplified structure of a TWT device (a) and the detailed structure of a TWT microwave amplifier (b).

IOP Conf. Series: Materials Science and Engineering **104** (2015) 012010 doi:10.1088/1757-899X/104/1/012010

The operation principle of TWT is based on an interaction of an electromagnetic wave, which is guided by the so-called delay line (e.g. helix), with the electron beam injected along its axis of a symmetry. The axial electromagnetic field component accelerates same electrons and at the same time decelerate others. As a result an electron beam density is modulated and the energy extracted from decelerating bunches of electrons is then transferred into the outer circuit. Microwave power generated in the delay line grows exponentially with a distance along its axis of symmetry. In fact the efficiency of the energy transfer of an electron beam into the energy of a microwave signal can be estimated up to 20% [5].

As stated above, one of the basic problems is a thermal behavior of TWT based devices, which directly influences its reliability parameters and performance. As the total energy should be conserved, in case of the analysed device the main sources of energy inputs and outputs can be detected as follows:

- input power of a microwave signal. e.g. 1 W - in fact could be neglected,
- output power of a microwave signal, e.g. 450 W,
- inlet power of an electron beam, e.g. 2350 W,
- outlet power of a collector, e.g. 800 W,
- dissipation power in the delay line, e.g. 70 W,
- other energy losses in the delay line, e.g. 80 W,
- recuperation of the input output power into the input power, e.g. 950 W.

An example of a power rough equilibrium estimation and a corresponding final TWT device used for experimental analysis are given in the figure 2.

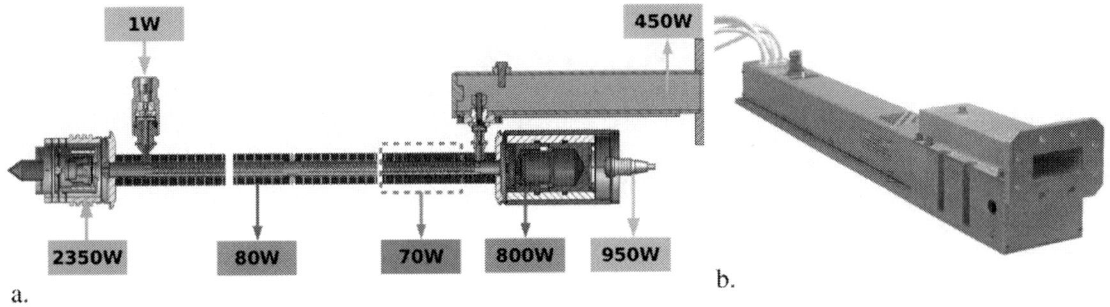

Figure 2. An example of a power equilibrium - rough estimation of a TWT microwave amplifier device (a) and a corresponding final TWT device used for experimental analysis (b).

As can be concluded from the above rough estimation, one of the key problems is an energy lost in the delay line, which causes temperature rise at the terminal part of the TWT device. In fact, the density of an electron beam power can be higher than 1500W/mm2. Actually, the energy lost in the delay line in a form of heat energy (e.g. 100W) is mainly due to the following phenomena:

- power associated to its resistivity p_{sf},
- power associated to the electron beam current intercepted by the helix structure through a bombarding phenomenon p_{sw}.

The above phenomena are a direct cause of a temperature rise of the delay line and can lead to early failure or performance deterioration of TWT device. In order to avoid the unexpected temperature rise, the generated heat energy over the delay line is to be dissipated to the ambient using a proper structural design of a TWT device. The heat is mainly dissipated through conduction and convection phenomena as the radiation could be neglected due to the typical working temperature of a TWT device below 100°C.

3. Analytical approach

Microwave power of an electron beam generated in the delay line grows exponentially with a distance along its axis of symmetry and the highest temperature is expected at the final section of the delay line. In case of an analytical approach the key problem was to determine the so-called linear density function p_c of a power loss that is dissipated along the symmetry axis z, as a result of an electron beam interaction with the delay line. Such relationship could be used in numerical analysis in order to evaluate and predict the temperature distribution in a TWT device. In case of the current research, such relationship was evaluated by taking into account the following assumptions [6]:

- all beam electrons in a point z inside an incremental length layer dz have the same energy,
- initial power of an electron beam coming entering the interaction zone equals to $U_0 I_0$
- electrons intercepted by the delay line in point z and due to an interaction with an electromagnetic wave and cause energy decrease by value of energy transfer to the wave from 0 to z - all its energy is converted into heat,
- density of a beam current intercepted by the delay line - $j(z)$ can be described by linear relationship of microwave power in point z

As stated above the linear power loss p_c along the delay line can be given as:

$$p_c = p_{sf} + p_{sw} \tag{1}$$

In case of the line resistivity according to the TWT's Pierce's theorem it is possible to assume the local power density P_f of a microwave signal along z axis as a function of a power at the end of the delay line P_{out} as :

$$P_f = P_{out}\, e^{B(z-L)} \tag{2}$$

where g is a gain constant, a is an attenuation constant ([g]=dB/m, [a]=dB/m), L is a delay line length and

$$B = \frac{(g-a)\ln(10)}{10} \tag{3}$$

The above assumption allows for evaluation of the linear power loss of a microwave signal p_{sf} due to the attenuation constant a as:

$$p_{sf}(z) = \frac{-dP_f(z)}{dz} = \frac{a\ln(10)}{10} \cdot P_f \tag{4}$$

On the other hand, the formula on the linear power dissipation of an electron beam intercepting with the delay line p_{sw} can be given as:

$$p_{sw}(z)\,dz = \frac{j(z)\,dz}{I_w(z)} P_w(z) \tag{5}$$

where $j(z)dz$ is a beam current intercepted at dz interval, $P_w(z)$ and $I_w(z)$ are correspondingly local power and current of an electron beam while the value of the intercepted current density $j(z)$ is given by:

$$j(z) = I_a \frac{Be^{Bz}}{e^{BL}-1} \tag{6}$$

where I_a is an overall delay line current due to an interaction with an electromagnetic wave:

$$I_a = \int_0^L j(z)\,dz \tag{7}$$

Thus the local current $I_w(z)$ function along the z axis would be given by:

$$I_w(z)=I_0-I_a\frac{e^{Bz}-1}{e^{BL}-1} \tag{8}$$

where I_0 is an initial current of an electron beam for $z=0$. The local power $P_w(z)$ along the z axis can be extracted by using the principle of energy conservation:

$$\frac{dP_w(z)}{dz}=-\frac{dP_f(z)}{dz}-\frac{dP_{cf}(z)}{dz}-p_{sw}(z) \tag{9}$$

Analytical solution of the above formula for the initial condition in point $z=0$: $P_f(0)=P_{in}$ and $P_w(0)=U_0I_0-P_{in}$ while using the notation $J_a/J_0=\chi$, and taking into account that $U_0J_0<<P_{in}$, $e^{Bz}>>1$ and $P_{in}=P_{out}e^{-BL}$ would be given by:

$$p_{sw}(z)=U_0I_0Be^{B(z-L)}\left(\chi+\frac{P_{out}}{U_0I_0}\ln\left(1-\chi e^{B(z-L)}\right)\right) \tag{10}$$

So, the final formula for linear power density $p_c(z)$ dissipated in the delay line can be described by formula 1, where components p_{sf} and p_{sw} are given by equations 4 and 10. Figure 3 show an example result of the linear power loss evaluated accordinglly.

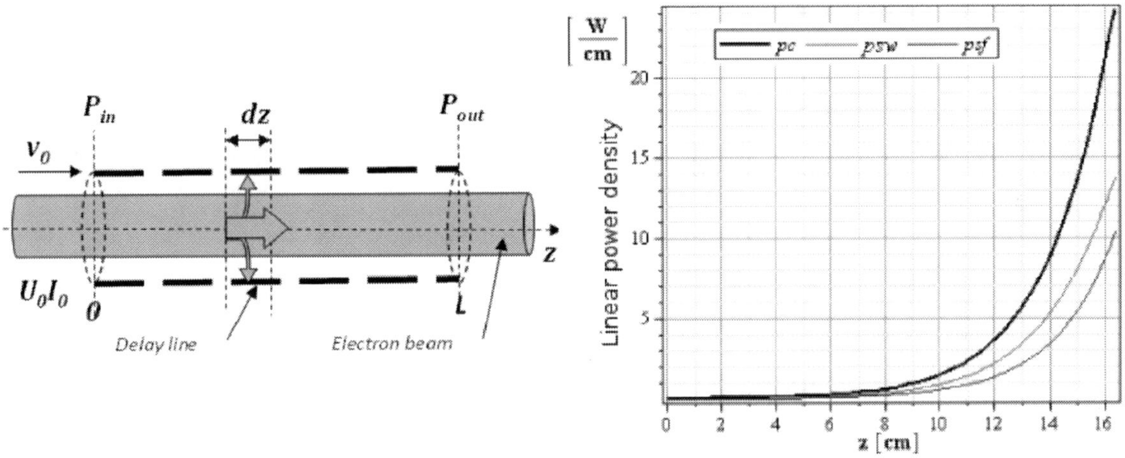

a. b.

Figure 3. Scheme of an interaction of an electron beam with the delay line (a) and an example of a linear power density $p_c(z)$ dissipated in the delay line according to the analytical analysis (b).

Finally, the above analytically evaluated dependence was assumed as the reference one in the followed up computer simulation of a temperature distribution in the TWT device structure.

4. Numerical approach

In case of numerical modelling of a temperature distribution, according to the presented analytical analysis, it was assumed that only the output section of the TWT device can be taken into account. The computer / numerical model, prepared in ANSYS package v.12, with the detailed scheme is given in figure 4a and the detailed structure of the delay line in the figure 4b. In order to simply the whole structure a couple of assumptions were taken into account [7]:

- only the final part of the delay line could be simulated due to the evaluated nonuniform linear power density dissipation,

- the microwave linear power density in the delay line can be represented by a nonuniform resistivity of the helix structure and the corresponding Joule heat,
- the collector structure can be represented by a simple metal block,
- only linear material properties could be used for simulations.

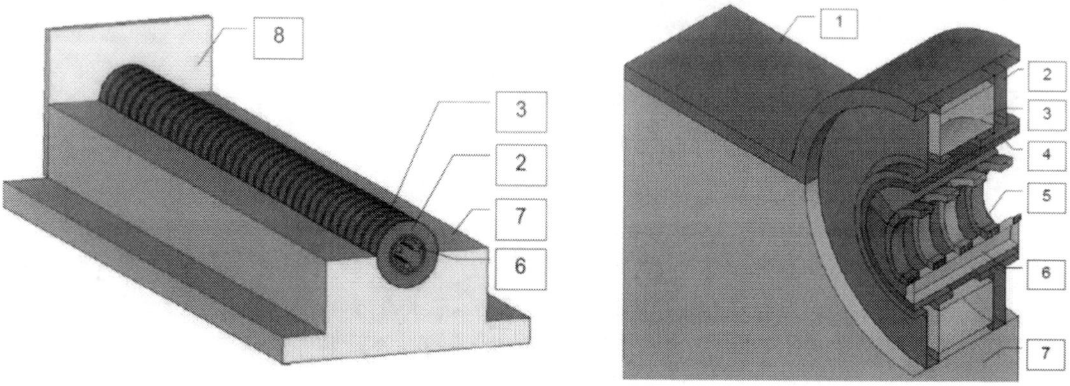

a. b.

Figure 4. Computer / numerical model of a final region of the analysed TWT device (a), and a detailed structure of an internal part the delay line including the helix structure (b), where: 1-clamp, 2-pole, 3-magnet, 4-vacuum shield, 5-helix, 6-ceramic rods, 7-base, 8-collector.

In order to implement the nonuniform linear power density distribution in the delay line it was assumed to use the segmented helix structure, which dissipates the total power P_n, according to the Joule heat:

$$P_n = I^2 R_n = I^2 \rho_n \frac{2\pi a}{d w} \tag{11}$$

where R_n is the resistance of each helix segment while ρ_n is its resistivity and a, d, w are its geometrical dimensions.

The main goal of the numerical analysis was to compare the temperature distribution of a TWT device $T(z)$ shield with the given distribution of a linear power dissipation in the delay line $p_c(z)$ given by equation 1. It was assumed that there should be a correlation between the both along z axis, which can be expressed in a form:

$$T(z) = k \cdot p_c(z) \tag{12}$$

where k is a constant proportional coefficient, that does not depend on z. In case when the heat transfer in radial direction is dominant then both $T(z)$ and $p_c(z)$, should have a similar distribution as given in the figure 5 and additionally in the figure 6 for two different linear power dissipation dependencies - smooth and steep. In fact, both curves when normalized seem to fulfill the equation 12, except the final region of the delay line (around *2cm*), which is mainly due to the collector influence. The same, it was concluded that using the presented analytical analysis and results of numerical simulations along with the so-called inverse engineering and formula 12, it should be possible to assess the linear power dissipation character along the delay line [8].

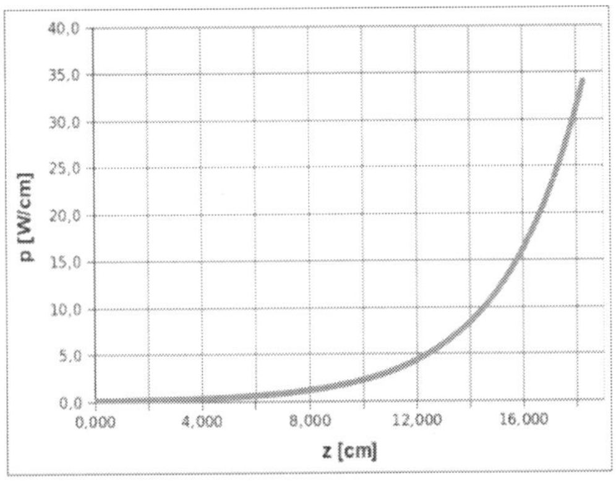

a.

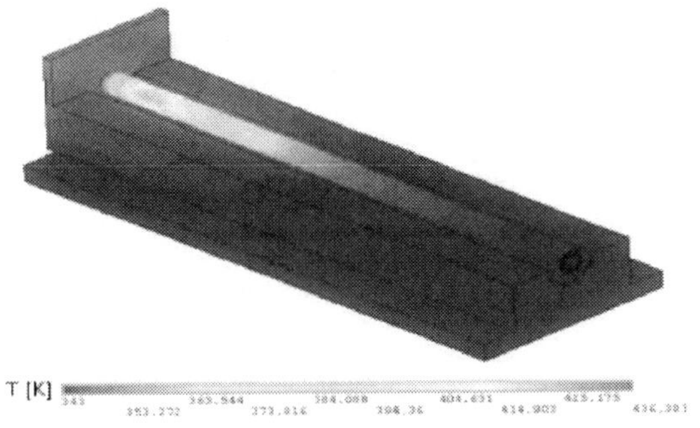

b.

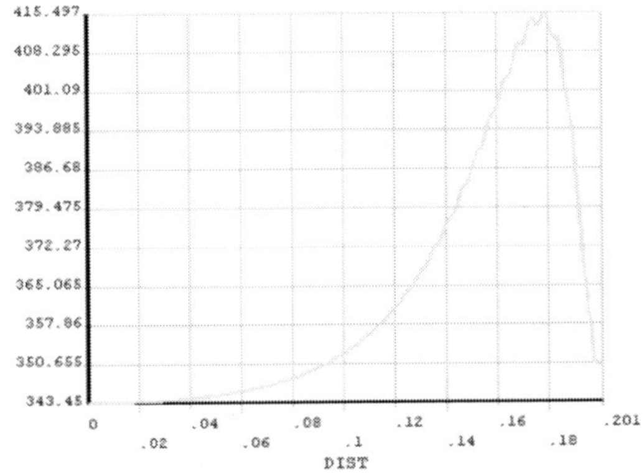

c.

Figure 5. Example of a numerical analysis of a TWT device: (a) assumed linear power dissipation $p_c(z)$, (b) corresponding temperature distribution, (c) temperature dependence $T(z)$ along the shield.

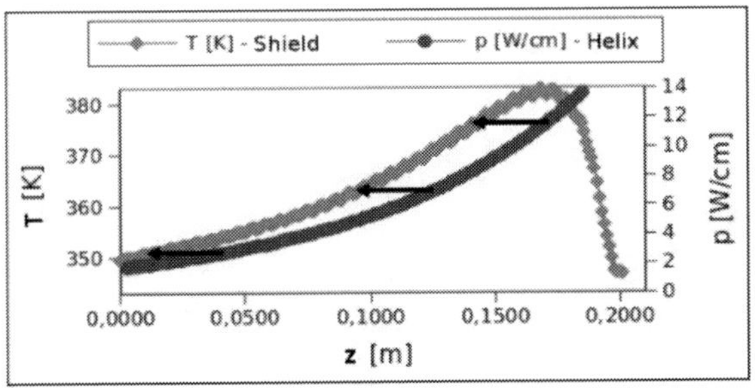

a.

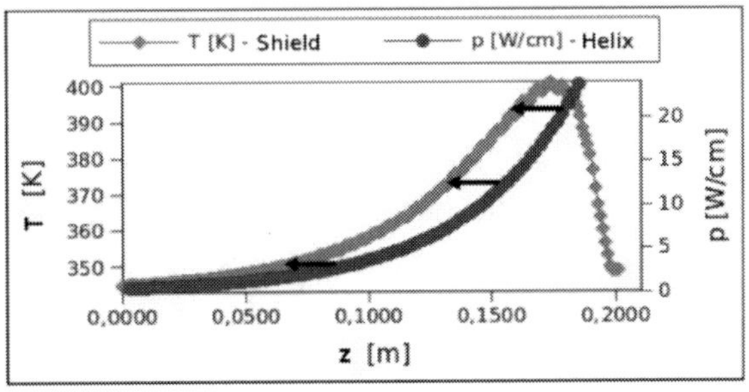

b.

Figure 6. The results of numerical analysis for two different linear power dissipation curves $p_c(z)$ (a) smooth (b) steep, where total p_c=107W.

5. Experimental validation

In order to validate the conclusions based on presented analytical analysis and numerical simulations, an experimental setup was designed and manufactured, which is shown in the figure 7. The objective was to compare the temperature distribution in different parts of the helix TWT assembly during the working conditions with the analytical and numerical results for corresponding power dissipation. The experimental setup consisted of eight thermocouples bounded to the delay line shield, which were

IOP Conf. Series: Materials Science and Engineering **104** (2015) 012010 doi:10.1088/1757-899X/104/1/012010

placed between the magnets and the whole system was cooled with water. The measurements were done for the selected power supply values (U_0=*7400V*, I_0=*310mA*) and for two microwave input powers *200mW* (P_{wy}=*412W*) and *300mW* (P_{wy}=*425W*). Figure 8 contains the comparison of a measured normalized linear power dissipation curves with three different analytical functions, which seem to fulfill the corresponding partial differential equations.

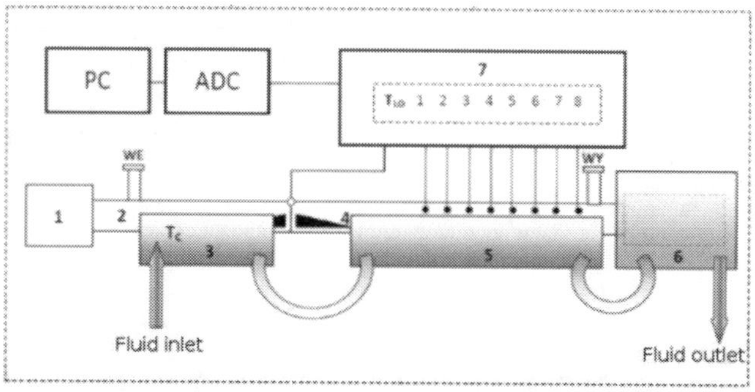

a.

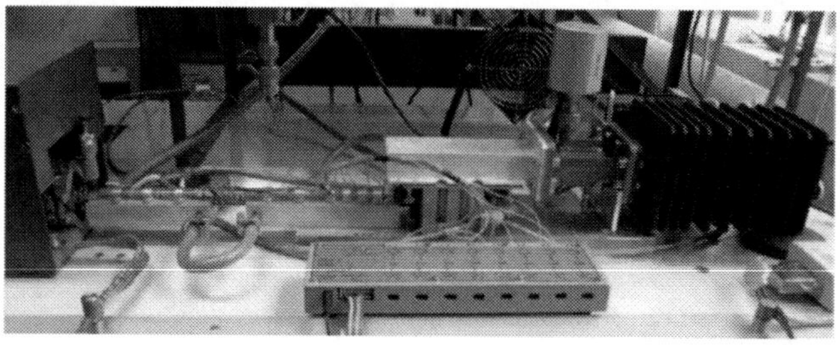

b.

Figure 7. Experimental setup for temperature distribution measurements: (a) diagram, where: 1-electron gun, 2-input, 3,5,6-corresponding coolers, 4-output section, 7-eight channels of DC amplifier, ADC- converter, PC-computer, (b) view of the manufactured system.

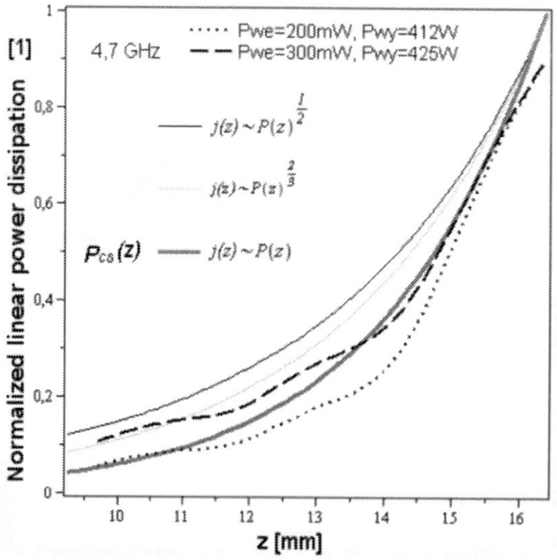

Figure 8. A comparison of a measured normalized linear power dissipation in the delay line versus three different analytical dependencies.

According to the presented results, it can be concluded that the best fit of the experimental results with the analytical functions are in case of the linear power dissipation that was given by the formula 1. In fact the formula was derived under the assumption that the density of a beam current intercepted by the delay line $j(z)$ is proportional to the local microwave power $P_f(z)$.

6. Conclusions

The above paper describes the thermal analysis of the TWT device by combined analytical and numerical, which were validated by experimental measurements. Analytical model was focused on an evaluation of the beam current power dissipation and microwave losses along the delay line, which is not uniform and most of the power is accumulated at its terminal part. High power results in a non-uniform temperature distribution, with very high temperatures at terminal part of the delay line. The analytically evaluated power dissipation function was then used in numerical simulation of a temperature distribution in the working TWT device. The experimental results validated the proposed methodology and through the inverse engineering it was possible to validate the undertaken assumptions on the dependence of a dissipated liner power density along the delay line.

The presented results focused mainly on a problem of the temperature distribution. In fact working TWT devices are exposed to high temperatures, which can influence their reliability. Therefore, the presented methodology would be implemented at the design stage and would hopefully lead to better products, which mainly means improved thermal properties and corresponding performance and, last but not the least, more reliable final products.

7. References

[1] W. Wiejak, A. Wymysłowski; 2009; "Thermal analysis of TWT delay line by combined theoretical and numerical approach", *33th International Conference and Exhibition, IMAPS–Poland Chapter*, pp. 359-362

[2] W. Wiejak, A. Wymysłowski; 2010; "Uproszczony analityczny model rozpraszania mocy w linii późniającej lampy fali bieżącej"; *Elektronika*; **vol. 51, nr 3**, pp. 114-119

[3] V.E. Cosslett; 1950; "Introduction to electron optics"; *Oxford University Press, London*

[4] B. Paszkiewicz; 1960; "Optyka elektronowa"; *PWN, Warszawa*

[5] Jr. A.S. Gilmour; *"Principles Of Traveling Wave Tubes"*; Artech House Boston, London, 1994

[6] W. Wiejak; 2013; PhD thesis: "Zastosowanie analitycznych i numerycznych metod wspomagania projektowania do lamp z falą bieżącą", *Wroclaw University of Technology, Poland*

[7] W. Wiejak, A. Wymysłowski; 2015; "Analytical, numerical and experimental approach to thermal analysis and design of a travelling wave tube"; *Proceedings of the 16th International Conference on Thermal, Mechanical and Multi-Physics Simulation and Experiments in Microelectronics and Microsystems, EuroSimE*, pp. 1-8

[8] W. Wiejak, A. Wymysłowski; 2011; "Zweryfikowany analityczny model rozpraszania mocy w linii opóźniającej lampy fali bieżącej."; *Elektronika*; **vol. 52, nr 8**; pp. 108-113

Acknowledgements

The project was supported by Ministry of Science and Higher Education of Poland as a development project R00 0010 06.

IOP Conf. Series: Materials Science and Engineering **104** (2015) 012011 doi:10.1088/1757-899X/104/1/012011

Modelling the pulse transformer in SPICE

Małgorzata Godlewska, Krzysztof Górecki, Krzysztof Górski

Gdynia Maritime University, Department of Marine Electronics, Morska 83, 81-225 Gdynia, Poland

E-mail: k.gorecki@we.am.gdynia.pl, m.godlewska@we.am.gdynia.pl

Abstract. The paper is devoted to modelling pulse transformers in SPICE. It shows the character of the selected models of this element, points out their advantages and disadvantages, and presents the results of experimental verification of the considered models. These models are characterized by varying degrees of complexity - from linearly coupled linear coils to nonlinear electrothermal models. The study was conducted for transformer with ring cores made of a variety of ferromagnetic materials, while exciting the sinusoidal signal of a frequency 100 kHz and different values of load resistance. The transformers operating conditions under which the considered models ensure the acceptable accuracy of calculations are indicated.

1. Introduction

Pulse transformers are widely used in switched-mode power supplies with galvanic isolation between the input and the output [1, 2]. In literature, there are a lot of papers concerning ferromagnetic material properties [3 – 10], as well as discussions concerning the impact of electromagnetic interactions occurring in transformers on their characteristics and operating parameters [11 - 16]. Few works in literature concern modeling pulse transformers, that take into account various physical phenomena [11, 12, 13, 16].

With a computer analysis of electronic circuits, the models of all the elements contained in these circuits are needed. These models should take into account all physical phenomena relevant to the circuits under consideration, while the character of these models should be simple enough to provide short duration calculations. These requirements are mutually contradictory. Therefore, it is important to choose the right model to solve every simulation problem.

Currently, a popular program for analyzing electronic circuits is SPICE [17], in which the basic models of electronic components are built-in [18]. However, it has no built-in model of the transformer in this program. Therefore, it is necessary to apply a macromodel of varying forms to describe the element.

An important physical phenomenon occurring in electronic components is self-heating, reflected by an increase in the internal component temperature due to conversion of electrical energy into heat by non-ideal cooling of the element [16, 19, 20]. Models which take into account the self-heating phenomenon are called electrothermal models and models which skip this phenomenon - isothermal models.

The paper describes the considered models and analyzes their suitability for modeling characteristics of pulse transformers containing cores made of different materials. The results of calculation are compared with the measurement results.

2. Models descriptions

In the paper five arbitrary selected literature transformer models, implemented in SPICE, are considered. These models can be divided into two groups: isothermal and electrothermal. The first of the considered models of the transformer consists of linearly coupled linear coils, and the second - linearly coupled linear coils with series resistors, which model losses in the windings. In turn, the model from paper [11] describes non-linear properties of the core, but does not take into account losses. The next two models are electrothermal ones. The fourth model (electrothermal), described in [11] takes into account non-linearity of core characteristics and an increase of the internal temperature of the transformer windings and core as a result of self-heating. The last model (also electrothermal) – elaborated by the authors [12] allows calculating core temperature and winding temperature taking into account self-heating and mutual thermal coupling between the core and windings.

2.1. Linearly coupled linear coils

Figure 1 shows a circuit representation of a transformer model composed of linearly coupled linear coils. The primary winding represents a coil L_1, while the coil L_2 represents a secondary winding. This model describes the ideal transformer and transmission of the transformer is equal to

$$n = \sqrt{L_2/L_1} \tag{1}$$

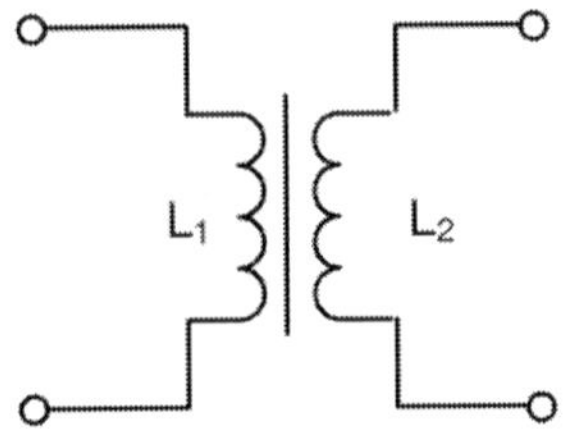

Figure 1. Circuit representation of the linearly coupled linear coils model

In this model we omit losses. It is impossible to include self-heating phenomena, either.

2.2 Linearly coupled linear coils with series resistances

Figure 2 shows a circuit representation of a transformer model composed of coupled linear coils and series resistors representing losses in the windings. The values of the resistance change depending on the parameters of the core used and the number of turns in the windings and their diameter.

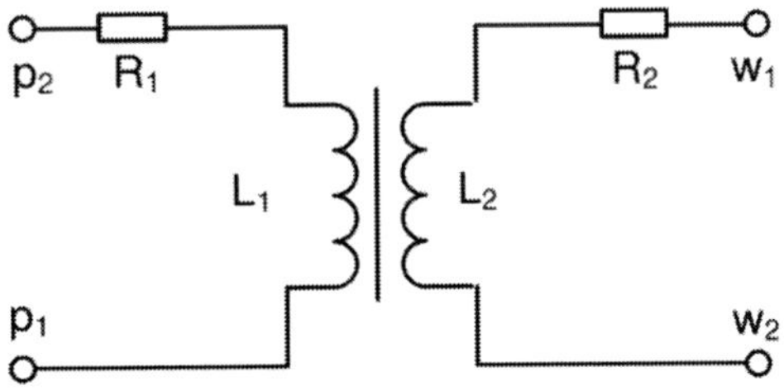

Figure 2. Circuit representation of the model with linearly coupled linear coils with series resistances.

2.3 Isothermal model of the transformer from paper [11]

In paper [11] an isothermal model of the transformer consisting of the ferromagnetic core and windings is proposed. The circuit representation of the transformer model for SPICE based on the text description presented in [11] is shown in Figure 3.

In the model of windings there are two pairs of clamps - electric clamps (respectively p1, p2 and w1, w2), corresponding to transformer outputs, and magnetic clamps (respectively p3, p4, and w3, w4), the voltage of which corresponds to the magnetomotive force, and the current – to the magnetic flux. The magnetic clamps are connected to the core clamps (r1, r2) between which there is a controlled current source G_1 representing reluctance of the core. In the core model, the controlled voltage source E_1 is used to determine the effective magnetic force H_e and the controlled voltage source E_2 is used to determine the value of irreversible magnetization M_{an}. The normalized reversible magnetization function M_{an} is approximated by the Langevin function $L(H_e)$, of the form:

$$M_{an} = \frac{1}{\tanh\left(\frac{H_e}{A}\right)} - \frac{A}{H_e} \qquad (2)$$

where A is the model parameter shaping a hysteresis loop.

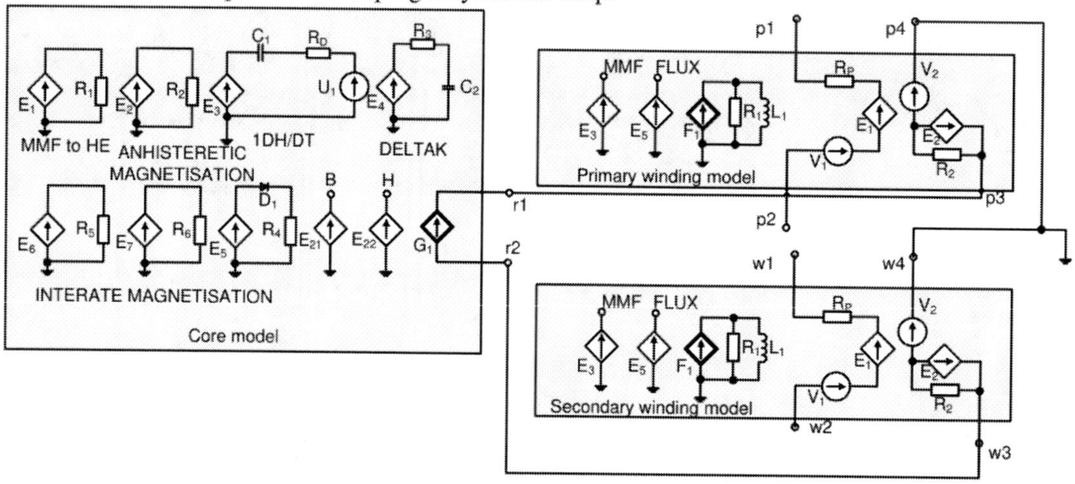

Figure 3. Circuit representation of the isothermal transformer model from paper [11]

Real reversible magnetization is obtained by multiplying the right side of equation (2) by saturation magnetization M_s. The value of magnetization M can be determined by solving a differential equation of the form:

$$\frac{dM}{dH} = \frac{1}{1+c}\frac{M_{am} - M}{\frac{\delta k}{\mu} - (M_{am} - M)} + \frac{c}{1+c}\frac{dM_{an}}{dH} \qquad (3)$$

where μ - magnetic permeability, α - coefficient characterizing the magnetic field δ - coefficient to determine the direction of a change of the magnetic field (δ = 1 for dH/dt>0, δ = -1 for dH/dt<0) k - coefficient of non-elastic deformation of domain walls c - coefficient of elastic deformation of domain walls.

The circuit consisting of the controlled voltage source E_3, the capacitor C_1, the resistor R_D and the voltage source V_1 is used to determine the time derivative of the magnetic force dH/dt. In turn, the circuit composed of the series connected controlled voltage source E_4, the resistor R_2 and the capacitor C_2 is responsible for the delay of the analyzed signal, slowing signal slopes or limiting the derivative dH/dt. Next four circuits containing the controlled voltage source E_6 and the resistor R_5, the circuit including the controlled voltage source E_7, the resistor R_6 and the controlled voltage source E_5 connected in series with the diode D_1 and the resistor R_4, and the controlled current source G_1, are used to determine total magnetization M. The two branches consisting of the controlled voltage sources E_{21} and E_{22} are used to designate the magnetization curve B(H) of the core.

The model of the windings contain the electromotive force corresponding to flux density (voltage source E_1), the winding resistance (R_p) and a circuit modeling the magnetomotive force generated by the current of each of the windings. The source voltage V_1 whose value is zero allows monitoring the output current of the winding. The voltage at the voltage sources E_3 and E_5 correspond to the magnetomotive force and magnetic flux. The circuit composed of the controlled current source F_1, the resistor R_1 and the coil L_1 allows determining the time derivative of the magnetic flux.

In the presented model core losses are skipped, but energy losses in the winding are taken into account.

2.4 Electrothermal model from paper [11]

Figure 4 shows the electrothermal model of the transformer from paper [11]. Resembling the model described in section 2.3, this model contains blocks modeling the characteristics of the core and windings, but in the equations describing the components of the transformer the influence of temperature and self-heating phenomena are taken into account .

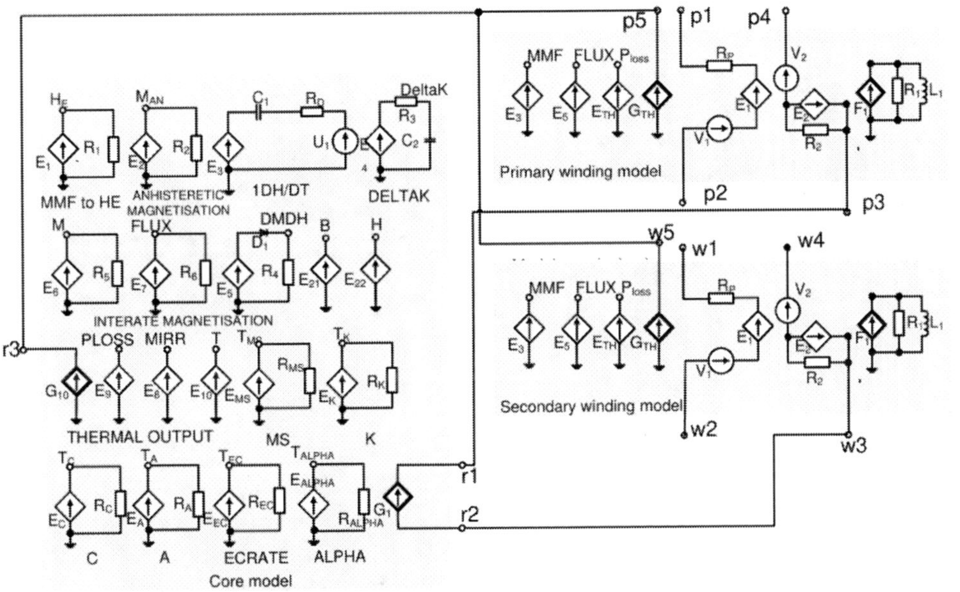

Figure 4. Circuit representation of the electrothermal model from paper [11]

In the models of the core and windings there are elements modeling energy losses. In the core model the voltage sources E_8, E_9, E_{10} and the current source G_{10} are used to determine power generated in the core. Similarly, in the models of windings the controlled current source G_{TH} representing losses in the windings is included. Clamps of these sources are connected to the external RC network, representing transient thermal impedance of the transformer. Node p5, w5 and r3 are connected, the voltage on these nodes corresponds to the excess of the transformer temperature above the ambient temperature.

Thus, in the considered model only one internal temperature of the whole transformer is used, i.e. in the considered model the temperature of the winding and the core are the same. In the model of the core circuits consisting of the series connected controlled voltage source E and the resistor R are used to determine the sequence of parameters of the core M_S, M, C, A, ECRATE ALPHA taking into account line influence on temperature. The temperature coefficients are designated A_{TC}, C_{TC}, α_{TC}, $ECRATE_{TC}$.

IOP Conf. Series: Materials Science and Engineering **104** (2015) 012011 doi:10.1088/1757-899X/104/1/012011

The power loss in the windings is described by

$$G_{TH} = I_w^2 \cdot R_w \quad (4)$$

where I_W is the winding current, R_W is internal resistance of the windings.

In turn, core losses are described by the equation:

$$P_{LOSS} = H \cdot LEN \cdot \mu_0 \cdot (M_S \cdot (1 + M_{STC} \cdot T) \cdot (M - M_{an}) + H) \quad (5)$$

where μ_0 – denotes magnetic permeability in free air, LEN – is the magnetic path length, MSTC – is the temperature coefficient of saturation magnetization, T – denotes the excess of core temperature over the ambient temperature determined from the thermal model.

In the formula (5) the ratio of the magnetic field strength H and the magnetic path length LEN is equal to the value of the magnetomotive force, observed between clamps of the core model (r1 and r2).

2.5. Model from paper [12]

Figure 5 is a circuit representation of the electrothermal model of a transformer for SPICE, presented by the authors in [12].

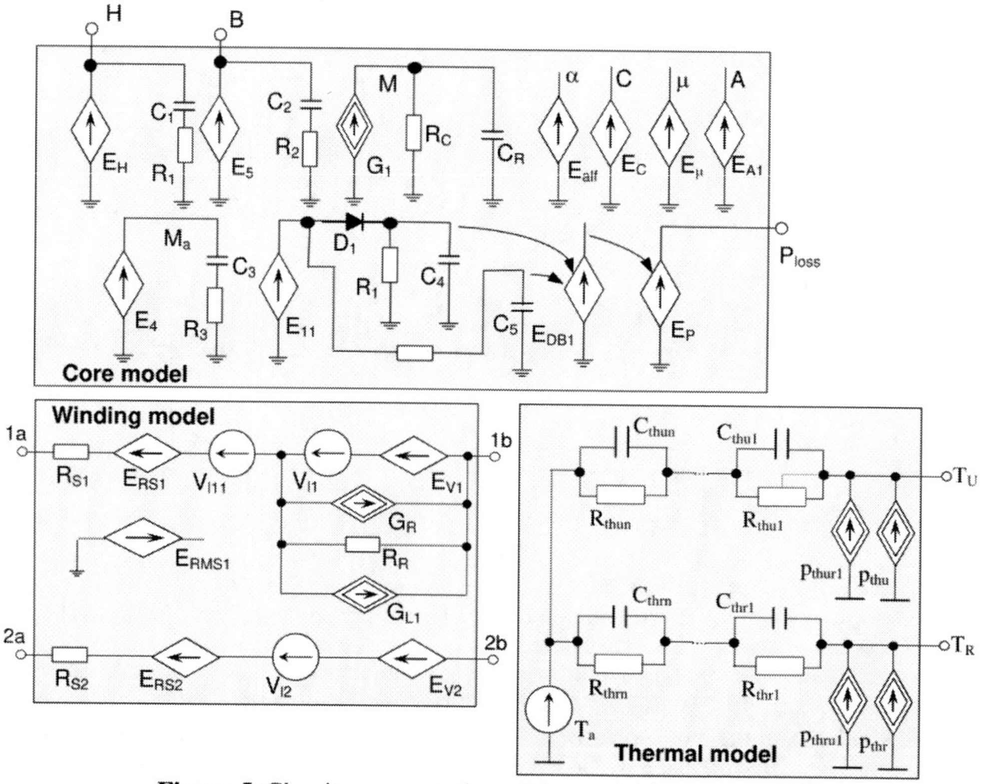

Figure 5. Circuit representation of the model from paper [12]

This model consists of three separate blocks: the core model, the windings model and the thermal model. The core model is based on a modified Jiles-Atherton model [3]. In this model the controlled voltage source E_H connected in series with the capacitor C_1 and the resistor R_1 is used to determine the magnetic force H. The flux density corresponds to the voltage at the output node B. Power losses in the core are equal to the voltage at the node P_{loss}.

The thermal model allows calculating the core temperature T_R and the winding temperature T_U taking into account self-heating and mutual thermal coupling between the core and windings. This

IOP Conf. Series: Materials Science and Engineering **104** (2015) 012011 doi:10.1088/1757-899X/104/1/012011

model is in the form of the RC Foster network stimulated by the source representing power dissipated in the core P_{THR} and in the windings P_{THU}. In order to take into account thermal coupling between the core and the windings the controlled current sources P_{thur1} and P_{thru1} are applied. The voltage source T_a is used to model the ambient temperature.

The detailed description of the core model contained in papers [8, 21], and the description of the thermal model in [22] show the performance of the controlled voltage sources included in the winding model. The voltage on the voltage source E_{V1} is given by

$$E_{V1} = z_1 \cdot S_{Fe} \cdot \frac{dB}{dt} \tag{6}$$

where z_1 - the number of primary winding turns, S_{Fe} - cross-sectional area of the winding wire. Similarly, the voltage value is determined at the voltage source E_{V2} taking into account the number of turns of the secondary windings.

In the core model the controlled voltage source E_H representing the magnetic force H is added. This magnetic force is described by the following formula

$$H = (z_1 \cdot i_1 + z_2 \cdot i_2)/l_{Fe} \tag{7}$$

where z_1 and z_2 are numbers of turns in primary and secondary windings, respectively, l_{Fe} is the magnetic path length, whereas i_1 and i_2 are currents of the windings.

Additionally, in order to improve the convergence of calculations in the differentiation networks used in this model the independent voltage sources are replaced by resistors, e.g. the network including the controlled voltage source E_H, capacitor C_1 and resistor R_1 is used to calculate the derivative dH/dt.

The flux density corresponds to the voltage on the terminal B. The power losses in the core are equal to the voltage at the terminal P_{loss}. The core losses P_{thr} are described by the formula

$$P_{loss} = V_E \cdot \left(\frac{B_{pp}}{2}\right)^{\beta-\alpha} \cdot \left(1 + \alpha_P \cdot (T_R - T_m)^2\right) \cdot \frac{P_{V0}}{T} \int_0^T \left|\frac{dB}{dt}\right|^{\alpha} dt \tag{8}$$

In Eq. (8), α_P denotes the square temperature coefficient of power losses, T_m – the temperature for minimum losses, T is a period of flux density and B_{pp} is the peak-to-peak value of flux density, V_E – the volume of the core, β and α are the Steinmetz's model parameters.

The windings model contains two networks representing primary winding, connected between the connectors 1a and 1b, and the secondary winding, connected between connectors 2a and 2b. In the model of the primary winding the resistor R_{S1} represents series resistance of this winding in the reference temperature T_0, the controlled voltage source E_{RS1} models a change of this resistance during temperature changes of the windings T_U and the skin effect. The controlled voltage source E_V represents the voltage induced in the primary winding, the controlled current source G_{L1} represents the magnetizing current by the formula:

$$G_{L1} = l_{Fe} \cdot \frac{(|H| + A)^2}{z_1^2 \cdot S_{Fe} \cdot B_{sat} \cdot A} \cdot \int_0^t U_P(x) \cdot dx \tag{9}$$

where U_p - voltage between clamps of the primary winding, l_{Fe} – magnetic path length, B_{sat} – core saturation flux density.

The controlled current source G_R models losses of energy in the core. The current of this source is given by the following equation

$$G_R = P_{loss} \cdot V_1 / V_{RMS1}^2 \tag{10}$$

where V_{RMS1} denotes the rms value of the voltage on the source G_R, while V_1 is the actual value of this voltage. The voltage V_{RMS1} is calculated by means of the controlled voltage source E_{RMS1}. Voltage sources V_{I1} and V_{II1} have a zero-value of the output voltage and they are used to monitor the value of their currents. In turn, the model of the secondary winding contains only elements modelling voltage induced on the winding (E_V) and series resistance of this winding (R_{S2} and E_{RS2}).

3 Results

In order to examine suitability of the transformer models described in section 2, calculations of the characteristics of the selected transformers containing cores made of different materials were carried out, while using them. Each test transformer had two windings of 20 turns of copper enamel wire having the diameter of 0.8 mm. The used toroidal cores respectively had the outer diameter of about 26 mm and were made of powdered iron (RTP), nanoperm (RTN) and ferrite (RTF).

Figure 6 shows a measuring circuit, where resistance of the resistor is $R_1 = 33\ \Omega$ and R_0 is load resistance of the circuit. The voltage source V_1 generates a sine wave with the amplitude of 67 V and frequency of 100 kHz.

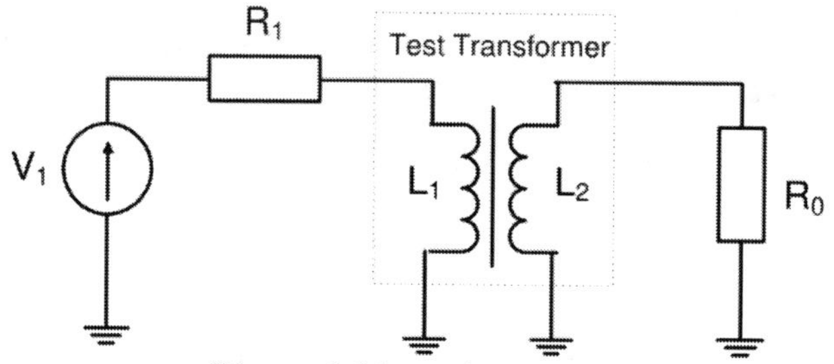

Figure 6. Measuring circuit.

The selected characteristics of this circuit are shown in the succesive figures. In these figures the results of measurements are denoted by points, whereas the results of calculations – by lines.

For each of the considered models the values of the parameters corresponding to each of the tested transformers based on datasheets or additional measurements and estimation algorithms described in [23, 24] are used. Figures 7 - 9 show the calculated and measured characteristics of the transformer considered in the steady state. The results presented in Figure 7 concerns the transformer core RTP, Figure 8 - the transformer core RTF and Figure 9 - the transformer core RTN. The lines in purple represent calculations for the model described in section 2.1, the green colour - for the model from section 2.2. In turn, the isothermal model of paper [11] is presented by lines in black, and the electrothermal model of the same work – by blue lines. The results obtained with the authors' model described in [12] are denoted by the red colour.

Figure 7 shows the results of the input voltage, output voltage, watt-hour efficiency and temperature on load resistance for the transformer with the RTP core. In this figure, solid lines represent the results of calculations obtained for the nominal values of parameters, and dotted lines - the results obtained with the adjusted parameter values, which better match of the results of calculations and measurements. To this end, value of MS parameter (for the models from [11]) was increased 3.2 times. In the case of the model from paper [12] B_{S0} parameter was increased by 5%. After these changes the model of paper [12] best fit measurement results, and the difference in terms of measurement of the input voltage as a function of load resistance did not exceed 2.7%.

Figure 7a and 7b show that for all the considered models the increasing function describing the dependence of the input and output voltages on load resistance is obtained. One can see that all the models under consideration, except models from [11] (up to three times lower values obtained from calculations than values obtained from measurements) provide a satisfactory agreement between the calculations and measurements results, wherein the difference between these results is higher for the smaller values of resistance R_0. After the change of model parameters from [11] the maximum difference between the results of calculations and measurements does not exceed 15%. The authors' model with the adjusted value of the B_{S0} parameter provides the best accuracy of calculations, but also the models of the transformer in the form of coupled coils ensure the correct calculation results.

On the other hand, watt-hour efficiency is typically a decreasing function of load resistance. The authors' model provides the most accurate description of the measured characteristics $\eta(R_0)$, and the used models from paper [11] obtained the understated values of watt-hour efficiency. Coupled coils have inflated the value of watt-hour efficiency, even several times.

Changes in the core temperature as a function of load resistance were only possible to examine for model from paper [12] and the electrothermal model from paper [11]. For both these models the increasing function $T_R(R_0)$ was obtained. The model from paper [12] fits satisfactorily the measured characteristics and the model from [11] has no convergence between the results of calculations and measurements. This proves the incorrect modeling power loss in this model. It is worth noting that the core temperature reaches about 120°C at high load current, which shows high energy loss in the core associated with the high value of the amplitude of the flux density in the core.

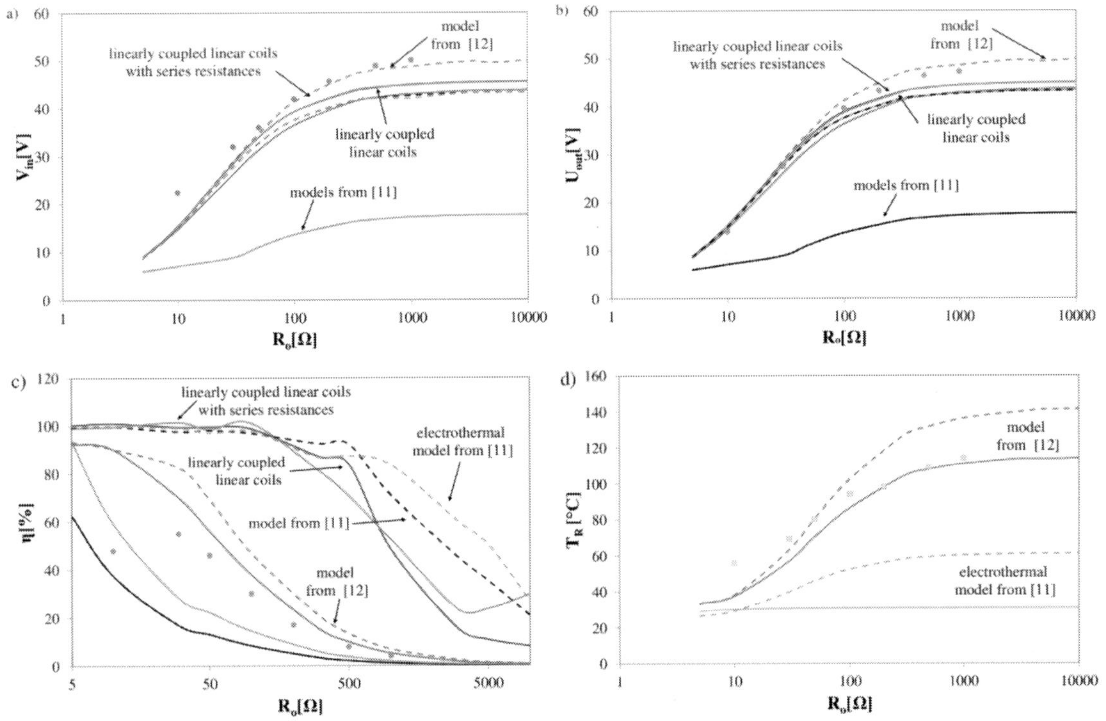

Figure 7. Calculated and measured characteristics of the input voltage (a), output voltage (b) watt-hour efficiency (c) and the temperature of the core (d) on load resistance for the transformer with the RTP core.

Figure 8 shows the dependence of watt-hour efficiency and the core temperature on load resistance of the RTF core transformer.

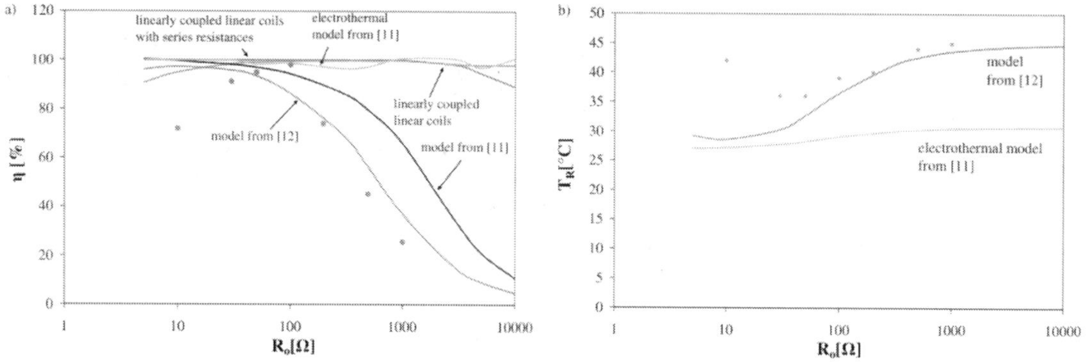

Figure 8. Calculated and measured characteristics of watt-hour efficiency (a) and the core temperature (b) on load resistance for transformer with the RTF core.

The shape of the considered characteristics is similar to the shape of the characteristics shown in Figure 6. It is worth noting that the transformer with the RTF core is characterized by higher watt-hour efficiency and lower core temperature than the transformer with the RTP core. The best compliance of the calculated and measured characteristics was obtained for the authors' model of the transformer. Unfortunately, significant deviation between the results of calculations and measurements in the range of small values of load resistance appeared.

Figure 9 shows the dependence of the output voltage and the core temperature on load resistance of the transformer with the RTN core.

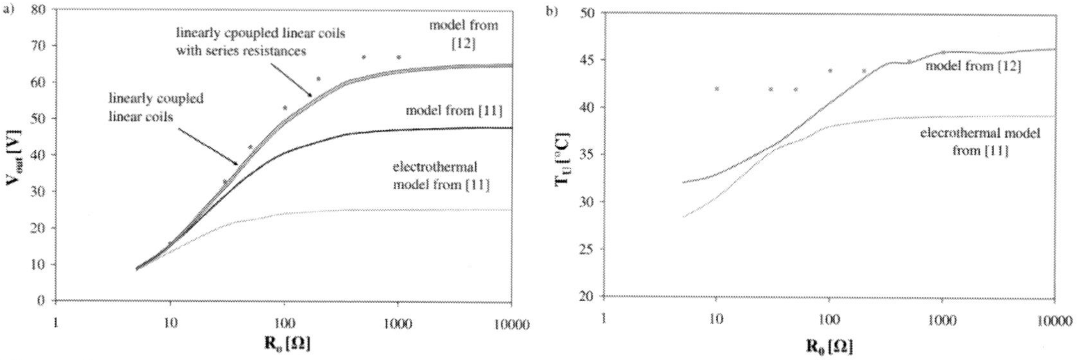

Figure 9. Calculated and measured characteristics of the output voltage (a) and the core temperature of the core (b) on load resistance for the transformer with the RTN core.

The shape of the characteristics of the presented considered dependences is similar to the characteristics of the transformer with the RTP core. Big deviation between the calculated output voltage obtained using models from paper [11] is comprising . On the other hand, the simplest model of the transformer in the form of coupled coils ensures the achievement of similar correspondence between the calc ulated and measured values of the output voltage as the electrothermal model from [12]. The results of calculations realized by the model from [12], differ from measurements by not more than 6%.

In turn, each of the considered models underestimates the value of the core temperature in a range of small values of load resistance, and for the resistance $R_0 > 100\ \Omega$ the authors' model assures nearly perfect agreement with the results of measurements.

4 Conclusions

In the paper, five arbitrarily selected models of transformers for SPICE were described and their suitability for modeling the characteristics of this class of elements was tested. The considered models are characterized by varying degrees of complexity of the description of phenomena occurring in the transformer. Experimental tests were carried out for capacitors containing toroidal cores made of different ferromagnetic materials. The results confirmed that with wide range of load resistance in order to correctly determine the value of the amplitudes of the input and output voltage of the transformer it is necessary to apply sufficient models of this element in the form of linearly coupled linear coils. On the other hand, reliable determination of watt-hour efficiency of the tested transformers requires the use of non-linear models of the element, taking into account energy losses both in the core and in the windings.

Models from papers [11] and [12] based on the same Jiles-Atherton model describing properties of the core, nevertheless achieved by means of the calculation results significantly differ from each other. The better matching results of calculations and measurements are assured by the authors' electrothermal transformer model from paper [12]. This model allows, in addition to the designation of voltage and current at transformer terminals and its watt-hour efficiency, calculation of the core and the windings temperature. The results described in the previous section show that this model provides the most accurate description of the characteristics of all transformers being considered. It should also be noted that under identical control and load conditions achieved significant differences in the values of parameters characterizing the considered transformers were obtained. For example, the greatest losses of energy and the highest core temperature values were obtained for the transformer with the RTP core. On the other hand, the highest watt-hour efficiency was obtained for the transformer with the RTF core.

The obtained results may be useful for designers of switched-mode power supplies. By using the presented models, the designers of this class of circuits will be able to make a realistic simulation of the designed equipment, which will reduce production costs of these circuits.

5. References

[1] Ericson R, Maksimovic D 2001 Fundamentals of Power Electronics *Kluwer Academic Publisher* (Norwell)

[2] Rashid MH 2007 Power Electronic Handbook *Academic Press Elsevier*

[3] Jiles DC, Atherton DL 1986 Theory of ferromagnetic hysteresis *Journal of Magnetism and Magnetic Materials* **61** 48-60

[4] Coit RH 2001 Designing with magnetic Cores at High temperatures *PCIM 2001 Europe Official Proc. of 43rd international Power Electronics Conference* (Nurnberg) 607-612

[5] 2001 Using Magnetic Cores at High Temperatures *Technical Bulletin, Magnetics* No. CG-06

[6] Tenant J, Rousseau J, Zegadi L 1995 Hysteresis modeling taking into account the temperature *Proceedings of European Power Electronics Conference* **1** 1.001-1.006

[7] Ladjimi A, Mekideche M R 2012 Modeling of thermal effects on magnetic hysteresis using the Jiles-Atherton model *Przegląd Elektrotechniczny* **88** (4a) 253-256

[8] Górecki K, Rogalska M, Zarębski J, Detka K 2014 Modelling characteristics of ferromagnetic cores with the influence of temperature *Journal of Physics: Conference Series* **494** 012016

[9] Raghunathan A, Melikhov Y, Snyder JE, Jiles DC 2009 Modeling the temperature dependence of hysteresis based on Jiles-Atherton theory *IEEE Transactions on Magnetics* **45** (10) 3954-3957

[10] Zirka SE, Moroz YI, Harrison RG, Chwastek K 2012 On physicl aspects of the Jiles-Atherton hysteresis models *Journal of Applied Physics* **112** 043916-1 – 043916-7

[11] Wilson PR, Ross JN, Brown A D 2002 Simulation of Magnetic Component Models in Electric Circuits Including Dynamic Thermal Effects *IEEE Transaction on Power Electronics* **17** (1) 55-65

[12] Górecki K, Górski K, Rogalska M 2014 Estimation of parameters of the electrothermal model of the impulse-transformer *38th International Conference of IMAPS-CPMT Poland* (Rzeszów-Czarna) paper 112
[13] Van den Bossche A, Valchev VC 2005 Inductors and transformers for Power Electronics *CRC Press, Taylor & Francis Group* (Boca Raton)
[14] Maxim A, Andreu D, Boucher J 1998 Novel behavioral method of SPICE macromodeling of magnetic components including the temperature and frequency dependencies *Proceedings of IEEE Application Power Electronics Conference and Exposition APEC'98*, **1** 393-399
[15] Maxim A, Andreu D, Boucher J 1999 New SPICE behavioural macromodelling method of magnetic components including the self-heating process *Proceedings of IEEE Annual Power Electronics Specialist Conference PESC* **2** 735-740
[16] Górecki K, Zarębski J 2009 Electrothermal analysis of the self-excited push-pull dc-dc converter *Microelectronics Reliability* **49** (4) 424-430
[17] Maksimovic D, Stankovic AM, Thottuvelil VJ, Verghese GC 2001 Modeling and simulation of power electronic converters *Proceedings of the IEEE* **89** (6) 898-912
[18] Wilamowski B, Jager RC 1997 Computerized circuit Analysis Using SPICE Programs *McGraw-Hill* (New York)
[19] Zarębski J, Górecki K 2010 The electrothermal large-signal model of power MOS transistors for SPICE *IEEE Transaction on Power Electronics* **25** (5-6) 1265 – 1274
[20] Górecki K, Detka K 2012 Electrothermal model of choking-coils for the analysis of dc-dc converters *Materials Science & Engineering B* **177** (15) 1248-1253
[21] Górecki K, Godlewska M 2015 Electrothermal model of ferromagnetic cores *Przegląd Elektrotechniczny* **91** (6) 161-165
[22] Górecki K, Rogalska M 2014 The Compact Thermal Model of the Pulse Transformer *Microelectronics Journal* **45** (12) 1795-1799
[23] Górecki K, Detka K 2015 The parameter estimation of the electrothermal model of inductors *Informacije MIDEM – Journal of Microelectronics, Electronic Components and Materials* **45** (1) 29-38
[24] Górecki K, Rogalska M, Zarębski J 2014 Parameter estimation of the electrothermal model of the ferromagnetic core *Microelectronics Reliability* **54** (5) 978-984

Modelling reverse characteristics of power LEDs with thermal phenomena taken into account

Przemysław Ptak, Krzysztof Górecki

Gdynia Maritime University, Department of Marine Electronics,
Morska 83, 81-225 Gdynia, Poland

E-mail: p.ptak@we.am.gdynia.pl, k.gorecki@we.am.gdynia.pl

Abstract. This paper refers to modelling characteristics of power LEDs with a particular reference to thermal phenomena. Special attention is paid to modelling characteristics of the circuit protecting the considered device against the excessive value of the reverse voltage and to the description of the temperature influence on optical power. The network form of the worked out model is presented and some results of experimental verification of this model for the selected diodes operating at different cooling conditions are described. The very good agreement between the calculated and measured characteristics is obtained.

1. Introduction

Power LEDs (Light Emitting Diode) are more and more commonly used in illumination technology [1 - 8]. They are also used in motorization [9]. In the last years one can observe LEDs operating parameters improvement and a significant decrease in purchasing costs of a single device [1].

From the point of view of a LED application in illumination technology the most significant is resistance of this devices at negative voltage polarization. In papers [5, 6] the circuit protection of LEDs against negative polarisation is discussed. It results from these papers that thyristors are more frequently used when the allowable reverse voltage obtains the level of above forty volts. When the allowable reverse voltage is in the range from several to tens volts typically Zener diodes are used as the circuit protection. In the analysis of characteristics of LEDs by CREE producer from the Xlamp family one can notice that the producer used the Zener diode as the circuit protection. The measured reverse characteristics of the selected power LEDs are shown in Figure 1.

On the other hand, in the stage of designing electronic circuits, computer simulations are made. In these simulations models of the devices occurring in the analysed system, which take into account all essential physical phenomena, are indispensable. The well-known from a literature models of power LEDs for SPICE, eg. [5, 10 - 13] skip the description of reverse characteristics of this device, because typically LEDs operate in the forward mode. However, in the case of the analysis of the power supply of light sources correct modelling of characteristics of power LEDs in all ranges of their operation is indispensible.

As it is commonly known, characteristics of diodes depend on their internal temperature, which is the sum of the ambient temperature and an excess resulting from thermal phenomena [14, 15]. Taking into account self-heating phenomena in the model of power LEDs demands the use of the electrothermal model of this device.

In the paper, the modified electro-thermal model of the LED for SPICE is proposed. This model takes into account the description of LED characteristics in the reverse range of the p-n junction.

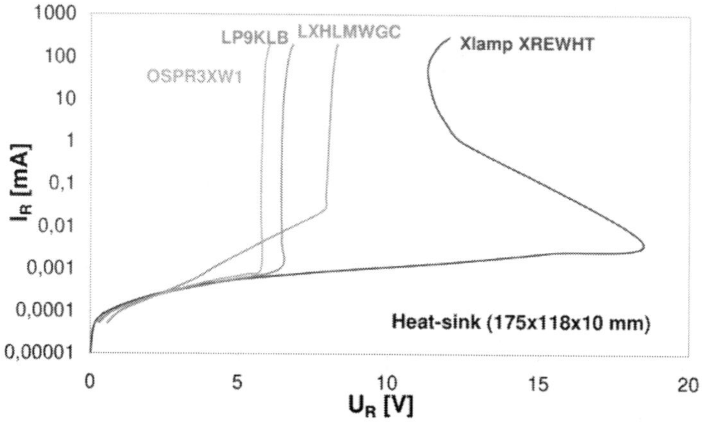

Figure 1. Reverse characteristics of the selected power LEDs.

In the following sections the form of the worked out model and the results of its experimental verification are presented.

2. The model form

The presented model belongs to a group of compact electrothermal models [10, 11, 14, 16, 17, 18]. Its network representation is shown in Figure 2.

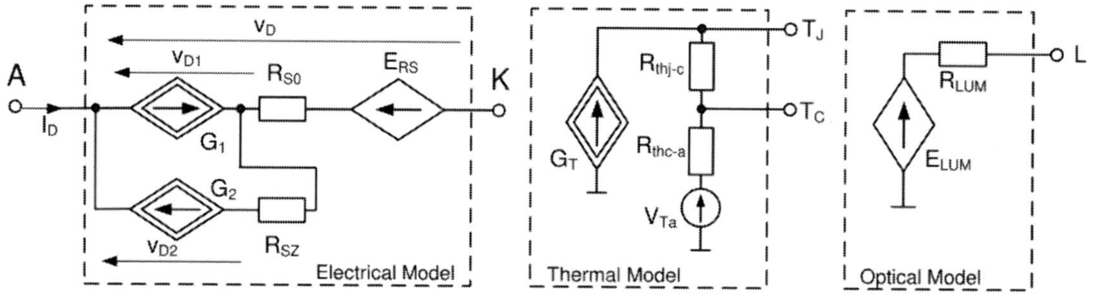

Figure 2. Network representation of the electrothermal model of the power LED.

This model contains 3 blocks representing the electrical model, optical model and thermal model of the power LED. The electrical model describes electrical characteristics of the considered device with regard to the influence of temperature on their course. The optical model describes the dependence of illuminance of lighted up area on temperature and the diode current. The thermal model makes possible to calculate the internal temperature T_j and the case temperatures T_C of the considered LED with self-heating phenomena taken into account.

The electrical model of the LED contains two branches. The first of them, containing the controlled current source G_1, the resistor R_{S0} and the controlled voltage source E_{RS} models the characteristics of the LED. In turn, the other branch, containing the controlled current source G_2 and the resistor R_{SZ} describes the characteristics of the protecting circuit of the LED against a breakdown. The controlled current source G_1 describes the diffusive component of the current of the diode, described with the classical equation contracted among other things in papers [10, 11, 14]. The resistor R_{S0} represents the series resistance of this diode in the reference temperature T_0, while the controlled voltage source E_{RS} models the influence of internal temperature of the diode on this resistance.

The current of the source G_1 is given by the formula [11]

$$I_{G1} = I_0 \cdot \left(\frac{T_j}{T_0}\right)^3 \cdot \exp\left(-\frac{U_{g0}}{n \cdot \frac{k}{q} \cdot T_j}\right) \cdot \left[\exp\left(\frac{v_{D1}}{n \cdot \frac{k}{q} \cdot T_j}\right) - 1\right] \tag{1}$$

where I_0 denotes a constant independent of temperature, T_j – internal device temperature, T_0 – reference temperature, U_{g0} – voltage corresponding to the band-gap of a semiconductor used to produce the considered device, n – emission coefficient, k – Boltzmann constant, q – electron charge, v_{D1} – voltage drop on the p-n junction of considered LED.

In turn, the voltage on the source E_{RS} is given by the formula [11]

$$E_{RS} = v_{RS0} \cdot \alpha_{RS} \cdot (T_j - T_0) \tag{2}$$

where v_{RS0} denotes the voltage drop on the resistor R_{S0}, α_{RS} - temperature coefficient of diode series resistance.

The reverse current of power LED is described by the controlled current source G_2, the output voltage of which is given by the formula

$$I_{G2} = I_Z \cdot \exp\left(\frac{-v_{D2} - U_{BR} \cdot \left(1 + \alpha_{UZ} \cdot \left(T_j - T_0\right)\right)}{\lambda \cdot \frac{k}{q} \cdot T_j}\right) \tag{3}$$

where I_Z denotes the breakdown current of the diode at its voltage equal to $v_D = - U_{BR}$ and temperature T_0, U_{BR} – breakdown voltage of the LED at the reference temperature T_0, α_{UZ} – temperature coefficient of the breakdown voltage of the LED, λ – coefficient modelling hardness of the LED characteristics in the breakdown mode, v_{D2} – the voltage on the current source G_2.

The resistor R_{SZ} models a series resistance in the branch representing the reverse current of the LED.

The optical model describes the dependence of illuminance on the forward current of the diode and on temperature with the use of the dependence presented in paper [11]. The voltage at the removal L corresponds to illuminance of the light emitted by the LED and is described by the following formula:

$$E_{LUM} = E_0 \cdot \left(1 - \exp\left(-\alpha_L \cdot I_D \cdot (1 + \alpha_T \cdot (T_j - T_0) + \alpha_{T2} \cdot (T_j - T_0)^2\right)\right) \tag{4}$$

where E_0 denotes the maximum value of illuminance at temperature T_0, α_L – current coefficient of illuminance, α_T, α_{T2} – linear and square temperature coefficients of illuminance.

The manner of estimating values of electrical and optical parameters occurring in the model of the power LED are described in paper [10] except parameters of the source G_2 describing the reverse current of the investigated LED.

The thermal model makes it possible to calculate the internal temperature T_j and case temperature T_C of the power LED. Values of these temperatures correspond to voltages in nodes T_J and T_C, respectively. The controlled current source G_T represents thermal power given off in the investigated device, and the voltage source V_{Ta} represents the ambient temperature T_a.

The thermal power given off in the LED shows the difference between the electric power received from the power source (the product of the voltage and the current on connectors of the investigated device) and the power of emitted radiation. In compliance with the well-founded dependence in papers [8, 11], optical power is equal to the product of electric power and efficiency of conversion of electrical energy to light η_L. This efficiency is proportional to the luminous efficiency of the LED, given by the producer. Of course, in the reverse mode of the investigated LED optical power is zero.

In papers [10, 11, 13] the constant value of the parameter η_L is used, because the modelled LEDs is characterised by the low value of luminous efficiency. However, the results of measurements, included in the quoted papers prove that power of radiation emitted by the LED is a decreasing function of the internal temperature T_j. In the presented model this dependence is taken into account and it is described by the current of the source G_T with the following equation

$$G_T = \begin{cases} v_D \cdot I_D \cdot \left(1 - \eta_L \cdot \dfrac{1 - \exp\left(-\alpha_L \cdot I_D \cdot (1 + \alpha_T \cdot (T_j - T_0) + \alpha_{T2} \cdot (T_j - T_0)^2)\right)}{1 - \exp\left(-\alpha_L \cdot I_D\right)}\right) & for \ \ I_D > 0 \\ v_D \cdot I_D \ \ for \ \ I_D \le 0 \end{cases} \quad (5)$$

where v_D and I_D are marked in Figure 2.

3. Results

The correctness of the model described in Section 2 was tested for the selected power LEDs offered by CREE. The research was conducted for 3 diodes: MCE4WT-A2-0000-00K01 (marked further as MCE), XPEWHT-L1-0000-00F01 (marked further as XPE) and XMLBWT-00-0000-0000U2051 (marked further as XML). The values of basic parameters of these diodes are given in Table 1. As it is visible, the admissible reverse voltage V_{Rmax} for every diode is 5V, the maximum forward current I_{Fmax} accepts values from 0.7 to 3 A, and the admissible internal temperature T_{jmax} for the investigated diodes is 150°C.

Table 1. Values of basic exploitive parameters of the investigated diodes.

Type of LED	I_{Fmax} [A]	T_{jmax} [°C]	V_{Rmax} [V]
XML	3	150	5
XPE	1	150	5
MCE	0.7	-	5

The investigated diodes operated at two cooling conditions: the operation of these devices without any heat-sink and while being mounted to the aluminium heat-sink, called the large heat-sink, of the dimensions 175x118x10 mm.

The values of parameters describing electrical and optical models of the considered devices are estimated with the use of the idea of local estimation [10, 19], whereas thermal parameters are measured with the method described in paper [20]. The obtained values of electrical and optical parameters for the considered diodes are assembled in Table 2, and the values of thermal parameters - in Table 3.

Table 2. Values of electrical and optical parameters of the investigated diodes.

Type of LED	E_0 [klx]	α_L [A^{-1}]	α_T [10^{-3}/K]	α_{T2} [10^{-3}/K^2]	λ	α_{UZ} [10^{-3}/K]	α_{RS} [10^{-3}/K]	η_L
XML	12.3	0.43	8	-0.04	1	0.5	1.5	0.373
XPE	9	0.5	2	-0.085	1	0.55	1.5	0.373
MCE	8.5	0.6	2	-0.085	2,5	0.5	1.5	0.373
	R_{S0} [Ω]	R_{SZ} [Ω]	I_0 [A]	U_{g0} [V]	n	I_Z [mA]	U_{BR} [V]	
XML	0.14	16	124	3.33	3.7	5	17.7	
XPE	0.1	18	220	3.6	3.7	5	17.8	
MCE	0.17	7	260	3.6	3.5	5	12.25	

Table 3. Values of thermal parameters of investigated diodes.

Type of LED	LED without any heat-sink		LED on the large heat-sink	
	R_{thJ-C} [K/W]	R_{thC-A} [K/W]	R_{thJ-C} [K/W]	R_{thC-A} [K/W]
XML	9.7	29.6	9.7	3.8
XPE	16.49	22	16.49	5.79
MCE	4.4	29	4.4	5

In Figures 3 - 7 the calculated and measured characteristics of the considered diodes are presented. In these figures points present the results of measurements, and lines – the results of calculations. Figure 3 presents the calculated and measured characteristics of the considered diodes in the forward mode, and Figure 4 - corresponding to them dependences of the case temperature (solid lines) or the internal temperature (dashed lines) on the forward current.

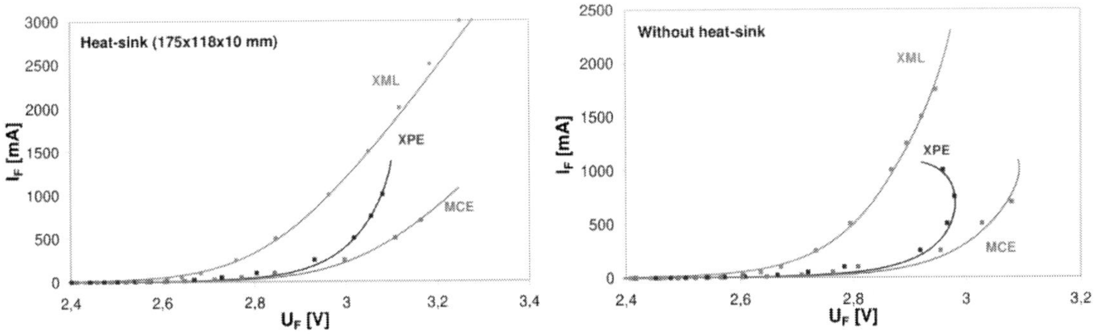

Figure 3. Current-voltage characteristics of the forward biased LEDs at different cooling conditions.

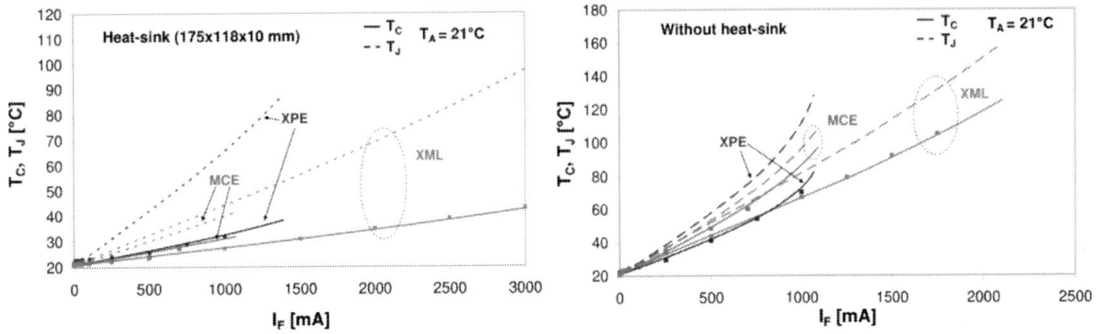

Figure 4. Dependences of the internal and case temperatures on the forward current at different cooling conditions.

As one can notice, the very good agreement between the calculated and measured characteristics of the investigated devices at both considered cooling conditions was obtained. As expected, for diodes operating without the heat-sink considerably higher values of the case and internal temperatures of the considered devices were obtained. Excesses of the values of these temperatures over the ambient temperature (T_a = 21°C) differ between each other even 5-times. It is worth noticing also that the difference of the values of the internal temperature and the case temperature of the investigated diodes is considerably higher for diodes situated on the heat-sink. This is due to the fact that values of thermal resistance R_{thJ-C} results from the construction of the investigated diodes and it does not depend on the cooling system, while values of thermal resistance R_{thC-A} for the diode without any heat-sink are even 7- times greater than for the diode situated on the heat-sink. The result of a rise of the internal temperature of the investigated diodes is a fall of the forward voltage of these devices. This fall is most strongly visible for the diode XML. For the diode XPE operating without any heat-sink the negative slope of the i(v) characteristics is observed for high values of the forward current. It is result of self-heating phenomenon.

In Figure 5 the dependence of illumination of LEDs on the forward current I_F is presented. The measurements and simulations were conducted for both considered cooling conditions. During the measurements the sensor was situated 19 cm from the surface of the investigated diode in its optical axis. In this figure solid lines refer to diodes operating without any heat-sink, whereas dashed lines - diodes situated on the heat-sink.

In this figure one can notice that typically an increase in the value of the forward current causes an increase in illuminance emitted by the investigated diodes. Simultaneously, worsening of cooling conditions of the diode and connected with this the internal temperature rise is effective with depreciation of illuminance. In the extreme case, for the diode MCE operating without any heat-sink, the considered dependence has the maximum.

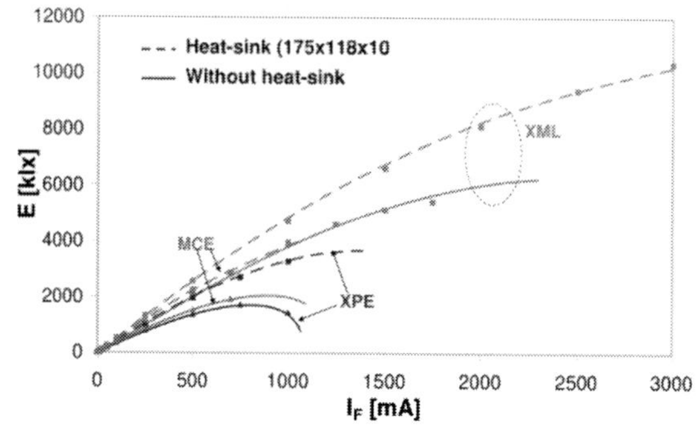

Figure 5. Calculated and measured dependences of illuminance of the investigated diodes on the forward current at different cooling conditions.

Figure 6 presents current-voltage characteristics of reverse biased LEDs mounted on the large heat-sink (a) and without heat-sink (b). As it can be observed the value of the breakdown voltage of the considered diodes changes in the range from 12 to 18 V, whereas self-heating causes the shift in the current-voltage characteristics to the right side.

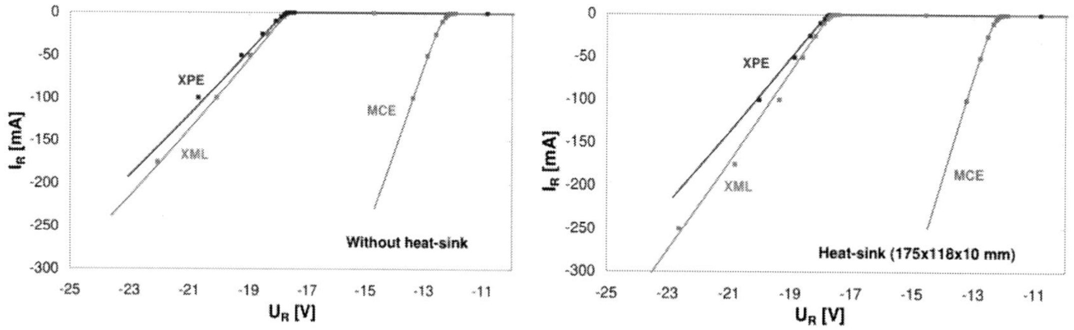

Figure 6. Reverse characteristics of the considered diodes situated on the heat-sink and operating without any heat-sink .

Figure 7 shows the dependence between the junction temperature of the LED on the total power in these devices. In this figure, solid lines correspond to the forward biased LEDs, whereas dashed lines – to the LEDs operating in the reverse mode.

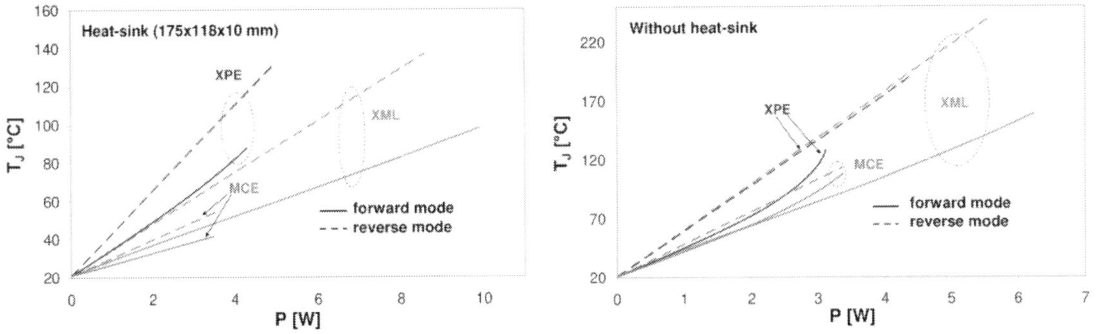

Figure 7. The dependence of internal temperature T_j of the investigated diodes on the total power P dissipated in the LED.

As expected, dependences $T_j(P)$ obtained for both ranges of the operation of the considered diodes differ between each other. At the same value of the total power for the diode operating in the forward mode the smaller value of the internal temperature than for the diode operating within the breakdown range is obtained. The difference of the value of the total power at which identical values of the internal temperature are obtained corresponds to the power of emitted light P_{opt}. On the basis of the obtained courses $T_j(P)$ optical efficiency of the considered diodes is calculated from the equation:

$$\eta_{opt} = \frac{P_{opt}}{P} \tag{6}$$

The obtained values of optical efficiency η_{opt} for the considered diodes, obtained at different cooling conditions, are collected in Table 4.

Table 4. Values of optical efficiency of the considered LEDs at maximum value of forward current and different cooling conditions

Type of LED	LED without any heat-sink		LED on the large heat-sink		I_{Fmax} [A]
	η_{opt}[%]	T_j [°C]	η_{opt}[%]	T_j [°C]	
XML	42	176.4	42	98.1	2.3
XPE	11	128.4	30	87.9	1
MCE	5	108.3	37	41.6	1

As one can notice, the influence of cooling conditions on the optical efficiency for considered LEDs is strongly different. Considered optical efficiency accepts the highest value, equal 42 % for the diode XML and practically does not change at the change of cooling conditions. On the other hand, the strongest influence of self-heating on the optical efficiency appears for the diode MCE in which the optical efficiency decreases even 7 times at operating of considered device without any heat-sink.

4. Conclusions

In the paper the electrothermal model of power LEDs for SPICE, taking into account the reverse characteristics of this device and influence of temperature on optical power is proposed. The investigations carried out for three different types of power LEDs produced by CREE show that this model assured the very good agreement between the calculated and measured electrical, optical and thermal characteristics of the considered devices operating at different cooling conditions. This testifies to the correctness of the proposed model.

Using the dependence of internal temperature of the investigated diodes on the total power, optical power of these diodes and efficiency of conversion of electrical energy to light are calculated. For the considered diodes this efficiency exceeds 30% and it decreases when its cooling conditions are worsening.

The simple form of the presented model and little duration time of calculations (up to the split second) allows anticipating that this model can be useful for designers of lighting systems with power LEDs.

5. References

[1] Craford MG 2007 Current State of the Art in High Brightness LEDs *CTO American Physical Society, Solid State Lighting Session*

[2] Czyżewski D 2012 LED substitutes of conventional incandescent lamps *Przegląd Elektrotechniczny* **88** (11a) 123-127

[3] Weir B 2012 Driving the 21st Century's Lights *IEEE Spectrum* **49** (3) 42-47

[4] Górecki K 2013 The influence of power supply voltage on exploitive parameters of the selected lamps *Informacije MIDEM- Journal of Microelectronics, Electronic Components and Materials* **43** (3) 193-198

[5] Lasance CJM, Poppe A 2014 Thermal Management for LED Applications *Springer Science+Business Media* (New York)

[6] Schubert EF 2008 Light-Emitting Diodes *Second Edition Cambridge University Press* (New York)

[7] Krejcar O, Frischer R 2013 Smart intelligent control of current source for high power LED diodes *Microelectronics Journal* **44** (4) 307-314

[8] Uddin S, Shareef H, Mohamed A, Hannan MA 2012 Harmonics and thermal characteristics of low wattage LED lamps *Przegląd Elektrotechniczny* **88** (11a) 266-271

[9] High power automotive forward lighting LED *Philips Lumileds* [11 Dec 2013] www.philipslumileds.com/uploads/15/AB35-pdf .

[10] Górecki K 2015 Modelling mutual thermal interactions between power LEDs in SPICE *Microelectronics Reliability* **55** (2) 389-395

[11] Górecki K, Ptak P 2015 Modelling power LEDs with thermal phenomena taken into account. *Proceedings of 22nd International Conference Mixed Design of Integrated Circuits and Systems Mixdes 2015* (Toruń) 431-435

[12] Poppe A and Lasance CJM 2009 On the standardization of thermal characterization of LEDs *25th IEEE Semi-Therm Symposium* 151-158

[13] Poppe A 2012 A step forward in multi-domain modeling of power LEDs *28th Annual IEEE Semiconductor Thermal Measurement and Management Symposium SEMI-THERM* (San Jose) 325–330

[14] Górecki K 2012 Electrothermal model of a power LED for SPICE *International Journal of Numerical Modelling Electronic Network, Devices and Fields* **25** (1) 39-45

[15] Górecki K, Górecka K 2011 Wpływ zjawisk cieplnych na właściwości diody LED mocy (The influence of thermal phenomena on properties of power LEDs) *Przegląd Elektrotechniczny* **87** (7) 144-147

[16] Mawby PA, Igic PM, Towers MS 2001 Physically based compact device models for circuit modelling applications *Microelectronics Journal* **32** 433–447

[17] Zarębski J, Górecki K 2010 The electrothermal large-signal model of power MOS transistors for SPICE *IEEE Transaction on Power Electronics* **25** (5-6) 1265 – 1274

[18] Starzak Ł, Zubert M, Janicki M, Torzewicz T, Napieralska M, Jabłoński G, Napieralski A 2013 Behavioral Approach to SiC MPS Diode Electrothermal Model Generation *IEEE Transactions on Electron Devices* **60** (2) 630-638

[19] Zarębski J, Górecki K 2002 Parameters Estimation of the D.C. Electrothermal Model of the Bipolar Transistor *International Journal of Numerical Modelling Electronic Networks, Devices and Fields* **15** (2) 181-194

[20] Górecki K 2014 Measurements of thermal resistance of power LEDs *Microelectronics International* **31** (3) 217-223

IOP Conf. Series: Materials Science and Engineering **104** (2015) 012013 doi:10.1088/1757-899X/104/1/012013

Modelling characteristics of photovoltaic panels with thermal phenomena taken into account

Ewa Krac and Krzysztof Górecki

Gdynia Maritime University, Department of Marine Electronics,
Morska 83, 81-225 Gdynia, Poland

E-mail: e.krac@we.am.gdynia, k.gorecki@we.am.gdynia.pl

Abstract. In the paper a new form of the electrothermal model of photovoltaic panels is proposed. This model takes into account the optical, electrical and thermal properties of the considered panels, as well as electrical and thermal properties of the protecting circuit and thermal inertia of the considered panels. The form of this model is described and some results of measurements and calculations of mono-crystalline and poly-crystalline panels are presented.

1. Introduction

Photovoltaic panels are important components of solar power stations and autonomous photovoltaic (PV) systems [1 - 7]. They consist of photovoltaic cells connected in series, in which there is conversion of solar energy into electrical energy. These devices contain a p-n junction, the properties of which strongly depend on its internal temperature T_j [8, 9, 10]. This temperature increases both as a result of absorption of heat generated by the radiation source and due to self-heating [1, 8, 9, 11, 12]. Therefore, these phenomena should be taken into account while designing and analyzing solar power stations [3, 13, 14, 15].

Currently, computer programs are commonly used to design and analyze characteristics of PV panels [3, 12, 13]. In order to calculate the characteristics of such a system before its construction the model of this device should be elaborated. Unfortunately, there is a lack of papers describing the operation of PV panels, which take a thermal phenomenon into account [16]. Some of the mentioned models treat phenomena in a simplified way [12]. In these models temperature is only a parameter.

In earlier papers of the authors [13, 16] electrothermal models of solar cells, which is the main component of photovoltaic panels are proposed. However, these models omit thermal inertia of the investigated devices. In paper [13] the use of the considered models to compute dc characteristics of the photovoltaic panels is proposed. The presented in quoted paper method of calculations demands uses the dc analysis in SPICE software and additional calculations on the basis of the obtained results of computations. In neither of the quoted papers the properties of the circuit protecting the photovoltaic panel against irregular connection to the photovoltaic system are taken into account.

In this paper, the new form of the electrothermal model of PV panels taking into account electrical, optical and thermal phenomena as well as thermal inertia in the panel, is proposed. In section two the detailed description of the proposed model of a photovoltaic panel is presented. The results of experimental verification of the presented model are shown and discussed in section three.

2. The model form

The elaborated by the authors electrothermal model of the photovoltaic panel belongs to the group of compact electrothermal models, presented for many semiconductor devices e.g. in papers [14, 17, 18]. This kind of models consists of two sub-models: the electrical model and the thermal model. The electrical model describes the current-voltage characteristics of the considered devices with the influence of temperature taken into account. In turn, in the thermal model the value of the device internal temperature T_j is calculated taking into account the changes of the ambient temperature T_a and self-heating phenomenon.

The network representation of the elaborated electrothermal model of a photovoltaic panel is shown in Figure 1.

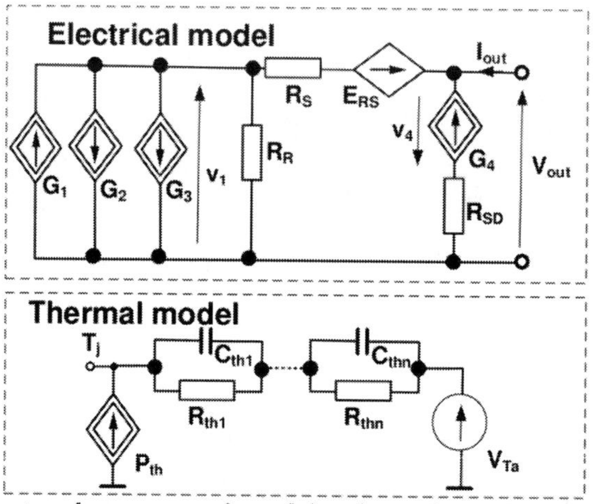

Figure 1. The network representation of the electrothermal model of PV panels.

In the electrical model, the controlled current source G_1 represents the photoelectric current, in turn, the controlled current source G_2 represents a component of the diffusion current of the p-n junctions, the controlled current source G_3 models the generation-recombination component of the p-n junctions current, whereas the controlled current source G_4 represents the current of the diode protecting the modelled panel against a reverse connection of this panel to the photovoltaic system. The resistor R_R models the leakage current of the panel, whereas the resistor R_S –series resistance of the panel at the reference temperature T_0. The controlled voltage source E_{RS} models the temperature dependence of series resistance. The resistor R_{SD} represents series resistance of the protecting diode.

In the thermal model, the controlled current source P_{th} represents the sum of power dissipated in the PV panel and power of absorbed radiation, whereas R_{th1}, …, R_{thn} and C_{th1}, …, R_{thn} are thermal resistances and thermal capacitances of the panel, respectively. The voltage source V_{Ta} represents the ambient temperature T_a.

The photoelectric current is described by the following formula [16]

$$I_1 = P \cdot S \cdot \eta \cdot x \cdot \left[1 + \alpha_T \cdot \left(T_j - T_0\right)\right] \tag{1}$$

where P denotes power density of lighting radiation, S - active area of the solar cell, η – efficiency of photovoltaic conversion, x - coefficient equal to 1 V^{-1}, α_T – temperature coefficient of changes of the photoelectric current, T_j – internal temperature of the solar cell, T_0 - reference temperature.

In turn, the diffusion current of the p-n junctions is described by the equation

$$I_2 = J_o \cdot S \cdot \left(\frac{T_j}{T_0}\right)^2 \cdot \exp\left(-\frac{U_{go}}{m \cdot n \cdot k/q \cdot T_j}\right) \cdot \left[\exp\left(\frac{v_1}{m \cdot n \cdot k/q \cdot T_j}\right) - 1\right] \tag{2}$$

where J_0 denotes the parameter of the saturation current density, n – emission factor of the p-n junction, m – number of solar cells in the panel, U_{go} – voltage corresponding to the band-gap of silicon, v_1 – voltage on the current source G_2, k is the Boltzmann constant, q - electron charge.

The generation-recombination current is described by the formula

$$I_3 = J_{02} \cdot S \cdot \frac{T_j}{T_0} \cdot \exp\left(-\frac{U_{go}}{m \cdot n_1 \cdot \frac{k}{q} \cdot T_j} \right) \cdot \left[\exp\left(\frac{v_1}{m \cdot n_1 \cdot \frac{k}{q} \cdot T_j} \right) - 1 \right] \tag{3}$$

where J_{02} denotes the parameter of the generation-recombination current density, n_1 – emission factor for the generation-recombination current.

The controlled voltage source E_{RS} models the influence of temperature on series resistance of the panel by linear dependences of the form

$$E_{RS} = V_{RSO} \cdot \alpha_{RS} \cdot (T_j - T_0) \tag{4}$$

where α_{RS} denotes the temperature coefficient of series resistance, whereas V_{RSO} – the voltage on the resistor R_S.

The current of the protecting diode is described by the formula

$$I_4 = I_{04} \cdot \left(\frac{T_j}{T_0} \right)^2 \cdot \exp\left(-\frac{U_{go}}{n_4 \cdot \frac{k}{q} \cdot T_j} \right) \cdot \left[\exp\left(\frac{v_4}{n_4 \cdot \frac{k}{q} \cdot T_j} \right) - 1 \right] \tag{5}$$

where I_{04} is the parameter of the saturation current of the protecting diode, whereas n_4 is the emission factor of this diode.

In the thermal model, the current source P_{th} represents power dissipated in the panel and power of radiation absorbed by the panel. The current of the controlled current source P_{th} is described by

$$P_{th} = P \cdot a \cdot S_1 \cdot m + (V_{out} - v_1) \cdot I_{out} + v_1 \cdot (I_2 + I_3) \tag{6}$$

where a denotes the ratio of conversion of radiation energy into heat, S_1 – area of the solar cell, while V_{out} and I_{out} are the output voltage and the output current of the panel, respectively.

3. Results

The correctness of the elaborated electrothermal model of photovoltaic panels is verified experimentally for two arbitrary selected panels. First of them is a mono-crystalline panel SGM -250D of the dimensions 1650 x 990 x 40 mm, and the other is a poly-crystalline panel Q.PRO-G3 250-265 of the dimensions 1670 x 1000 x 35 mm. Both the panels consist of 60 solar cells and their nominal output power is equal to 250 W.

Using the method described in paper [16] the electrical and optical parameters of the considered panels are estimated, whereas the thermal parameters are measured using the method described in [19]. The values of parameters of the considered panels are collected in Table 1.

In Figures 2 – 7 the results of measurements (points) and calculations (lines) of the considered panels are presented. In these Figures the results obtained for the poly-crystalline panel are denoted with blue colour, whereas the results obtained for the mono-crystalline panel are denoted with red colour. All measurements are realized using the measurement system described in [13] and calculations are realized with the use of SPICE software.

Figure 2 illustrates the output characteristics of the considered panels obtained at two values of power density of lighting radiation equal to 168 W/m^2 and 335 W/m^2, respectively. Figure 3 illustrates the dependence of temperature of the panels on the output current.

As it is visible, the considered v(i) characteristics are decreasing functions. An increase in the power density of lighting radiation causes an increase in the value of the output current. For the poly-crystalline panel, the obtained values of the panel output voltage are higher than for the monocrystalline panel by even about 1.5 V. Analysing the dependence $T_j(i)$, it is visible, that the output current very weakly influences the value of the panel temperature. This proves that the observed excess of the panel temperature over the ambient temperature results mostly from conversion of lighting energy into heat. This excess is even equal to about 70°C for power density of lighting radiation equal to 335 W/m^2.

Table 1. The estimated values of model parameters of the considered panels

parameters	S [m^2]	S$_1$ [m^2]	η	α_T [K^{-1}]	J$_0$ [A/m^2]	J$_{02}$ [A/m^2]	I$_{04}$ [A]
poly-crystalline	0.0231	0.024	0.5952	2.51x10^{-3}	9.44x10^{12}	1.51x10^5	10^4
mono-crystalline	0.0231	0.024	0.588	2.34x10^{-3}	9.44x10^{12}	1.51x10^5	70
parameters	n	n$_1$	n$_4$	m	U$_{go}$ [V]	α_{RS} [K^{-1}]	R$_S$ [Ω]
poly-crystalline	0.95	2.72	1.3	60	1.206	0.00279	0.186
mono-crystalline	0.98	2.82	1.8	60	1.206	0.00293	0.234
parameters	R$_{SD}$ [Ω]	R$_R$ [Ω]	R$_{th1}$ [K/W]	C$_{th1}$ [J/W]	R$_{th2}$ [K/W]	C$_{th2}$ [J/W]	a
poly-crystalline	0.08	10^4	0.0106	3.18x10^5	0.03083	3.474x10^4	3.25
mono-crystalline	0.135	10^4	0.0166	1.57x10^5	0.0234	3.54x10^4	3.41

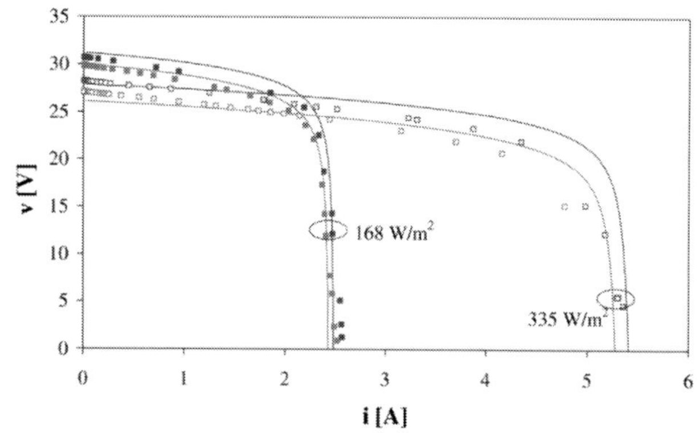

Figure 2. The calculated and measured output characteristics of the lighted up panels.

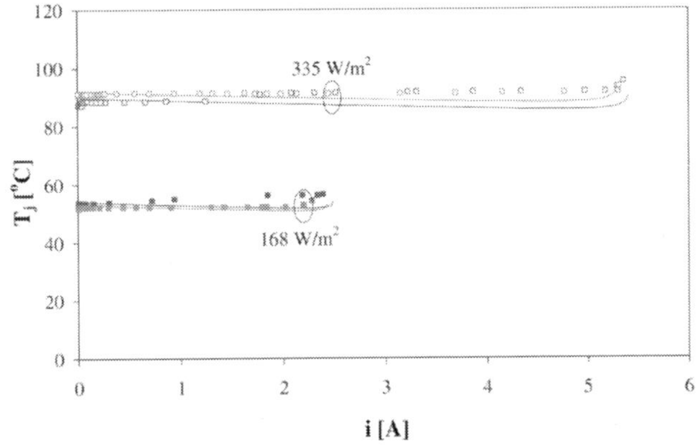

Figure 3. The calculated and measured dependences of the temperature of the considered panels on their output current.

In Figure 4 the output characteristics of the considered panels operating without lighting up are presented, whereas in Figure 5 – corresponding to them dependences of temperature of these panels on the panel output current.

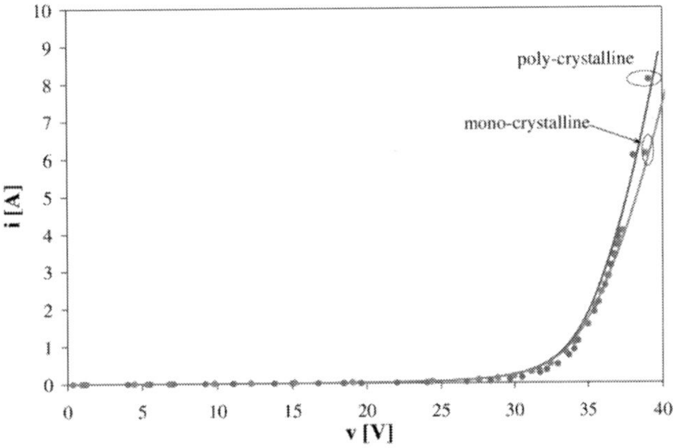

Figure 4. The calculated and measured output characteristics of the non lighted up panels.

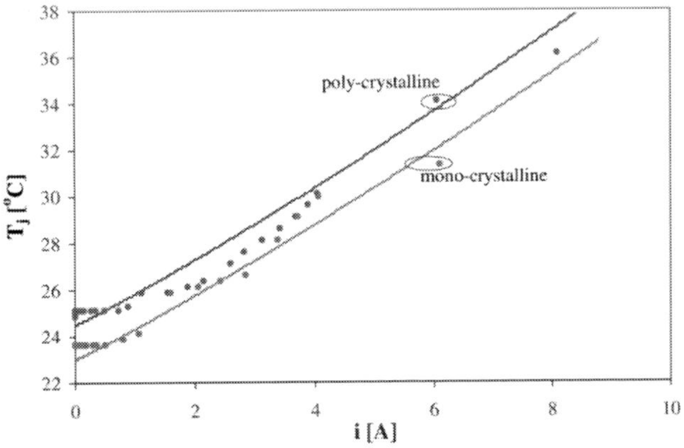

Figure 5. The calculated and measured dependences of the temperature of non lighted up panels on the output current.

As it is seen, the forward current of the considered panels is an exponentialy increasing function of the forward voltage. This voltage assumes higher values for the mono-crystalline panel. In turn, the temperature of the panel linearly increases with an increase of the forward current. It is worth noticing, that for the maximum allowable value of the panel forward current (equal to 8 A), a self-heating phenomenon causes an increase in the panel temperature value equal only by 12°C.

The characteristics presented in Figures 2-5 correspond to operation of the considered panels at the steady state. In Figure 6 a waveform of temperature of non lighted up panels obtained at rectangular pulses train of their output current of the low value equal to 0 and the high value equal to 8 A is presented. The pulse width and period are selected in order to obtain the steady state in the panel during each pulse of the current. It is worth noticing that the steady state is obtained in the considered panel after about 3 hours (10000 s).

Figure 7 illustrates the characteristics of the non lighted up panels operating at the reverse biasing voltage.

It is easy to observe that the reverse voltage of the poly-crystalline panel is higher than this voltage of the mono-crystalline panel. Power dissipated in the protecting diodes is very small and therefore, an increase of the panel temperature is not observed.

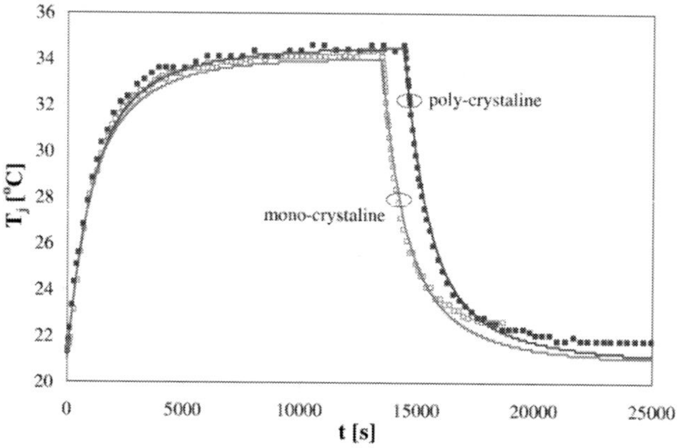

Figure 6. The calculated and measured waveforms of panels temperature by stimulation with the rectangular current pulses train.

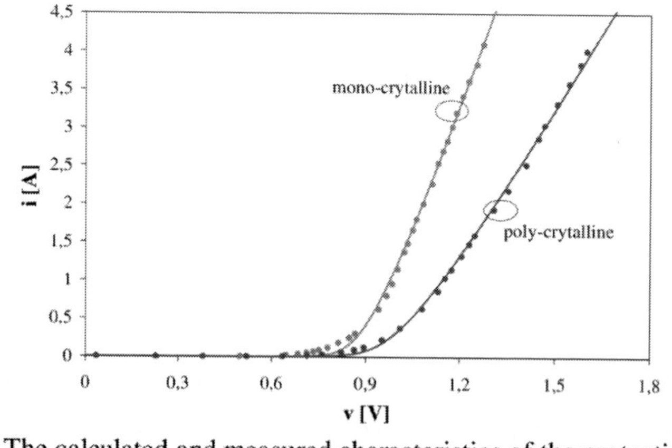

Figure 7. The calculated and measured characteristics of the protecting diodes.

4. Conclusions

In the paper the new electrothermal model of photovoltaic panels is proposed. This model takes into account optical, electrical and thermal properties of the considered panels. Apart from properties of solar cells included in the considered panels, the characteristics of protecting diodes are taken into account in this model. In the thermal model of the panel, thermal inertia is included.

The correctness of the model was tested by examining characteristics of mono-crystalline and poly-crystalline solar panels for different values of power of lighting radiation and different biasing conditions. For all the considered operating conditions the good agreement between the results of calculations and measurements was obtained.

The presented model can be useful for designers of solar energy generation systems, as well as for designers of electronics systems supplied with solar panels.

5. References

[1] Rashid MH 2007 Power Electronic Handbook *Academic Press Elsevier*

[2] Zdanowski M 2015 Wielogałęziowy wysokoczęstotliwościowy przekształtnik DC/DC z łącznikami z SiC i elementami magnetycznymi o zredukowanej pojemności pasożytniczej *Doctor Thesis Politechnika Warszawska*

[3] Górecki P 2015 Voltage Regulators for the Laptop's Power Supply Station with Photovoltaic Modules *Proceedings of 22nd International Conference Mixed Design of Integrated Circuits and Systems Mixdes 2015* (Toruń) 571-575

[4] Lee WM, Infield DG and Gottschalg R 2001 Thermal modelling of building integrated PV systems *17th European Photovoltaic Solar Energy Conference and Exhibition* Munchen 1447

[5] Klugmann-Radziemska E 2010 Fotowoltaika w teorii i praktyce *Wydawnictwo BTC* (Legionowo)

[6] Ersam T, Kimball J, Krein P, Chapman P and Midya P 2006 Dynamic Maximum Power Point Tracking of Photovoltaic Arrays Using Ripple Correlation Control *IEEE Transactions on Power Electronics* **21** (5) 1282-1291

[7] Mulvanet D 2014 Solar's Green Dilemma *IEEE Spectrum* **51** (9) 26-29

[8] Blackburn DL, Oettinger FF 1976 Transient Thermal Response Measurements of Power Transistors *IEEE Transactions on Industrial Electronics and Control Instrumentation* **IECI- 22** (2) 134-141

[9] Zarębski J and Górecki K 2008 A Method of the Thermal Resistance Measurements of Semiconductor Devices with P-N Junction *Measurement* **41** (3) 259-265

[10] Singh J 2000 Semiconductor devices: basic principles *John Wiley&Sons*

[11] Krac E, Górecki K 2014 Measurements of thermal resistance of solar cells *Zeszyty Naukowe AMG* **84** 56-65

[12] Castaner L, Silvestre S 2002 Modelling photovoltaic systems using Pspice *John Wiley&Sons*

[13] Górecki K, Krac E, Zarębski J 2015 Photo-electro-thermal characteristics of photovoltaic panels *Springer Proceedings in Energy, 2nd International Congress on energy Efficiency and Energy Related Materials ENEFM 2014* (Oludeniz) 45-51 doi: 10.1007/978-3-319-16901-9_6.

[14] Gadjeva E and Hristov M 2013 Behavioral Parametrized SPICE Models of Photovoltaic Modules *Proceedings of the 20th International Conference Mixed Design of Integrated Circuits and Systems MIXDES* (Gdynia) 355-359

[15] Villalva MG, Gazoli JR and Filbo ER 2009 Comprehensive Approach to Modeling and Simulation of Photovoltaic Arrays *IEEE Transactions on Power Electronics* **24** (5) 1198-1208

[16] Górecki K, Górecki P, Paduch K 2014 Modelling solar cells with thermal phenomena taken into account *Journal of Physics: Conference Series* **494** 012007 doi:10.1088/1742- 6596/494/1/012007

[17] Mawby PA, Igic PM and Towers MS 2001 Physically based compact device models for circuit modelling applications *Microelectronics Journal* **32** 433 – 447

[18] Zarębski J and Górecki K 2010 The electrothermal large-signal model of power MOS transistors for SPICE *IEEE Transactions on Power Electronics* **25** (5-6) 1265 – 1274

[19] Górecki K, Krac E 2015 Measurements of thermal parameters of a solar module *Proceedings of Microtherm 2015, Microtechnology and Thermal Problems in Electronics* (Łódź) 165-170

IOP Conf. Series: Materials Science and Engineering **104** (2015) 012014 doi:10.1088/1757-899X/104/1/012014

SPICE-aided modeling of high-voltage silicon carbide JFETs

Kamil Bargieł, Janusz Zarębski, Damian Bisewski

Gdynia Maritime University, Dept. of Marine Electronics, Morska 83, 81-225 Gdynia, Poland

E-mail: k.bargiel@we.am.gdynia.pl,

Abstract. The paper presents the static characteristics of the SiC transistor SJEP170R550 offered by SemiSouth obtained from simulations using JFET model built-in in PSPICE. Values of the model parameters were estimated using MODEL EDITOR, as well as procedure described in the literature. Simulation results were verified experimentally by comparison of results of measurements.

1. Introduction

For many years, JFETs (Junction Field Effect Transistors) have been applied in the power electronic circuits and systems, especially in switching circuits [1], uninterruptible power supplies [2] and high voltage switching power supply systems [3]. The major drawbacks of silicon JFET is low electric strength and operating temperature. In general, the maximum value of the drain-source voltage of such transistors does not exceed 150 V, whereas a typical value of theirs operating temperature is up to 150°C.

Since 2008, power JFETs made of silicon carbide have been commercially available on the market. Properties of this material allow to get the drain-source voltage of considered device above 1 kV and a higher operating temperature - typically 175°C.

The models implemented in computer programs for analysis and designs of electronic systems are of a great importance for engineers and designers of electronic circuits. One of the most popular computer program is PSPICE, containing built-in models of semiconductor devices, including the JFET model. Apart of this, PSPICE allows to implement a various models offered i.e. by manufacturers of semiconductor devices.

An important problem when using the built-in JFET model, is to have a credible values of the model parameters, estimated by appropriate methods based on the results of measurements of transistor characteristics or catalogue characteristics available in the datasheet of the semiconductor device.

In the evaluation process of the JFET model parameters, literature estimation algorithms [4-7] or MODEL EDITOR program, which is a component of PSpice software package, can be used.

The paper presents and discusses results of calculations of static characteristics of the SJEP170R550 JFET made of silicon carbide fabricated by SemiSouth. The calculations were performed using the PSPICE built-in model of the considered transistor [8]. The values of the model parameters are determined by MODEL EDITOR, as well as using the estimation method described in the literature. The influence of the ambient temperature on the static characteristics and operating parameters of the JFET were investigated, as well.

2. The model of JFET

The network form of built-in in PSPICE JFET model is presented in Figure 1. Resistors R_D and R_S (Figure 1) represent parasitic serial resistances of the drain and the source of the transistor, respectively. Diodes D_1 and D_2 model currents flowing between the gate and the source (I_{gs}) as well as the gate and the drain (I_{gd}) of the transistor.

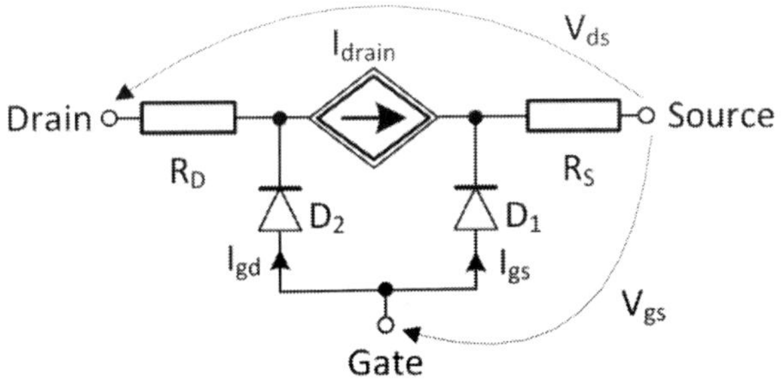

Figure 1. The network form of the JFET model.

The main element of the presented model is the controlled current source I_{drain} described for three operational regions of transistor, according to the equations [8]:

- in the cut-off region (for $V_{gs} - VTO(T) < 0$):

$$I_{drain} = 0 \tag{1}$$

- in the linear region (for $V_{ds} \le V_{gs} - VTO(T)$):

$$I_{drain} = BETA(T) \cdot (1 + LAMBDA \cdot V_{ds}) \cdot V_{ds} \cdot (2 \cdot (V_{gs} - VTO(T)) - V_{ds}) \tag{2}$$

- in the saturation region (for $0 < V_{gs} - VTO(T) < V_{ds}$):

$$I_{drain} = BETA(T) \cdot (1 + LAMBDA \cdot V_{ds}) \cdot (V_{gs} - VTO(T))^2 \tag{3}$$

where: V_{gs} – gate-source voltage, V_{ds} – drain-source voltage, $BETA(T)$ – temperature dependence of the transconductance coefficient, $LAMBDA$ – channel-length modulation coefficient, $VTO(T)$ – temperature dependence of the threshold voltage.

The currents I_{gs} and I_{gd} are of the form [8]:

$$I_{gs} = I_S(T) \cdot \left[\exp\left(\frac{V_{gs}}{N \cdot V_t(T)} \right) - 1 \right] + I_{SR}(T) \cdot \left[\left(\frac{V_{gs}}{N_R \cdot V_t(T)} \right) - 1 \right] \cdot \left[\left(1 - \frac{V_{gs}}{P_B(T)} \right)^2 + 0.005 \right]^{\frac{M}{2}} \tag{4}$$

$$I_{gd} = I_S(T) \cdot \left[\exp\left(\frac{V_{ds}}{N \cdot V_t(T)} \right) - 1 \right] + I_{SR}(T) \cdot \left[\left(\frac{V_{ds}}{N_R \cdot V_t(T)} \right) - 1 \right] \cdot \left[\left(1 - \frac{V_{ds}}{P_B(T)} \right)^2 + 0.005 \right]^{\frac{M}{2}} + I_i \quad (5)$$

where $I_S(T)$ – temperature dependence of the saturation current, $I_{SR}(T)$ – temperature dependence of the recombination current, $P_B(T)$ – temperature dependence of the diffusion potential, $V_t(T)$ – the thermal potential, N, N_R, M – model parameters. The impact ionization current I_i for the forward saturation region is described as follows [8]:

$$I_i = I_{drain} \cdot ALPHA \cdot vdif \cdot \left[\exp\left(-\frac{V_K}{vdif} \right) \right] \quad \text{if } 0 < V_{gs} - VTO(T) < V_{ds} \quad (6)$$

$$\text{else } I_i = 0 \quad (7)$$

where $vdif = V_{ds} - (V_{gs} - VTO(T))$ (8)

Several parameters of the JFET model described by Eqs. (1) – (8) are dependent on the temperature. The temperature dependences of: the thermal potential $V_t(T)$, the saturation current $I_S(T)$, the diffusion potential $P_B(T)$, the threshold voltage $VTO(T)$, the transconductance coefficient $BETA(T)$ and the recombination current $I_{SR}(T)$ are expressed by the formulas [8]:

$$V_t(T) = \frac{k \cdot T}{q} \quad (9)$$

$$I_S(T) = I_S \cdot e^{(T/T_{NOM}-1) \cdot E_G/(N \cdot V_t(T))} \cdot (T/T_{NOM})^{XTI/N} \quad (10)$$

$$P_B(T) = P_B \cdot T/T_{NOM} - 3 \cdot V_t(T) \cdot \ln(T/T_{NOM}) - E_g(T_{NOM}) \cdot T/T_{NOM} + E_g(T) \quad (11)$$

$$VTO(T) = VTO + VTOTC \cdot (T - T_{NOM}) \quad (12)$$

$$BETA(T) = BETA \cdot 1.01^{BETATC \cdot (T - T_{NOM})} \quad (13)$$

$$I_{SR}(T) = I_{SR} \cdot e^{(T/T_{NOM}-1) \cdot E_G/(N_R \cdot V_t(T))} \cdot (T/T_{NOM})^{XTI/N_R} \quad (14)$$

where the temperature dependence of energy band-gap is of the form [8]:

$$E_g(T) = 1.16 - .000702 \cdot T^2/(T + 1108) \quad (15)$$

In equations (6-14), I_S represents the saturation current parameter at the reference temperature, E_G is the energy band-gap at the reference temperature, q represents the elementary charge, k is the Boltzmann's constant, I_{SR} is the recombination current parameter at the reference temperature, $BETA$ is the transconductance coefficient, T_{NOM} is the reference temperature, whereas: XTI, $BETATC$, $ALPHA$, V_K, $VTOTC$ – are the model parameters.

3. Results of simulations and measurements

The results of simulations of selected characteristics of SiC JFET using built-in in PSPICE model of the considered transistor are presented and discussed. The results of calculations are compared with the results of measurements in the wide range of temperature change. Measurements of static characteristics were performed using pulse measuring source Keithley 2602.

Estimation of JFET transistor model parameters were carried out using MODEL EDITOR. Usefulness of this program in the estimation of parameters of semiconductor devices are described in detail, e.g. in the papers [4, 5]. Values of the considered model parameters obtained from MODEL EDITOR are presented in Table 1. In addition, the default values [8] of the model parameters are given in Table 1, as well.

Calculations of the static characteristics using both mentioned sets of parameters values (Table 1), were performed. As an example, the results of calculations (dashed lines and solid lines) of the current-voltage transfer characteristics of the transistor at two temperatures are shown in Figure 2. In comparison, the points in this figure represent results of measurements.

As seen, the use in the calculations both: parameter values obtained from the MODEL EDITOR, as well as the default values of model parameter provides to the very large discrepancy between the results of calculations and measurements. The difference between the simulation results and measurements in case of the first and the second calculation variant, expressed as the average square error, reached approximately 40% and 60%, respectively.

Table 1. Values of the model parameters obtained by MODEL EDITOR and default values

Parameter name	Parameters values: Parameter value by MODEL EDITOR	Parameters values: Default value	Parameter name	Parameters values: Parameter value by MODEL EDITOR	Parameters values: Default value
VTO [V]	0.67	-2.0	ISR [A]	0	0
BETA [A/V^2]	2.2	10^4	M [-]	0.5	0.5
LAMBDA [V^{-1}]	0.26	0	PB [V]	1.0	1.0
RS [Ω]	0.004	0	VK [V]	0	0
RD [Ω]	0	0	NR [-]	2.0	2.0
IS [A]	10^{-14}	10^{-14}	ALPHA [V^{-1}].	0.01	0
VTOTC [V/°C]	0	0	N [-]	2	1
BETATCE [%/°C]	0.5	0			

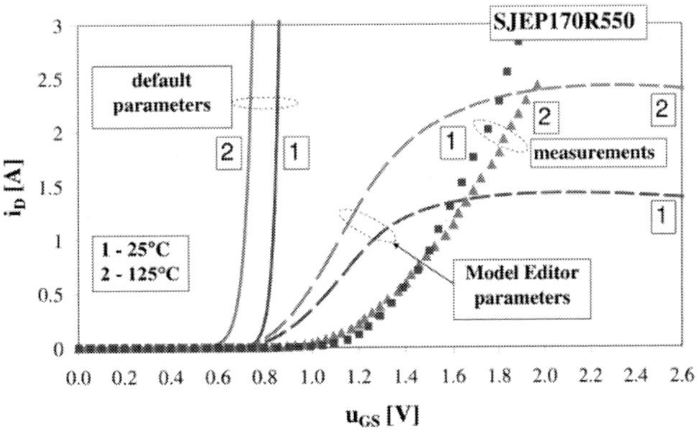

Figure 2. Calculated and measured transfer characteristics at two values of temperature

Due to the observed significant differences between the results of simulations and measurements (Figure 2), an additional estimation process of the JFET model parameters, using a direct algorithm described in [7], were performed. The direct algorithm is suitable for the considered transistor model (the Shichman-Hodges model) and allows to determine a value of a single parameter or a group of parameters directly from the results of measurements.

The values of the model parameters of JFET transistor obtained by the direct algorithm are presented in Table 2.

Table 2. Values of the model parameters of JFET obtained from the direct algorithm

Parameter name	Parameter value	Parameter name	Parameter value
VTO [V]	1.01	ISR [A]	0
BETA [A/V^2]	3.65	M [-]	0.5
LAMBDA [V^{-1}]	0.1	PB [V]	1.0
RS [Ω]	0.04	VK [V]	0
RD [Ω]	0.55	NR [-]	2.0
IS [A]	10^{-14}	ALPHA [V^{-1}]	0.01
VTOTC [V/°C]	0.0002	N [-]	6
BETATCE [%/°C]	0.2		

The calculations of the static characteristics of SiC JFET using the parameters set listed in Table 2, were performed. Figure 3 illustrates the results of calculations (solid lines) in comparison with the measurements results (points connected by dashed lines).

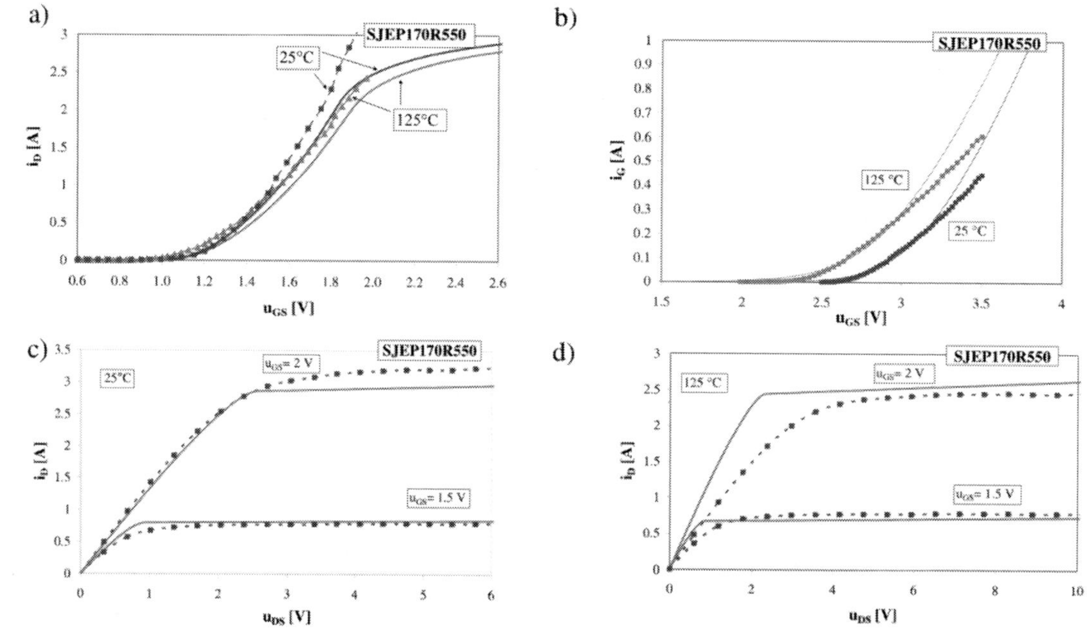

Figure 3. Calculated and measured transfer characteristics (a), input characteristics (b) and output characteristics (c),(d) of the SiC JFET at two values of the temperature: 25°C and 125°C.

Figure 3a shows the transient characteristics of the transistor at two temperatures. As seen from Figure 3a and from the results shown in Figure 2, the accuracy of the considered model, in case of using the model parameters obtained from the direct method was improved.

The calculated and measured input characteristics at two values of temperature (25°C and 125°C) are presented in Figure 3b. As seen, very good agreement between results of simulations and measurements at the range of up to about 3.3 V at temperature equal to 25 ° C and up to about 3 V at temperature equal to 125°C, are observed only. The increase of the temperature results in a reduction of the gate-source voltage u_{GS} at fixed gate current i_G, and the temperature coefficient of the gate-source voltage is equal to 2.6 mV/°C.

Figure 3c and 3d shows the output characteristics of the transistor for two values of the gate-source voltage and at two values of temperature. As seen, very good agreement between simulated and measured results are observed in the case of the output characteristics at U_{GS}= 1,5 V for the full range of the drain-source voltage, as well as for the output characteristics at U_{GS}= 2 V (25°C) - in linear region, only. The increase of the ambient temperature results in a decrease of the drain current at a fixed value of the control voltages. For example, the measured value of the drain current at U_{GS}= 2 V and U_{DS}= 6 V at 25°C equals approximately to 3 A (Figure 3c), whereas at the same operating point at 125°C value of the drain current is about 20% lower.

Value of the R_{ON} resistance of the considered transistor, determined using the results of measurements (Figures 3c, 3d) as the slope of the output characteristics of the transistor in the linear region, in arbitrarily selected operating point of coordinates U_{GS}= 2 V, U_{DS}= 2 V at ambient temperatures of 25 °C and 125 °C equals to: 0,8 Ω and 1,3 Ω, respectively. On the other hand, the R_{ON} resistance calculated by considered model at mentioned operating point and temperatures is substantially constant and is equal to about 0,9 Ω. It should be noted, the considered model does not include the influence of the temperature on the serial parasitic resistances R_D and R_S (see Section 2), which probably results in significant discrepancies between the calculated and measured values of the resistance R_{ON} at different ambient temperatures.

4. Conclusions

The paper presents the results of simulations of selected static characteristics of SiC JFET using built-in in PSPICE transistor model in comparison with the results of measurements. It has been shown that the use of the considered model in relation to the transistor made of silicon carbide provides in some cases to large differences between the results of simulations and measurements. Results presented in this paper will be used by the authors to formulate more accurate model of JFET transistor made of silicon carbide.

5. References

[1] Veliadis V et al 2013 Reliable operation of SiC JFET subjected to over 2.4 Million 1200-V/115-A hard switching events at 150°C *IEEE Electron Device Letters* vol. **34**(3)

[2] Katoh K et al 2014 Study on low-loss gate drive circuit for high efficiency server power supply using normally-off SiC-JFET *IEEE Conference Publications, Hiroshima.* 2285-2289

[3] Josifović I et al 2011 Power sandwich industrial drive with SiC JFETs *Power Electronics and Applications (EPE 2011) IEEE Conference Publications*(Birmingham) pp. 1-10

[4] Górecki K, Stepowicz W J, Zarębski J 2005 Estymacja parametrów modelu tranzystora JFET za pomocą programu PARTS *XXVIII Międzynarodowa Konferencja z Podstaw Elektrotechniki i Teorii Obwodów IC-SPETO'2005* Vol **2** 301

[5] Zarębski J, Bisewski D 2009 Estymacja parametrów modelu Danga tranzystora MOS *Elektronika – konstrukcje, technologie, zastosowania Sigma-Not* **6** 87-90

[6] Bisewski D 2004 Stałoprądowy model Shichmana-Hodgesa tranzystora MOS w programie SPICE II Konferencja Naukowa Studentów, *Oficyna Wydawnicza Politechniki Wrocławskiej* (Wrocław) t II, 397-403

[7] Bisewski D. 2009 Estimation of parameters of GaAs and SiC MESFETs using genetic algorithm 2009 *45th International Conference on Microelectronics, Devices and Materials, MIDEM* 81-84

[8] Design Systems, Inc, PSpice, manual ver 10

DC characteristics and parameters of silicon carbide high-voltage power BJTs

Joanna Patrzyk, Janusz Zarębski, Damian Bisewski

Gdynia Maritime University, Dept. of Marine Electronics, Morska 83, 81-225 Gdynia, Poland

E-mail: j.patrzyk@we.am.gdynia.pl

Abstract. The paper shows the static characteristics and operating parameters of the bipolar power transistors made of silicon carbide and for comparison their equivalents made of classical silicon technology. The characteristics and values of selected operating parameters with special emphasis on the effect of temperature and operating point of considered devices are discussed. Quantitative as well as qualitative differences between the characteristics of the transistor made of silicon and silicon carbide are indicated as well.

1. Introduction

High-voltage silicon bipolar power transistors (BJTs) have been known for decades and are still evolving. In the mid-nineties of the last century in the world's technological laboratories has begun intensive work on getting power BJT transistors made of silicon carbide (SiC) - a material which properties make many operating parameters and characteristics of SiC devices have features better than silicon counterparts of these devices.

Silicon carbide, compared with silicon, has about 2,5 times larger width of the band-gap, a much smaller value of the intrinsic carrier concentration at room temperature, more than three times larger thermal conductivity and the SiC junction has 10 times larger value of the critical electric field strength than silicon one. Favorable properties of silicon carbide make the SiC components, in comparison with the silicon devices, have better values of operating parameters, for example: greater voltage strength, better thermal properties and shorter switching times [1, 4].

In the literature, a lot of information about the characteristics and parameters of the silicon carbide power devices, which appeared previously on the market for example: Schottky diodes and transistors: MESFET, MOSFET and JFET can be found and are presented in [8, 9, 10, 11, 12].

In contrast, literature information relating to the characteristics and parameters of SiC power BJTs are rather poor. Static and dynamic characteristics of these devices are presented and discussed in [1-3], characteristics of SiC BJTs working at high temperatures are presented and discussed in [4] and selected operating parameters are considered in [5, 6, 7].

The market offers in terms of the considered class of semiconductor devices has been changing rapidly. The new generation of bipolar power transistors made in the technology based on silicon carbide appeared on the market in 2005 (TranSiC), whereas the company GeneSiC introduced a new technology of super junction bipolar transistors SJT made of silicon carbide, and the United Sillicon Carbide company promises starting the production of SiC BJT transistors at the turn of the year 2016/2017.

Producers of semiconductor devices do not provide in the catalog data of considered class of SiC devices a lot of characteristics and parameters useful for engineers-designers of electronic circuits and systems.

It is worth to notice that the temperature dependences of certain characteristics and parameters differ not only quantitatively but also qualitatively from similar characteristics and parameters of the silicon power bipolar transistors.

In the paper the authors present the results of their own measurements of current-voltage characteristics and parameters of the BT1206AC SiC bipolar power transistor made by TranSiC. The influence of the ambient temperature, the collector current and the collector-emitter voltage on the leakage current, the saturation voltage and the current gain factor (β) is presented and discussed. In particular, the influence of the self-heating phenomenon on characteristics of the investigated transistors are considered. For comparison, the characteristics of the 2SC5294 Panasonic silicon transistor is presented as well.

2. Results of investigations

The research subject is power BJTs made of silicon and silicon carbide mounted in a standard TO-247 package. The values of the most important parameters of these transistors are shown in Table 1.

Table 1. Values of the selected parameters of the SiC BJT and Si BJT [14, 15]

Manufacturer	Type	V_{CEO}	I_{Cmax}	P_{max}	T_{jmax}	β
TranSiC	**SiC BT1206AC**	1200 V	6 A	-	175°C	35 (V_{CE}=2,5 V, I_C=4 A)
Panasonic	**Si 2SC5294**	600 V	20 A	120 W (T_c=25°C)	150°C	12 (V_{CE}=5 V, I_C=10 A)

As seen from Table 1, the SiC BJT has a 2 times larger value of the permissible collector-emitter voltage V_{CE0}, as well as 25°C higher value of permissible junction temperature T_{jmax}, in comparison with the Si BJT. The current gain factor β of the SiC BJT has 3 times higher value than the silicon transistor, however, it should be noted that this parameter is determined for different operating points of the considered transistors under. Unfortunately, the manufacturer of transistor made of silicon carbide, did not disclose the value of permissible power loss P_{max} for the semiconductor device.

The following section presents author's results of measurements of selected parameters and characteristics of both considered transistors obtained under isothermal conditions, it means, when the junction temperature T_j of transistor approximately equals to the ambient temperature T_a, as well as under non-isothermal conditions in which the junction temperature of the semiconductor device may exceed, even significantly, the ambient temperature by self-heating phenomenon [13]. Isothermal measurements of the characteristics of the considered transistors were performed using a Keithley pulse measuring source type 2602, which minimized the influence of self-heating phenomenon on the measured characteristics of the transistor. In turn, the non-isothermal characteristics were made by "point by point" measurement for the transistors without heatsink in the thermal steady state conditions, which extended the measuring procedure, and a thermal steady state condition obtained even after 20 minutes. In addition, the temperature T_C of the transistors cases was measured using a temperature meter TM-2000 equipped with a platinum probe Pt-100. This sensor was mounted on the transistors cases using a thermally conductive glue type AG Termoglue 10G. Figures 1-5 present the author's results of measurements of the characteristics and parameters of considered transistors. The results of isothermal current-voltage i(u) characteristics were compared with the catalog data.

Figure 1 presents the isothermal dependence of the current gain factor β of the considered devices on the collector current at fixed value of the ambient temperature (Figure 1a) and on the ambient temperature at fixed value of the collector-emitter voltage and the base current (Figure 1b).

a) b)

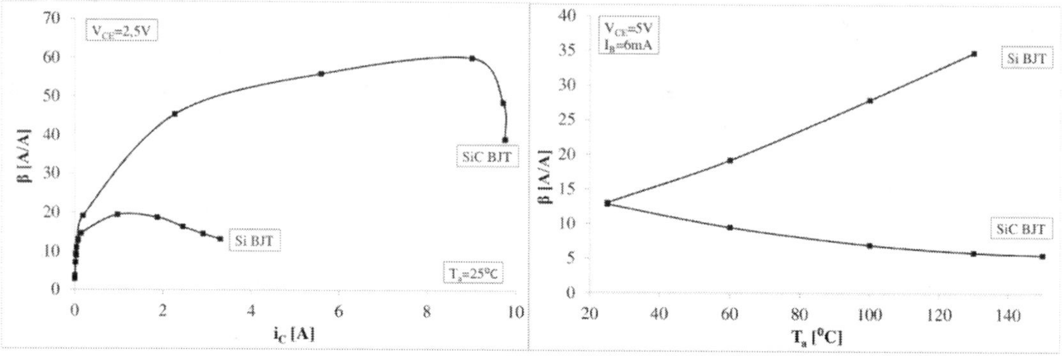

Figure 1. Dependence of the $\beta(i_C)$ (a) and $\beta(T_a)$ (b) characteristics of Si BJT and SiC BJTs.

As seen in Figure 1a, in the transistor made of silicon, for the collector current value range up to approximately 1 A the dependence of $\beta(i_C)$ is a monotonically increasing function, while for higher values of this current the considered parameter decreases. On the other hand, in the SiC BJT transistor, extreme value of $\beta(ic)$ exists at a value of I_C equal to 9 A. The increase of the ambient temperature (Figure 1b) results in, increasing of the β parameter value in the silicon transistor, while in the transistor made of silicon carbide – decreasing of the β parameter value at the fixed value of both the collector-emitter voltage and the base current. Temperature coefficients of β parameter for the Si BJT and SiC BJT are equal to 1.59 %/K and -0.52 %/K, respectively. Values of the temperature coefficient were determined using equation:

$$\gamma_A = \frac{A(T_2)-A(T_1)}{A(T_1)\cdot(T_2-T_1)} \cdot 100\% \qquad (1)$$

where $A(T_1)$ and $A(T_2)$ represent values of β parameter at temperature T_1=25°C and T_2=130°C, respectively.

a) b)

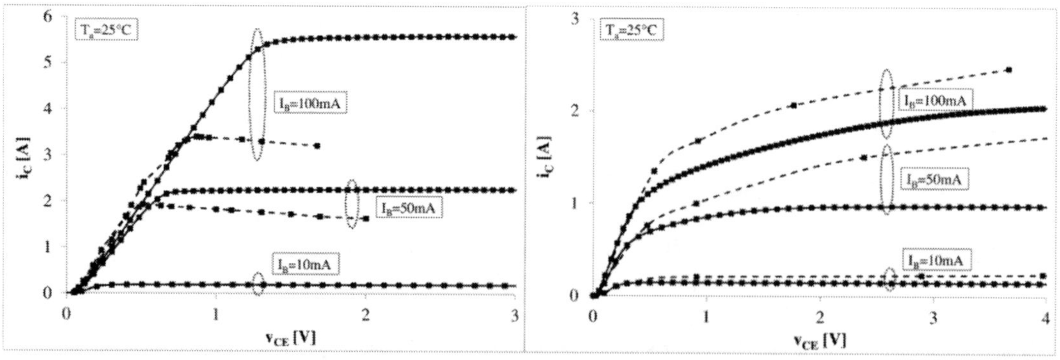

Figure 2 Output characteristics of SiC BJT transistor (a) and Si BJT transistor (b).

Figure 2a presents the output characteristics $i_C(v_{CE})$ of the SiC BJT, and Figure 2b - the output characteristics $i_C(v_{CE})$ of the Si BJT measured under isothermal conditions (points connected by solid lines) and under non-isothermal (points connected by dashed lines) at the fixed value of the base current and the ambient temperature.

Position of non-isothermal characteristics in comparison with their isothermal counterparts (considering the same value of the base current i_B) is qualitatively different for each of the investigated transistors. The junction temperature measured along the considered non-isothermal characteristics is growing, because of the device active power growth. However, because of the dependence of β(T) presented in Figure 1, the non-isothermal characteristics of SiC BJT lie under their isothermal counterparts (current gain factor β is a decreasing function of temperature), while in the silicon transistor, increasing temperature results in β drop (Figure 1a) which causes that the non-isothermal characteristics $i_C(v_{CE})$ of the silicon transistor are located above their isothermal counterparts (Figure 2a).

In addition, the non-isothermal output characteristics of SiC BJT transistor also differ qualitatively from their isothermal counterparts. In the literature, such shape of the characteristics is called N-type, having regard to the phenomenon of avalanche breakdown, not illustrated in the Figure 2a.

Figure 3 shows the isothermal input characteristics of the both transistors measured in a wide range of the temperature at the fixed value of the collector-emitter voltage equal to 10 V.

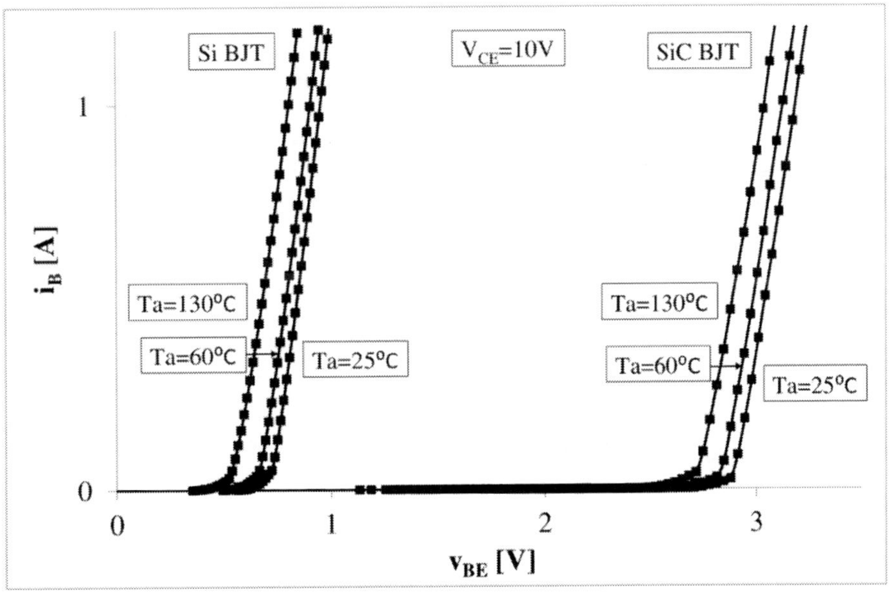

Figure 3 $i_B(v_{BE})$ characteristics of SiC and Si BJTs.

As seen, the transistor made of silicon carbide has more than three times higher value of the base-emitter voltage than the silicon transistor, at the same fixed value of the base current.

In the case of both considered transistors, as it was expected, the increase of the temperature causes a shifting of the input characteristics towards lower values of the base-emitter voltage. The temperature coefficient of base-emitter voltage at the constant base current (equal to 1A) for transistor made of silicon and silicon carbide is equal to -0.15 %/K and -0.05 %/K, respectively. Values of the temperature coefficient of base-emitter voltage were determined using Eq. 1, where $A(T_1)$ and $A(T_2)$ represent values of base-emitter voltage at temperature T_1 and T_2, respectively.

An important operating parameter of the transistor is the collector-emitter voltage in an ON-state (V_{ON}). Figure 4 presents the isothermal dependence of $V_{ON}(i_B)$ for the both transistors measured at two ambient temperatures.

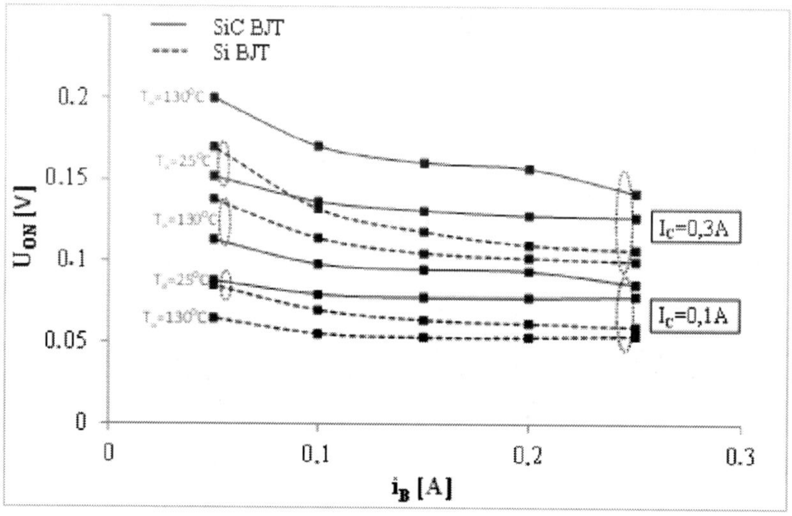

Figure 4 $V_{ON}(i_B)$ characteristics of SiC BJT and Si BJTs.

As seen, in the both considered transistors the V_{ON} voltage decreases with an increase of base current and increases with an increase of collector current. On the other hand, the temperature increase results in a decrease of the voltage V_{ON} value in the silicon transistor (increase - in the silicon carbide transistor) at a constant value of the base and collector current.

An important operating parameter of the BJT transistor operating in the cut-off state is leakage current flowing in the collector circuit referred to the terms of the established polarization of a control electrode. The most commonly presented by the manufacturer leakage current is I_{CB0} current measured in a common base measuring set with an opened emitter.

Figure 5 presents the non-isothermal dependence of $I_{CB0}(v_{CB})$ of considered transistors measured for different ambient temperatures.

a) b)

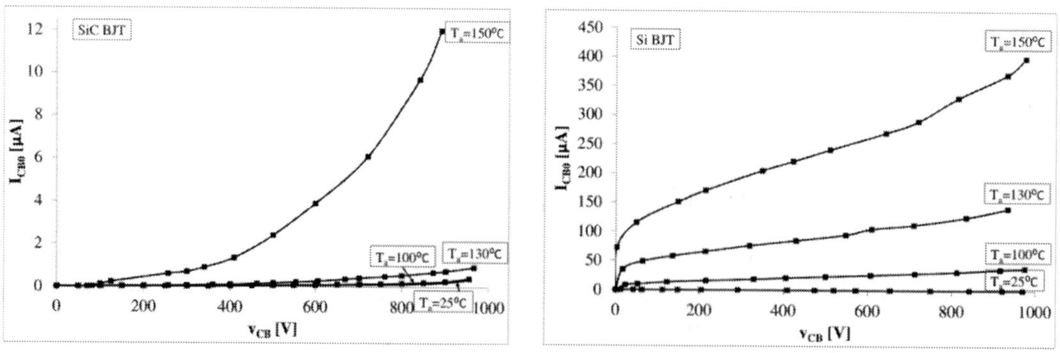

Figure 5 Dependence of the leakage current I_{CB0} on the collector-emitter voltage of SiC BJT (a) and Si BJT (b).

As seen, the transistor made of silicon carbide is characterized by several dozen times lower value of I_{CBO} current than the silicon transistor at the same, fixed values of collector-emitter voltage and ambient temperature. For example, the value of the considered parameter at 150°C and the voltage V_{CB} = 600 V in the Si BJT and SiC BJT is equal to 260 µA and 3.9 µA, respectively. Furthermore, it is presented that in the both considered transistors, the I_{CBO} current strongly increases with an increase of the temperature.

3. Conclusions

The paper presents the results of author's measurements of the selected static characteristics and operating parameters of the bipolar power transistor made of silicon carbide and for comparison the mentioned characteristics and parameters of the silicon counterpart of the considered device. In the case of considered transistors change both the junction and the ambient temperature due to self-heating phenomenon has an important impact on the characteristics of these semiconductor device.

The most important qualitative differences in properties of the BJT transistor made of silicon and silicon carbide are mainly due to the fact that the temperature dependence of the current gain factor in the first-mentioned transistor is an monotonically increasing function, and in the second - a monotonically decreasing function. As a result, the output characteristics presents a decrease of the collector current as the temperature increases only in the transistor made of silicon carbide. The occurrence of such a phenomenon may improve the thermal stability of the device and can also improve the reliability of electronic circuits containing such transistors.

The results of the measurements shown in the paper will be used by the authors in modeling SiC BJTs, and next – in a computer analyzes of electronic circuits containing power SiC BJTs.

4. References

[1] Sundaresan S, Jeliazkov S, Grummel B, Singh R 2013 10 kV SiC BJTs static, switching and reliability characteristics *25th International Symposium in Power Semiconductor Devices and ICs (ISPSD)* 303–306

[2] DiMarino C, Chen Z, Boroyevich D, Burgos R, Mattavelli P 2013 Characterization and comparison of 1.2 kV SiC power semiconductor devices *IEEE EPE* 1-10

[3] Ahmed M M R, Parker A, Mawby N A, Nawaz P A, Zaring M 2008 Characterization of the static and dynamic behavior of a SiC BJT *Power Electronics and Motion Control Conference* 2472-2477

[4] Niedra J M, Schwarze G E 2006 Static and Switching Characteristics of a 4H-SiC Based BJT to 200°C *International Conference on High Temperature Electronics (HiTEC 2006)*

[5] Zhang Y, Zhang B, Li Z, Lai C 2008 Analysis of thermal characteristics of current gain for BJT-BSIT compound device *International Conference on Communications, Circuits and Systems* 1278 – 1281

[6] Buono B, Ghandi R, Domeij M, Malm B G, Zetterling C M, Östling M 2010 Modeling and characterization of current gain versus temperature of 4H-SiC power BJTs *IEEE Trans. Electron Devices* **vol 57** 704-711

[7] Chailloux T, Calvez C, Planson D, Tournier D 2014 Etude de differents transistors de puissance SiC 1.2kV des temperatures cryogeniques aux hautes temperatures *Symposium de Genie Electrique (SGE'14)*

[8] Zarębski J, Dąbrowski J 2011 Investigations of SiC Merged PiN Schottky Diodes Under Isothermal and Non-isothermal Conditions *International Journal of Numerical Modelling: Electronic Networks, Devices and Fields* **vol 24** 207-217

[9] Zarębski J, Bisewski D 2008 Modifications of the DC Raytheon-Statz model for SiC MESFETs *International Journal of Numerical Modelling: Electronic Networks, Devices and Fields, John Wiley&Sons* **vol 21** 583-590

[10] Zarębski J, Bisewski D 2009 Modelowanie tranzystorów SiC-MOS *Elektronika – konstrukcje, technologie, zastosowania* Nr 7 177-180.

[11] Gorecki K, Zarebski J 2008 The nonlinear compact thermal model of power MOS transistors *International Conference of Microelectronics* 196 – 199

[12] Na H J 2006 Fabrication and characterization of 4H-SiC planar MESFETs *Microelectronic Engineering* **vol 83** 160-164

IOP Conf. Series: Materials Science and Engineering **104** (2015) 012015 doi:10.1088/1757-899X/104/1/012015

[13] La Spina L, d'Alessandro V, Russo S, Rinaldi N, Nanver L K 2009 Electrothermal behavior of highly-symmetric three-finger bipolar transistors *Bipolar/BiCMOS Circuits and Technology Meeting (BCTM 2009)* 21 – 24

[14] http://www.dacpol.eu/pl/elementy-polprzewodnikowe-z-weglika-krzemu/product/elementy-polprzewodnikowe-z-weglika-krzemu-1224

[15] http://www.alldatasheet.com/datasheet-pdf/pdf/108707/PANASONIC/2SC5294.html

IOP Conf. Series: Materials Science and Engineering **104** (2015) 012016 doi:10.1088/1757-899X/104/1/012016

The influence of losses in the core of an inductor on characteristics of the boost converter

Krzysztof Górecki, Kalina Detka

Gdynia Maritime University, Department of Marine Electronics, Morska 83, 81-225 Gdynia, Poland

E-mail: k.gorecki@we.am.gdynia.pl, k.detka@we.am.gdynia.pl

Abstract. In the paper the influence of core losses on characteristics of the boost converter is considered. In calculations, the average electrothermal models of the diode-transistor switch and of the inductor are used. The applied electrothermal models and the obtained results of calculations are presented. The selected results of calculations are compared with the results of measurements. The influence of such factors as converter load resistance and frequency of the control signal and lossiness of the core on characteristics of the considered converter and the loss of energy in the inductor are discussed.

1. Introduction

Boost converters are widely used sub-circuits of switched-mode power supplies, and they are characterized by high watt-hour efficiency [1, 2, 3]. To miniaturize dimensions of these devices an increase in switching frequency of electronic components is required.

The inductor included in boost converters is the basic element for energy storage and its properties influence characteristics of the considered circuit [4, 5, 6]. During the inductor operation, self-heating phenomena and mutual thermal interaction between the core and the winding of the inductor are observed [7, 8, 9]. The source of heat generated in the inductor are losses in the core and in the winding. As presented in paper [7, 8, 10] the inductor core characteristics strongly depend on magnetic material used to build this core and it well describes its dependence of the magnetic flux density B on the magnetic force H, which takes the shape of a hysteresis loop.

For the analysis and design of electronic circuits computer programs such as SPICE are used [11, 12, 13]. In many papers concerning the analysis of dc-dc converters only energy losses in semiconductor devices are taken into account [11, 12, 14 - 17], whereas losses in magnetic elements are omitted. In paper [5] the electrothermal model of an inductor dedicated for SPICE, which takes into account power losses in the core and a self-heating phenomenon is proposed. This model is dedicated to simulate dc-dc converters using the transient analysis, which is time-consuming [14]. Reduction in the duration time of calculations is ensured for the use of an average model of the inductor proposed in papers [18, 19]. These models allows calculating the non-isothermal characteristics of non-isolated dc-dc converters using the DC sweep analysis in SPICE.

In this paper, the influence of power losses of the inductor core on characteristics of the boost converter is analyzed. The calculations are made using the average electrothermal model of the diode-

IOP Conf. Series: Materials Science and Engineering **104** (2015) 012016 doi:10.1088/1757-899X/104/1/012016

transistor switch, presented in paper [15], and the average electrothermal model of the inductor, presented in paper [18], for converters containing cores made of different magnetic materials. The calculations are performed for different frequency of operation and converter's load resistance. The selected results of calculations are compared with the results of measurements.

In the second section the average models of components of the considered converters are shortly described, whereas in the third section the results of calculations and measurements are presented.

2. The used models

The analysis is performed using average electrothermal models of the diode-transistor switch and of the inductor, elaborated previously by the authors. In paper [19], usefulness of such a model for the analysis of boost converters at the steady state is proved. The network representation of these models are presented in Figs. 1 and 2, respectively.

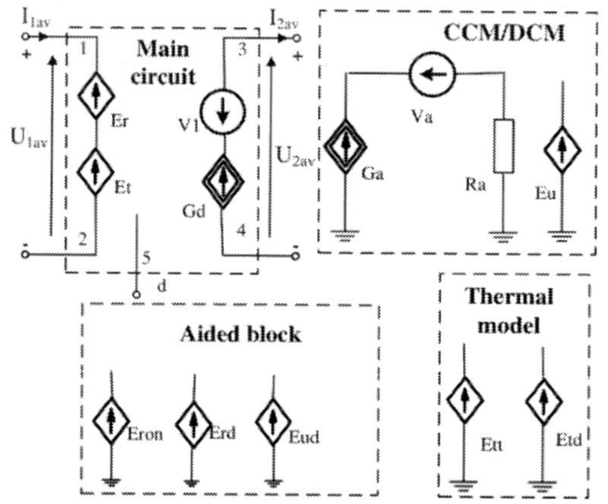

Figure 1. Network representation of the average electrothermal diode-transistor switch model [15].

The average electrothermal model of the diode-transistor switch takes into account energy losses for the switch-on semiconductor devices and a self-heating phenomenon. Additionally, this model allows determining characteristics of dc-dc converters operating in the CCM and DCM modes.

In Figure 1, the series connected control voltage sources Et and Er represent the unipolar transistor, whose output terminals are connected to terminals 1 and 2 and the control terminal is connected to the terminal 5, while the diode (between the terminals 3 and 4) is represented by the controlled current source Gd. The independent voltage source V1 of zero output voltage is used to control the diode average current, elements Ga, Ra, Va, Eu are the auxiliary circuit, which is used to determine the mode of dc – dc converter (CCM or DCM) including the considered switch. The controlled voltage sources Eron, Erd, Eud model temperature changes of transistor resistance R_{ON}, series diode resistance R_D and voltage U_D on the forward biased p-n junction. The value of the internal diode and transistor temperature are calculated using the voltage sources Etd and Ett. The detailed description of the controlled voltage sources efficiency contained in the considered model is presented in [15].

In turn, in the average electrothermal model of the inductor there are three blocks: the main circuit, auxiliary circuits and the thermal model. Terminals A and B are the terminals of the inductor. Additionally, the terminal L provides the voltage corresponding to the inductor inductance, terminal T_U - the temperature of the inductor winding and terminal T_R - the temperature of the inductor core.

In the main circuit the voltage source V_L with zero output voltage monitors the current of the inductor, the resistor R_{S0} represents the winding resistance of the inductor for the direct current at the reference temperature T_0 and the controlled voltage source E_{RS} describes the influence of temperature and losses of the inductor series resistance. In turn, the controlled voltage source E_{RN} models the skin effect phenomenon.

In the auxiliary block the controlled voltage sources are used to calculate the value of the magnetic force H, flux density B, saturation flux density B_{sat} and the auxiliary value C. The description of efficiency of these sources is given in [9]. The controlled voltage sources E_{LS} model the dependences of the inductor inductance L of the current, frequency and temperature.

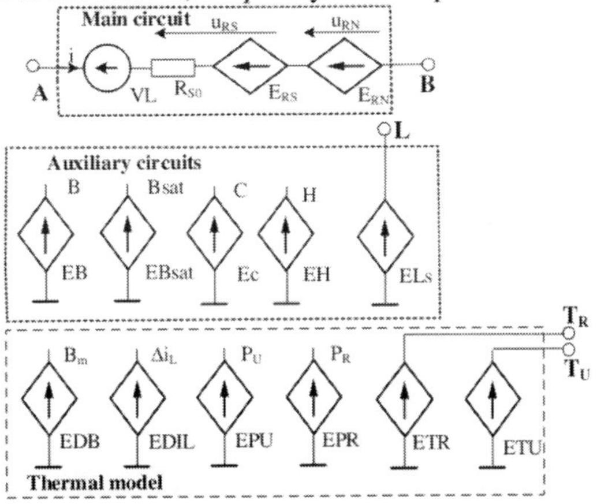

Figure 2. Network representation of the average electrothermal model of the inductor.

The thermal model allows calculating an excess of the core temperature T_R and winding T_U above the ambient temperature T_a. To determine these excesses the compact thermal model is used, in which the controlled voltage sources E_{TU} and E_{TR} are used to calculate the value of these temperatures taking into account a self-heating phenomenon in the core and winding and mutual thermal interactions between the core and the winding.

The total power losses occurring in the ferromagnetic core (representing by the voltage source E_{PR}) are the sum of losses connected with hysteresis losses, eddy current losses and remaining losses. The kind of losses which are dominant is determined by the type of magnetic material, the peak value of flux density, frequency and the shape of flux density [5]. Losses in the core are also characterized by power losses of the core per unit of volume or mass [7]. For a sine wave of induction, core losses are described by the formula given in [18]. In dc-dc converters the inductor current has the shape of a triangle wave. In this case, core losses can be described by the formula

$$P_R = V_e \cdot P_{V0} \cdot f^{\alpha} B_m^{\beta} \cdot 2^{\alpha} \cdot \left(1 + \alpha_P \cdot (T_R - T_m)^2\right) \cdot \left(d^{1-\alpha} + (1-d)^{1-\alpha}\right) \tag{1}$$

where V_e denotes the equivalent volume of the core , P_{v0} - volumetric power losses density in the core, α, β are parameters depending on magnetic material, B_m – the amplitude of flux density, α_P represents the temperature coefficient of power dissipated in the core, T_m is the value of the core temperature, at which power losses in the core have the minimum, f denotes frequency and d is the control signal duty cycle of the converter.

On the other hand, powers dissipated in the core and winding P_U is determined by means of the controlled voltage source E_{PU}. This power is described by the formula

$$P_U = u_{RS} \cdot i + u_{RN} \cdot \frac{\Delta i_L}{2} \tag{2}$$

The voltage U_{RS} and U_{RN} occurring in the formula (2) are shown in Figure 2. The peak to peak value of the current Δi_L and flux density amplitude B_m correspond to the voltage value on the controlled voltage sources EDIL and EDB, respectively. The values of these quantities are described by the formulas

$$\Delta i_L = \frac{u_{we} - u_{RS} - u_{RN} - V_{on}}{L \cdot f} \cdot d \tag{3}$$

$$B_m = \frac{L \cdot \mu_0 \cdot l_{Fe} \cdot z \cdot \Delta i_L}{2 \cdot (l_{Fe} + l_p) \cdot (\mu_0 \cdot z^2 \cdot S_{Fe} - L \cdot l_p)} \tag{4}$$

where u_{we} denotes the converter input voltage, V_{on} – the voltage on the active transistor contained in the converter (voltage at the source Et of Figure 1), z denotes the number of turns in the inductor winding, S_{Fe} –effective cross-section area of the core, l_{Fe} – the magnetic path in the core, l_p – the air gap length, μ_0 – magnetic permeability of free air.

3. The results of calculations and measurements

In order to investigate the influence of losses in the core of the inductor on characteristics of the boost converter some measurements and calculations of the considered converter, whose diagram is presented in Figure 3, are performed. In the considered network the transistor MOSFET of the type IRF540, the Schottky diode 1N5822 and passive elements of the following values: R_G = 30 Ω, C_1 = 47 μF are applied. The resistor R_0 represents load resistance of the converter, and the voltage source U_{we} - the power supply of the output value equal to 12 V.

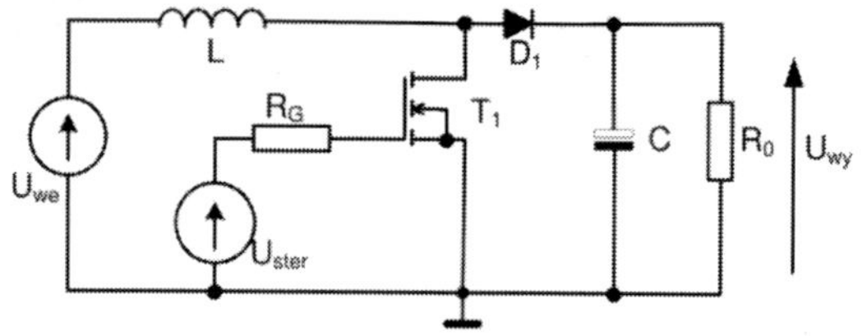

Figure 3. Diagram of the considered boost converter.

In the considered converter, the investigations are performed for inductors with ring cores of the diameter equal 26 mm made of ferrite (the core RTF), of powdered iron (the core RTP) and of nanoperm (the core RTN). On each core 27 turns of the copper wire in the enamel of the diameter 0.8 mm are wound. In Figure 4 the calculated dependences L(i) of the considered inductors are shown.

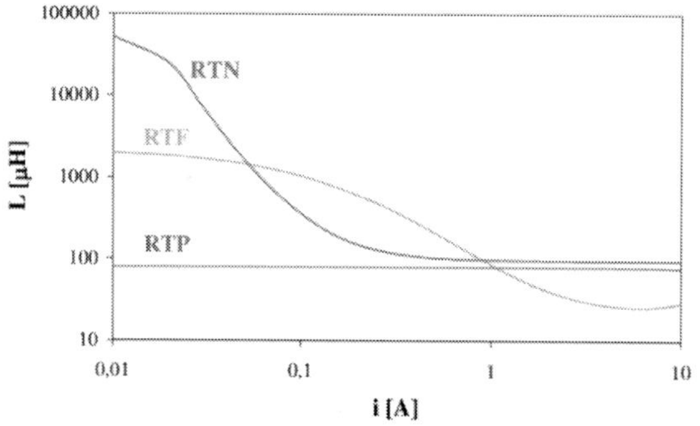

Figure 4. Calculated dependences L(i) for the investigated inductors.

As it is visible, depending on material of the applied core very different courses of the dependence L(i) for the considered inductors are obtained. Inductance of the inductor with the core RTP is practically constant in the considered range of changes of the current and is about 80 μH. In turn, inductance of inductors with the cores RTF and RTN strongly depends on the current. For example, for the inductor with the core RTN, in the considered range of changes of the current, a fall in the value of inductance from above 50 mH for i = 10 mA to 95 μH for i > 1 A is visible.

The considered inductors differ between each other not only is the course of the dependence L(i), but also in the values of parameters occurring in equation (1) and describing a power loss in the core. The values of these parameters are collected in Table 1.

Table 1. Values of parameters characterizing power losses in the cores of the considered inductors.

Core	P_{V0} [$Ws^{\alpha}/(m^3T^{\beta})$]	α	β	α_P [K^{-2}]	T_m [K]	V_e [m^3]
RTF	2.76	1.24	2.28	$2x10^{-4}$	343	$3.14x10^{-6}$
RTN	0.05	1.69	2.2	0	358	$3.95x10^{-6}$
RTP	103	1.15	2.07	0	361	$4.43x10^{-6}$

In Table 1 the considerable differentiation of the value of lossiness P_{V0} for the considered cores is visible. It accepts values from the section from 0.05 to 103 $Ws^{\alpha}/(m^3T^{\beta})$. It should be expected that this differentiation will influence also courses of characteristics of converters containing the considered inductors.

In Figure 5 the calculated (lines) and measured (points) characteristics of the boost converter containing in turn each of the investigated inductors are presented. The calculations and measurements are performed at frequency of the control signal f = 50 kHz and the duty cycle d = 0.5.

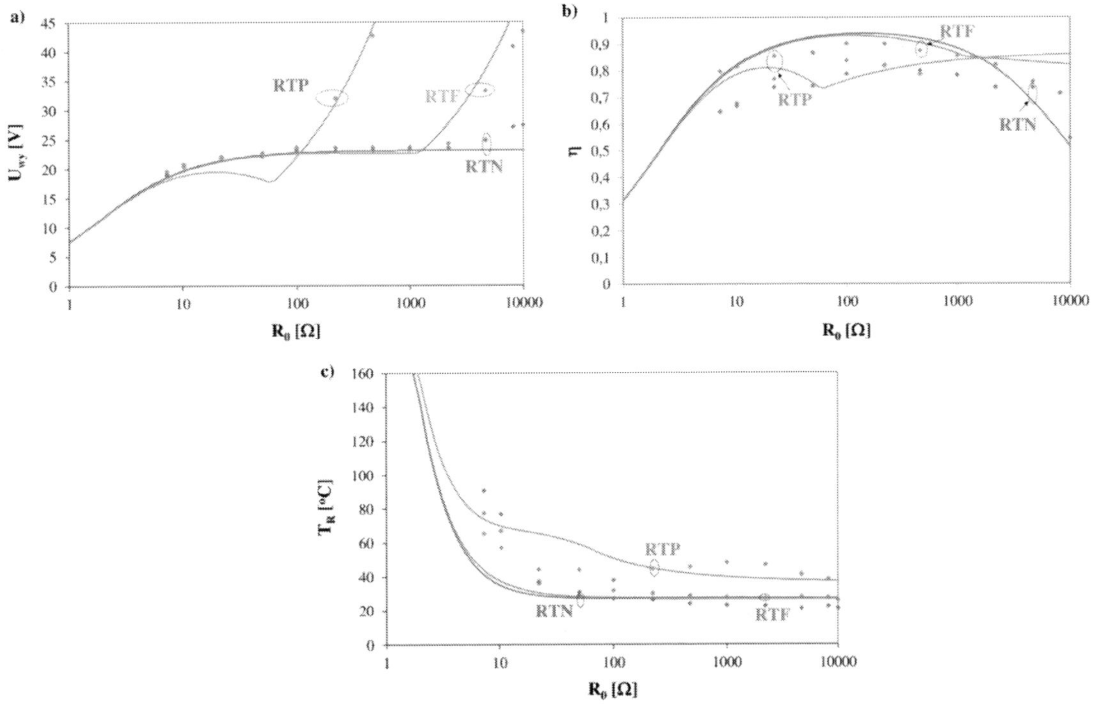

Figure 5. Calculated and measured dependences of the output voltage (a), watt-hour efficiency (b) of the boost converter and core temperature of the inductor (c) on load resistance.

As one can notice, the dependence $U_{wy}(R_0)$ within the range of small values of load resistance (when the converter operates in CCM) has practically the identical course for all the inductors. In turn, material of the core decides about the value of resistance R_0 at which the converter passes from the mode CCM to DCM. For the converter with the inductor containing the core RTP this passage appears at resistance R_0 equal to about 100 Ω, and for the converter with the inductor containing the core RTN - at resistance R_0 equal to about 3 kΩ. Unfortunately, the obtained by means of the presented models calculated values of watt-hour efficiency are overestimated with regard to the measured values of this parameter, and the calculated values of temperature of the core are typically understated.

The dependences of the converter output voltage, watt-hour efficiency and core temperature on load resistance of the considered converter including the inductor with the RTF core at different values of the parameter P_{v0} equal to values 0.01, 2.7, 25, 80 and 150 $Ws^{\alpha}/(m^3T^{\beta})$, respectively are presented in Figure 6. Research is conducted in the function of load resistance, at the constant value of frequency f = 50 kHz and the duty cycle of the control signal d = 0.5.

As it is visible for load resistance R_0 in the range from 10 Ω to about 1 kΩ (when the converter operates in the CCM mode) an increase in P_{v0} loss causes a decrease in the output voltage up to 40% and a decrease in watt-hour efficiency - about 30% for load resistance higher than 10 Ω, and up to a three-fold increase of the core temperature T_R for load resistance R_0 < 100 Ω. Power dissipated in the core does not exceed 1.5 W, and its value is the biggest in the range of small load resistance. As expected, an increase in the value of the parameter P_{V0} causes an increase in the value of power dissipated in the core of the inductor.

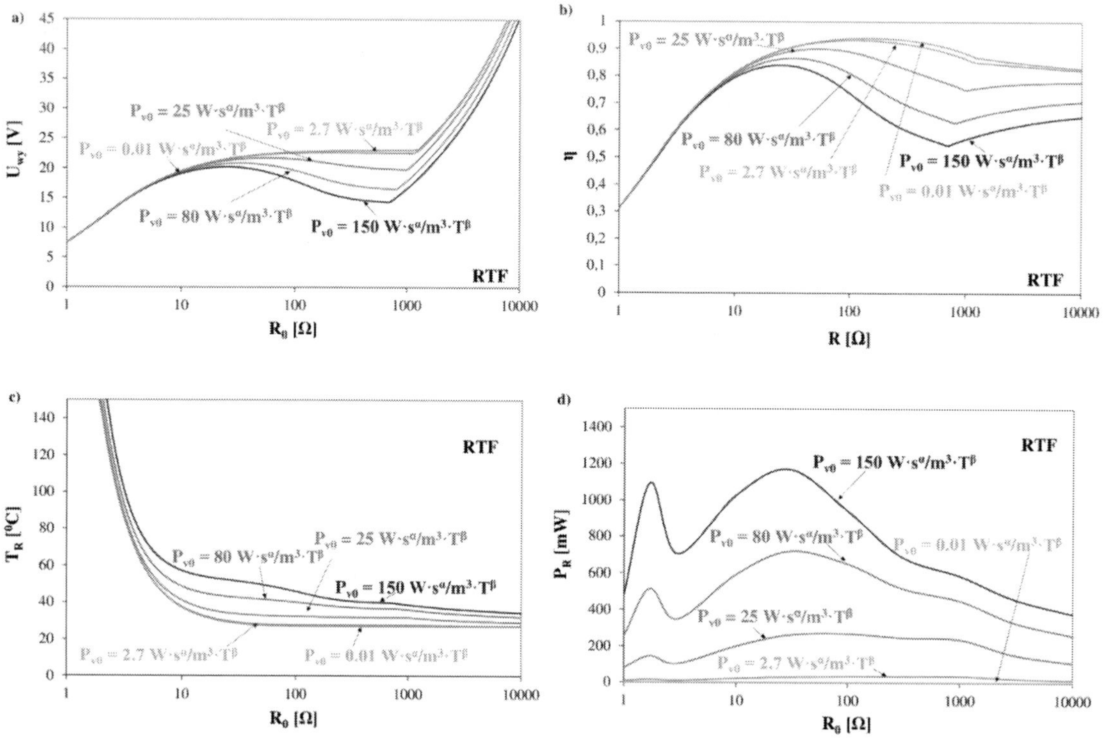

Figure 6. Dependence of the converter output voltage (a), watt-hour efficiency (b) and inductor core temperature of the considered boost converter (c) and power dissipated in the core (d) on load resistance for different values of parameter P_{V0}.

The dependences of the converter output voltage, watt-hour efficiency and core temperature as a function of load resistance of the considered converter with the ferrite core at different frequency equal to 50, 100 and 400 kHz, respectively are presented in Figure 7. As one can notice, with an increase of frequency of the control signal the point of passages of the considered converter into DCM moves to higher values of load resistance. It is proper to draw attention to the fact that only at frequency of the control signal f = 400 kHz the boost converter in the whole considered range of load resistance operates in CCM. One observes also not a large influence of frequency of the control signal on watt-hour efficiency. Whereas, temperature of the core is a decreasing function of load resistance and it is weakly depends on frequency.

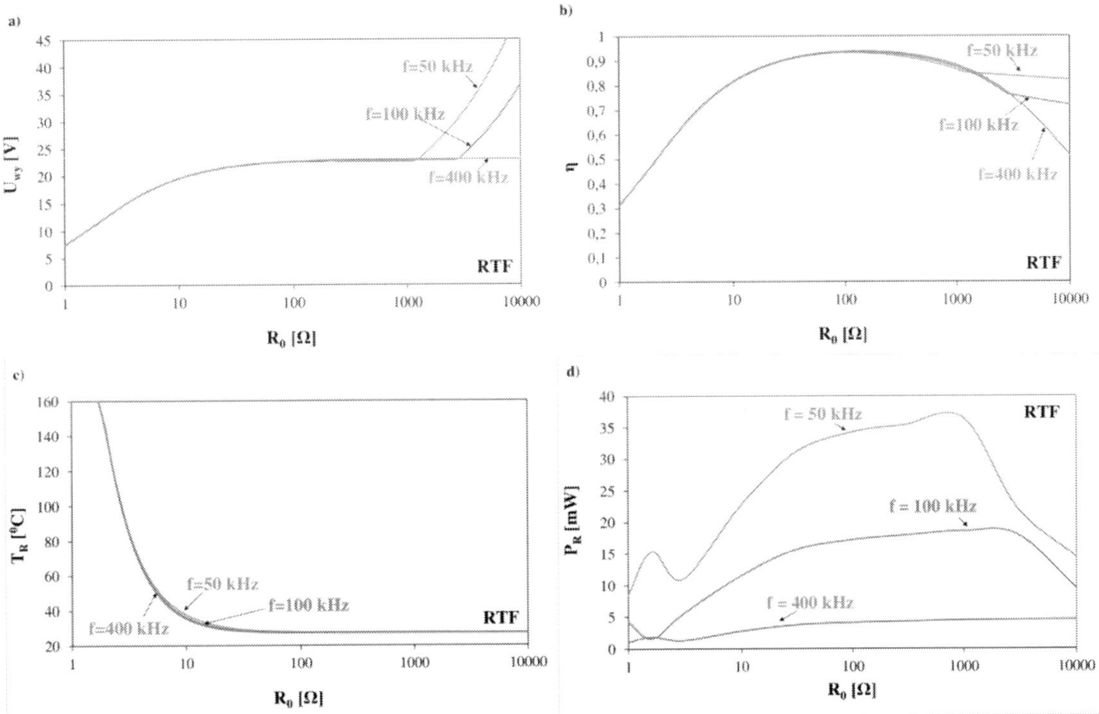

Figure 7. Dependence of the converter output voltage (a), energy efficiency (b) and inductor core temperature of the considered boost converter (c) of load resistance for different values of signal frequency.

4. Conclusions

The paper presents the influence of power losses in the core inductor on characteristics of the boost converter. This effect is studied on the basis of the proposed earlier averaged electrothermal models of a diode-transistor switch and the modified average electrothermal model of an inductor. The results show that the inductor core losses significantly affect characteristics of the considered boost converter operating in the CCM mode. An increase of loss causes a significant reduction of the output voltage and shifts the boundary between the CCM and DCM modes towards smaller values of load resistance. On the other hand, in the range of large and very small currents, the influence of losses on the output voltage of the boost converter is not observed, which means that in this range losses in the winding are dominant.

Investigations performed for the converter containing cores made of different materials show that the material of the core caused essential changes of inductance of the inductor and in a decisive manner influenced the value of load resistance at which the border between the modes CCM and DCM appear.

Additionally, one observes the essential influence of frequency of the control signal on the dependence $U_{wy}(R_0)$ over load resistance $R_0 = 1$ k Ω. With an increase of frequency follows the shift of the point of passage of the converter in the DCM mode of operation. Whereas, one does not observe influence of frequency on watt-hour efficiency of the considered converter.

5. References

[1] Borkowski A 1990 Zasilanie urządzeń elektronicznych *WKiŁ* (Warszawa)
[2] Rashid M H 2007 Power Electronic Handbook *Academic Press, Elsevier*
[3] Kazimierczuk M K 2008 Pulse-width Modulated DC-DC Power Converters *John Wiley & Sons*
[4] Detka K, Górecki K 2014 Wpływ samonagrzewania w dławiku na charakterystyki przetwornicy typu boost *Przegląd Elektrotechniczny* **90** (9) 19-21.

[5] Górecki K, Detka K 2015 The parameter estimation of the electrothermal model of inductors *Informacije MIDEM* **45** (1) 29 – 38.
[6] Górecki K, Stepowicz WJ 2008 Comparison of Inductor Models Used in Analysis of the Buck and Boost Converters *Informacije MIDEM* **38** (1) 20-25.
[7] Van den Bossche A, Valchev V C 2005 Inductors and transformers for Power Electronics *CRC Press, Taylor & Francis Group Boca Raton*
[8] Wilson P R, Ross J N, Brown A D 2002 Simulation of magnetics components models in electrics circuits including dynamic thermal effects *IEEE Transactions on Power Electronics* **17** (1) 55 – 65
[9] Górecki K, Detka K 2012 Electrothermal model of choking-coils for the analysis of dc-dc converters *Materials Science & Engineering B* **177** (15) 1248-1253
[10] Barlik R, Nowak M 2014 Energoelektronika – elementy, podzespoły, układy *Oficyna Wydawnicza Politechniki Warszawskiej* (Warszawa)
[11] Ericson R, Maksimovic D 2001 Fundamentals of Power Electronics *Norwell, Kluwer Academic Publisher*
[12] Basso C 2001 Switch – Mode Power Supply SPICE Cookbook *McGraw – Hill* (New York)
[13] Maksimovic D, Stankovic A M, Thottuvelil V J, Verghese G C 2001 Modeling and simulation of power electronics converters *Proceedings of the IEEE* **89** (6) 75 – 84.
[14] Górecki K, Zarębski J 2012 The method of a fast electrothermal transient analysis of single-inductance dc-dc converters *IEEE Transactions on Power Electronics* **27** (9) 4005-4012.
[15] Górecki K 2008 A New Electrothermal Average Model of the Diode-Transistor Switch. Microelectronics Reliability **48** (1) 51-58.
[16] Ben-Yaakov S, Glozman S, Rabinovici R 2000 Envelope simulation by SPICE-compatible models of electric circuits driven by modulated signals *IEEE Transactions on Industrial Electronics* **47** (1) 222-225.
[17] Górecki K, Zarębski J 2008 Modeling Nonisothermal Characteristics of Switch-Mode Voltage Regulators *IEEE Transactions on Power Electronics* **23** (4) 1848 – 1858.
[18] Górecki K 2015 Uśredniony elektrotermiczny model przetwornicy boost uwzględniający nieliniowość dławika *Przegląd Elektrotechniczny* **91** (5) 131-134.
[19] Górecki K, Zarębski J, Detka K 2015 Application of the electrothermal average inductor model for analyses of boost converters *Proceedings of 22nd International Conference Mixed Design of Integrated Circuits and Systems Mixdes* (Toruń) 417-421.

Assesment of $(Mn,Co)_3O_4$ powders for possible coating material for SOFC/SOEC interconnects

D. Szymczewska[1], S. Molin[2], V. Venkatachalam[2], M. Chen[2], P. Jasinski[1], P.V. Hendriksen[2]

[1] Faculty of Electronics, Telecommunications and Informatics, Gdansk University of Technology, ul. Narutowicza 11/12, 80-233 Gdansk, Poland
[2] Department of Energy Conversion and Storage, Technical University of Denmark, Frederiksborgvej 399, 4000 Roskilde, Denmark

E-mail: author: sebmo@dtu.dk

Abstract. In this work $(Mn,Co)_3O_4$ spinel powders with different Mn:Co ratio (1:1 and 1:2) and from different commercial suppliers are evaluated for possible powder for production of interconnect coatings. Sinterability of the powders is evaluated on pressed pellets sintered in oxidizing and in reducing/oxidizing atmospheres. For selected powder, coatings are then prepared by the electrophoretic deposition method on Crofer 22 APU stainless steel coupons. Effects of dispersant/iodine content and deposition voltage and times are evaluated. Thickness as a function of deposition parameters is described. Results show that with appropriate powder it is possible to produce adherent protective coating with a well-controlled thickness.

1. Introduction

High temperature corrosion of interconnects in Solid Oxide Fuel/Electrolysis stacks is one the most important degradation phenomena for operating stacks [1,2]. Oxides formed on steel interconnects have low electrical conductivity so over time they contribute more and more to the ohmic area specific resistance of the interconnect and thus the full stack [3–5]. Additionally, chromia, formed on the interconnect, can at high pO_2 and in the presence of steam evaporate and poison the oxygen electrode causing more degradation [6–8].

In order to mitigate high temperature corrosion, protective coatings are being developed [9–13]. Coatings can be in general divided into two categories: ones lowering corrosion rates and ones lowering chromia evaporation. For lowering corrosion rates reactive elements (e.g. Y_2O_3, CeO_2) were proven to be very successful whereas for mitigation of chromia evaporation one of the most promising materials reported up to date is the manganese cobalt spinel $(Mn,Co)_3O_4$ (abbreviated as MCO) [14–17]. It offers good electrical conductivity at working temperatures, good thermal expansion coefficient match to the steel and other cell components. Additionally it is very protective towards chromium diffusion – it blocks evaporation of chromia. One of the most important issues for fabrication of a well performing coating is its deposition method and sintering/thermal treatment.

In the available literature several different methods have been reported: electrophoretic deposition [18], physical vapor deposition [19], plasma spraying [20] and other [21–23]. In this work the powders are evaluated for possible use in the electrophoretic deposition [24,25] method developed in our group [26]. Electrophoretic deposition (EPD) technique is quite fragile in terms of powder requirements and its high quality. Additionally, as the volatility of powders is not certain due to producer's changes, it seems vital to evaluate different available powders to find possibly the best one and its alternatives.

2. Experimental

2.1. MCO powders.

Five different powders produced by four suppliers were tested. Main difference between powders was the Mn:Co elements ratio, as is shown in Table 1. Ratio 1:1 has been purchased from American Elements and Fuel Cell Materials. Ratio 1:2 from Fuel Cell Materials, Marion Technologies and CerPoTech companies. ID of particular powders will be used throughout the text.

Scanning electron microscopy (SEM) pictures and Energy Dispersive X-ray Spectroscopy (EDS) data of the powders were obtained using Hitachi TM3000 with Energy Dispersive X-ray Spectrophotometer Bruker Quantax70. X-ray diffractometry (XRD) measurements of the powders were made using X-ray diffractometer Bruker D8.

Table 1. Studied MCO powders.

Powder source	Molecular formula	Mn:Co ratio	ID
American Elements	$Mn_{1.5}Co_{1.5}O_4$	1:1	AE
Fuel Cell Materials	$Mn_{1.5}Co_{1.5}O_4$	1:1	FCM11
Fuel Cell Materials	$MnCo_2O_4$	1:2	FCM12
Marion Technologies	$MnCo_2O_4$	1:2	MT
CerPoTech	$MnCo_2O_4$	1:2	CPT

2.2. Sinterability.

From AE, FCM11, FCM12 and MT powders pellets were prepared and sintered in four different conditions. Temperatures of 900°C and 1000°C were used. For each temperature two conditions were tested (listed in Table 2). First sintering procedure consisted of only one step – sintering in air. Second procedure had two steps – in the beginning a reducing atmosphere (H_2/N_2) atmosphere was used. After furnace cooling reduced pellets were subjected to sintering in air as a second step. This kind of sintering, first in reducing and then in oxidizing atmosphere is often used for the spinel powder and is termed "redox" sintering.

Table 2. Sintering conditions.

Temperature [°C]	Cooling/heating rate [°C/h]	Sintering time [h]	Atmosphere
900	120	2	Air
1000	120	2	Air
900	120	2	H_2/N_2
	120	2	Air
1000	120	2	H_2/N_2
	120	2	Air

Sintered pellets were measured geometrically to calculate shrinkage and the density of each pellet. Information about density was received also from Archimedes measurements. SEM and EDX examinations of pellets polished cross-sections were made using Hitachi TM3000 with Energy Dispersive X-ray Spectrophotometer Quantax70.

2.3. Electrophoretic deposition of American Elements powder.
For EPD only American Elements powder was used. For substrates for the deposition, stainless steel coupons were prepared. These were cut from a 0.3 mm thick Crofer 22 APU steel sheet and were subsequently cleaned in ethanol. For the deposition, ethanol and isopropanol in volume ratio 1:1 were mixed as solvents for each suspensions. Three different suspensions were made with varying I_2 content. I_2 amount was different in each suspension: 0.1 wt.%, 0.2 wt.% and 0.3 wt.% of I_2 was used (wt.% in relation to powder weight). Powder content in the suspension was 2 wt.%. Influence of deposition time on layer thickness was checked. Layers were deposited for 1 min, 1.5 min and 2 min. A voltage of 60 V was used for deposition of this set. Additionally, the influence of the voltage during the deposition was also examinated. For this purpose solution with 0.2 wt.% I_2 was used and 2 min deposition time was chosen. Deposition voltages of 30 V, 40 V, 60 V, 80 V and 120 V were evaluated.

Deposited layers were sintered in two steps (redox sintering): 900°C in H_2/N_2 atmosphere for 2 hours, and after furnace cooling again at 900°C for 2 hours in air. Sintered layers were examined using Hitachi TM3000 with Energy Dispersive X-ray Spectrophotometer Quantax70. XRD measurements were made using X-ray diffractometer Bruker D8.

3. Results
In the first step the powder morphology and composition was checked by means of SEM and EDS analysis. This was done in order to ensure powder composition and quality.
SEM analysis of loose powder (attached to a carbon tape) is presented in Figure 1. All powders have fine, less than 1 micrometer grains. Powders produced by CerPoTech, coming from a spray drying process have a core-shell structure and to be suitable for the electrophoretic deposition would require an additional grinding/milling step. Therefore this powder has been excluded from further (sintering) analyses.
As provided by the manufacturers, Mn:Co ratio in the powders should be either 1:1 or 1:2 depending on the composition. All tested powders roughly fall in the expected range with a small and constant deviation, that all powders are slightly more Co rich then the original ratio. About 5-10 at.% enrichment in Co content is measured for all powders.

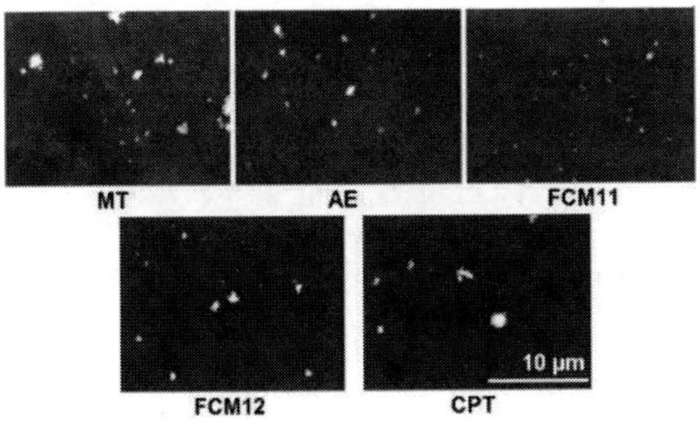

Figure 1. MCO powders

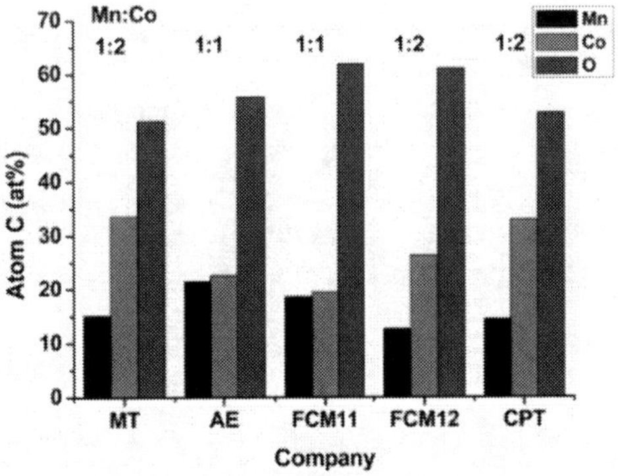

Figure 2. Atomic composition of different as-delivered powders as determined by EDS analysis.

After the SEM/EDS analysis, X-ray diffractometry was performed to check phase purity of the powders. Although no impurities (no foreign elements) were detected by the EDS analysis, still it is interesting to see phases present. Spinel with a nominal cation ratio of 1:1 is at room temperatures a mixture of a cubic $MnCo_2O_4$ and tetragonal Mn_2CoO_4 phases. This has been confirmed by XRD analysis of powders FCM11 and AE shown in Figure 3. Spectra are quite noisy with broad peaks which might indicate a small crystalline size and/or only partial crystallization of the powders that might require further thermal processing upon heating of the spinel powder, only single cubic phase remain at 800°C. On the other hand, XRD analysis of CPT, FCM12 and MT (Figure 4) powders revealed only a single phase of cubic spinel. In all cases peaks are narrower. No impurity phases were found in any of powders.

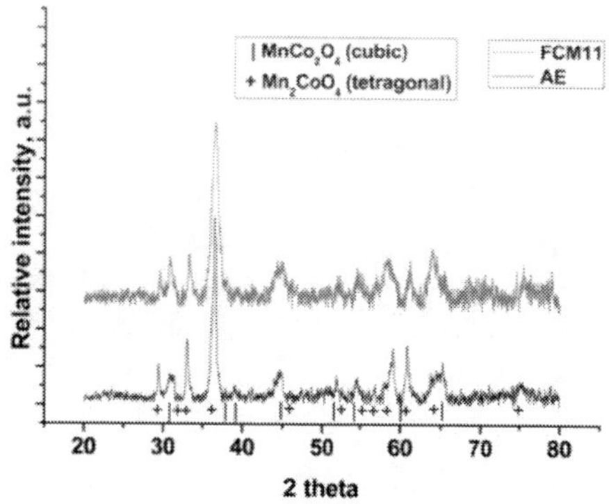

Figure 3. XRD spectra of powders with Mn:Co ratio of 1:1: AE and FCM11.

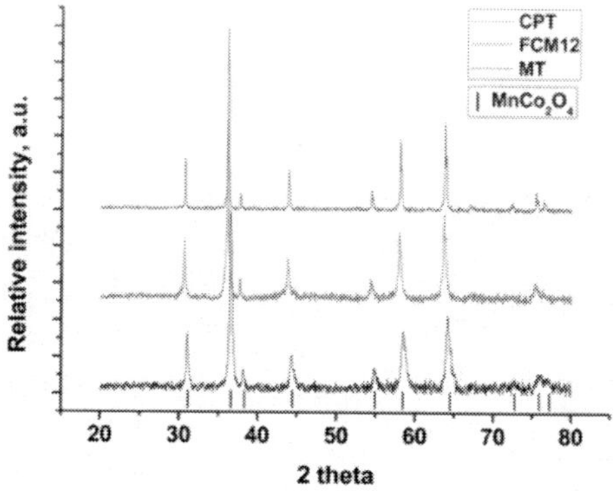

Figure 4. XRD spectra of powders with Mn:Co ratio of 1:2: CPT, MT and FCM12.

3.1. Sinterability

After ensuring powders quality, its sinterability was evaluated. Pellets were pressed from MT, AE, FCM11 and FCM12 powders. Pellets were then sintered in 4 different conditions (two temperatures and two sets of atmospheres). As a measure of sinterability, linear shrinkage, calculated density (based on pellet geometrical shrinkage) and measured density (by Archimedes principle) were used. Linear shrinkage of the pellets is presented in Figure 5. Geometrical measurement can be influenced by several factors: cracks or voids inside, parts of the pellet cracked away etc, so more reliable are the values from the Archimedes measurement and these are presented in Figure 6.

For sintering in air, there is a clear difference between the temperatures of 900°C and 1000°C. MT and AE powders have shown little sintering at 900°C and FCM 11 and FCM12 powders have shown visibly higher shrinkage (~5 %). For sintering at 1000°C, all powders sintered better than at 900°C with shrinkages in the range 5-12 %. Powders FCM11 and FCM12 again have shown superior shrinkage.

Sintering in reducing and then oxidizing atmosphere provided samples with much lower porosities. Even at 900°C samples are less porous then samples sintered in air at 1000°C. After sintering with the redox procedure, density of MT, AE and FCM11 pellets were very similar. Shrinkage was ~25 %. Only pellet prepared from the FCM12 powder reported density and shrinkage lower than expected both for 900°C and 1000°C.

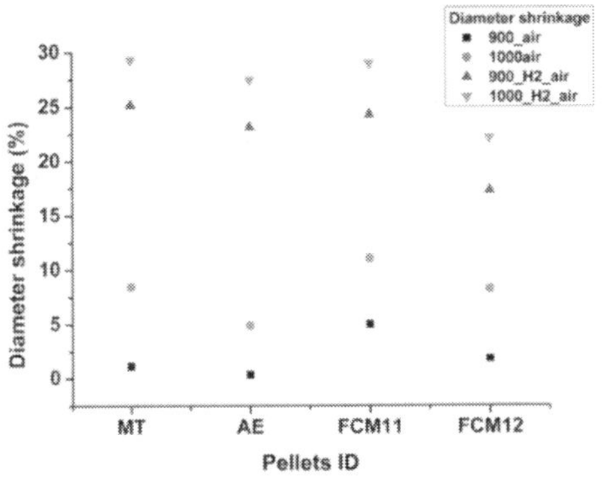

Figure 5. Measured shrinkage of pellets prepared from different powders and sintered at different conditions.

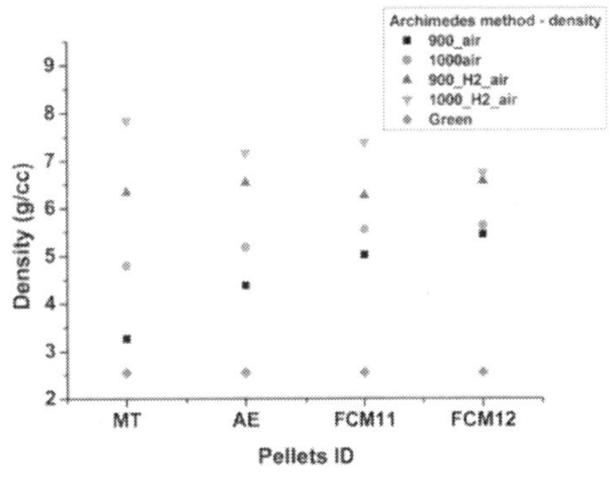

Figure 6. Measured density (by Archimedes method) of pellets prepared from different powders and sintered at different conditions.

Results of shrinkage and density analysis of sintered pellets were further compared with SEM images of pellets cross sections. Cross section images from MT, AE, FCM11 and FCM12 pellets sintered in all 4 conditions are shown in Figure 7. For samples sintered only in air, there is a visible difference in porosities of samples sintered at 900°C and 1000°C. Also grain size increase is noticed, especially from both FCM powders.

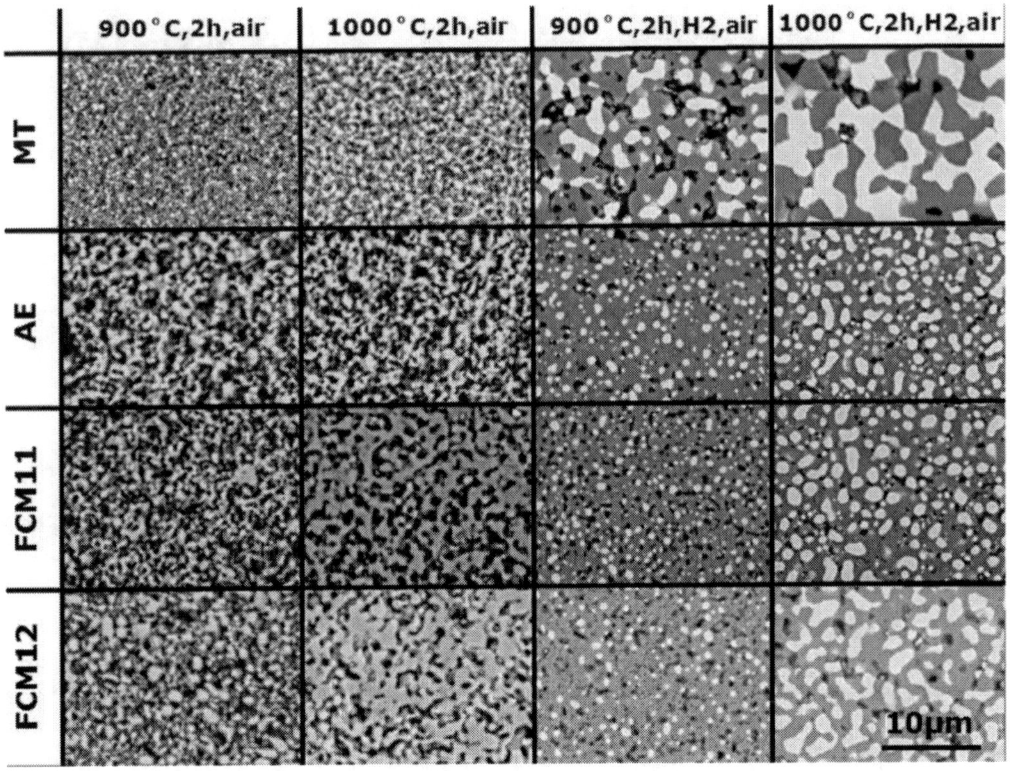

Figure 7. SEM cross sections of polished pellets prepared from different powders and sintered at different conditions.

SEM images of samples sintered using the "redox" procedure show an interesting microstructure consisting of two phases (bright and dark phases). Apparently, the darker phase is the spinel, whereas the brighter phase is a still reduce powder. After sintering in reducing conditions a mix of Co and MnO is formed that upon reoxidation should transform back to the oxide. However, due to kinetic limitations and porous structure of the samples, reoxidation did not occur fast enough. Grain size in MT and FCM12 powders (both powders with a pure cubic phase composition) was higher than for the other two powders with a mixed composition. It seems that at 900°C there is less metallic phase left after oxidation than at 1000°C. This might also indicate that the reduction conditions were not sufficient to fully reduce the powder. To fully oxidize samples longer exposure times would be needed. Due to importance of the sintering step and possible influence of the sintering atmosphere, temperature and time these studies will be continued in the coming future.

3.2. EPD of American Elements powder

3.2.1. Different I_2 contents and deposition time

In order to deposit powders via electrophoretic deposition, particles in the suspension must be given some electric charge. For this purpose in this study I_2 is added to the suspension. Different amount of I_2 has to be added to powders with different surface area etc. For the AE powder amounts of 0.1 wt.%, 0.2 wt.% and 0.3 wt.% were tested. Samples were deposited at 1 min, 1.5 min and 2 min. After deposition coatings were sintered using a "redox" procedure at 900°C. SEM of the as-sintered coatings are shown in Figure 8.

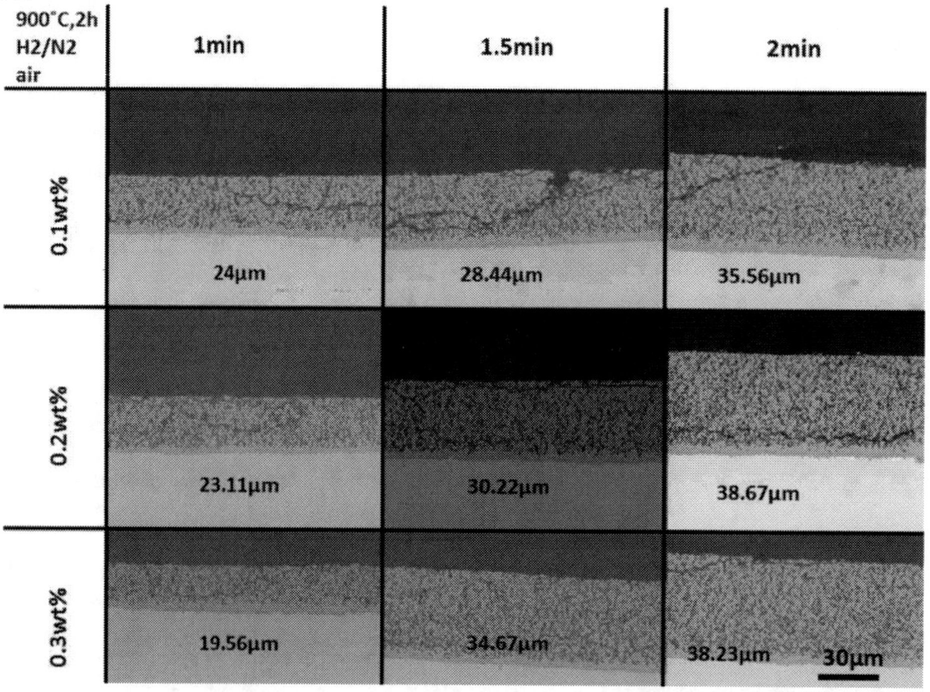

Figure 8. SEM polished cross sections of EPD coatings prepared from AE powder with different iodine content and varying deposition time. Thickness values measured from SEM pictures.

For coatings prepared on steel, there seems to be no unreacted reduced phase present after the oxidation stage of the sintering. It might be due to a relatively low thickness of the coating (~30 µm instead of few mm for pellets) which would facilitate gas exchange due to diffusion.

Thickness of the produced coatings, has been evaluated from the SEM images (Figure 9) and additionally, just after the deposition from the weight gain of sample coupons (Figure 10). For this a 30% porosity was included into calculations. Thickness calculations from the weight gain seems to be representative for real coating thickness and therefore can be used as an assessment of sample quality without the need to destroy the sample.

For the amount of dispersant of 0.1 wt.% and 0.2 wt.% the thickness vs. deposition time is very similar. Whereas for 0.3 wt.% for 1 minute coating thickness is lower than anticipated. One possible reason is that it has been too much of I_2 and some of it reacted at the electrodes and then the amount was reduced and for longer times results were similar for all studied concentrations. It seems from the images, that for higher iodine content, produced coating is less cracked and of higher quality. For 0.3 wt.% of iodine almost no cracks are visible. Looking at the results, using 0.2 wt.% or 0.3 wt.% with the AE powders seems sufficient to ensure homogenous coating.

By controlling the time of deposition coatings with thicknesses ranging from 20 μm to 40 μm might be produced.

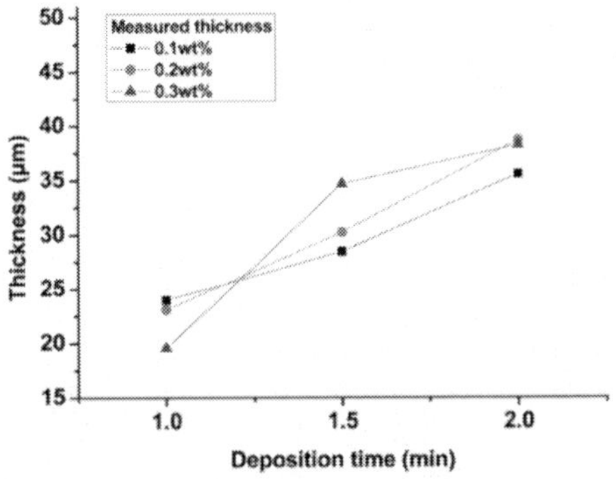

Figure 9. Thickness of the coatings as measured from the SEM cross section images.

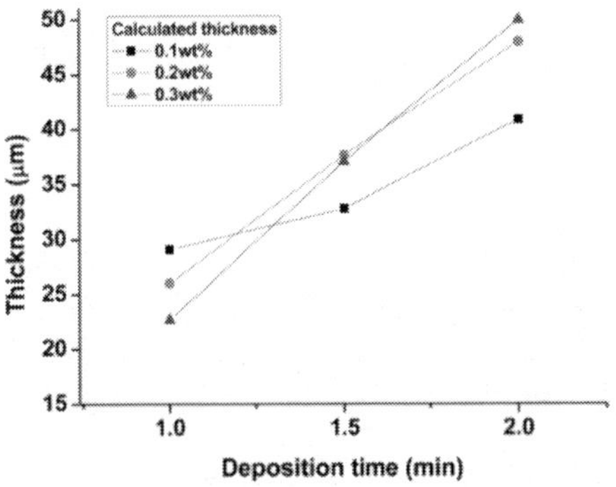

Figure 10. Thickness of the coatings as measured from weight gain after the EPD deposition.

After evaluation of coatings fabrication parameters, cross sections were analysed in more details by SEM/EDS analysis to check chromium diffusion and oxide layer thickness formed on steel during the "redox" sintering procedure. Samples analysed were samples prepared with 0.1, 0.2, 0.3 wt.% iodine and the deposition time was 1 minutes. Results of the analysis are presented in Figure 11. For all samples 3 regions of interest were analysed. "1" is the alloy, analysed just to ensure proper calibration of the EDS. "2" is the interface of the alloy and the coating, where the oxide (Cr_2O_3 and $(Mn,Cr)_3O_4$ spinel should occur). "3" is the bulk of the coating, far away from the interface. The composition of points number "1" is for all samples quite similar and close to the expected composition (between 21 and 24 wt.% as given by the producer of steel). At the interfaces of all samples, Cr rich phase is detected. Due to spatial limitation of the EDS detector it is not possible to clearly distinct the chromium oxide from the bulk of the alloy. Also Mn,Co spinel is detected in point "2". In points "3", major elements present are Mn and Co in the 1:1 ratio as expected for the AE powder. For all samples also Fe and Cr are detected in the coating. Up to 6 wt.% of Cr is present in the coating after sintering. This amount seems quite high. One possibility is that Fe and Cr diffused into the coating during sintering. However, the time was quite short, in total 4 hours at 900°C so maybe metallographic preparation of the samples was not careful enough and some alloy particles were embedded in the coating during polishing. Certainly this will be studied further.

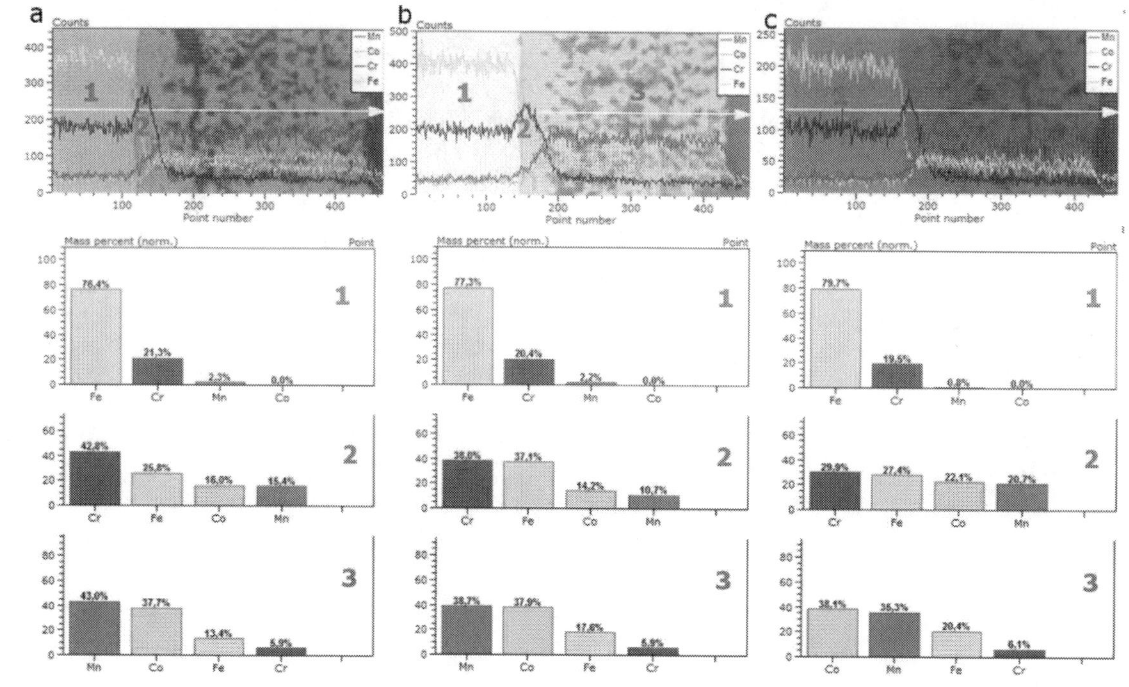

Figure 11. SEM and EDS line profiles of Mn, Co, Cr and Fe on the sintered coatings. Samples with a) 0.1 wt%, b) 0.2wt%, c) 0.3wt% of I_2.

4. Conclusions

In this study several commercial spinel powders were evaluated for their phase purity, composition and sinterability. Four different sintering conditions were used. Results show that sintering using a "redox" procedure results in much denser pellets. For pellets, reduction and/or reoxidation step seems to be kinetically/diffusion limited so that large amount of the reduced phase remain in the pellet after "redox" sintering. This is not present in the coatings prepared on stainless steels.

Electrophoretic deposition method has been evaluated for producing coatings with different dispersant content. For iodine content >0.2 wt.% crack free and homogenous layers were prepared. By finding suitable solvent, dispersant content and deposition parameters (voltage, time) protective coatings for steel interconnects can be produced reproducibly by the EPD method.

5. References

[1] Shaigan N, Qu W, Ivey D G and Chen W 2010 A review of recent progress in coatings, surface modifications and alloy developments for solid oxide fuel cell ferritic stainless steel interconnects *J. Power Sources* **195** 1529–42

[2] Fergus J W J W 2005 Metallic interconnects for solid oxide fuel cells *Mater. Sci. Eng. A* **397** 271–83

[3] Żurek Z, Brylewski T, Jaroń A and Chmura E 2013 Area specific resistance of the scale formed on Crofer 22APU ferritic steel in atmospheres containing SO2 *Solid State Ionics* **234** 33–9

[4] Brylewski T, Nanko M, Maruyama T and Przybylski K 2001 Application of Fe–16Cr ferritic alloy to interconnector for a solid oxide fuel cell *Solid State Ionics* **143** 131–50

[5] Fergus J W 2005 Metallic interconnects for solid oxide fuel cells *Mater. Sci. Eng. A* **397** 271–83

[6] Jiang S P and Chen X 2014 Chromium deposition and poisoning of cathodes of solid oxide fuel cells – A review *Int. J. Hydrogen Energy* **39** 505–31

[7] Wang K, Liu Y and Fergus J W 2011 Interactions Between SOFC Interconnect Coating Materials and Chromia *J. Am. Ceram. Soc.* **94** 4490–5

[8] Gindorf C, Singheiser L and Hilpert K 2005 Vaporisation of chromia in humid air *J. Phys. Chem. Solids* **66** 384–7

[9] Molin S, Chen M and Hendriksen P V 2013 Oxidation study of coated Crofer 22 APU steel in dry oxygen *J. Power Sources* **251** 488–95
[10] Szymczewska D, Molin S, Chen M, Hendriksen P V and Jasinski P 2014 Ceria Based Protective Coatings for Steel Interconnects Prepared by Spray Pyrolysis *Procedia Eng.* **98** 93–100
[11] Molin S, Kusz B, Gazda M and Jasinski P 2008 Protective coatings for stainless steel for SOFC applications *J. Solid State Electrochem.* **13** 1695–700
[12] Yang Z, Walker M S, Singh P, Stevenson J W and Norby T 2004 Oxidation Behavior of Ferritic Stainless Steels under SOFC Interconnect Exposure Conditions *J. Electrochem. Soc.* **151** B669–78
[13] Yang Z, Xia G-G, Li X-H and Stevenson J W 2007 (Mn,Co)3O4 spinel coatings on ferritic stainless steels for SOFC interconnect applications *Int. J. Hydrogen Energy* **32** 3648–54
[14] Larring Y and Norby T 2000 Spinel and Perovskite Functional Layers Between Plansee Metallic Interconnect (Cr-5 wt % Fe-1 wt % Y[sub 2]O[sub 3]) and Ceramic (La[sub 0.85]Sr[sub 0.15])[sub 0.91]MnO[sub 3] Cathode Materials for Solid Oxide Fuel Cells *J. Electrochem. Soc.* **147** 3251–6
[15] Larring Y, Haugsrud R and Norby T 2003 HT Corrosion of a Cr-5 wt % Fe-1 wt % Y[sub 2]O[sub 3] Alloy and Conductivity of the Oxide Scale *J. Electrochem. Soc.* **150** B374–9
[16] Brylewski T, Kucza W, Adamczyk A, Kruk A, Stygar M, Bobruk M and Dąbrowa J 2014 Microstructure and electrical properties of Mn1+xCo2−xO4 (0≤x≤1.5) spinels synthesized using EDTA-gel processes *Ceram. Int.* **40** 13873–82
[17] Liu Y, Kumar C J D and Fergus J W 2012 Electrical Properties of Transition Metal-Doped (Mn,Co) pp 421–7
[18] Smeacetto F, De Miranda A, Cabanas Polo S, Molin S, Boccaccini D, Salvo M and Boccaccini A R 2015 Electrophoretic Deposition of Mn1.5 Co1.5 O4 on Metallic Interconnect and Interaction with Glass-ceramic Sealant for Solid Oxide Fuel Cells Application *J. Power Sources* **280** 379–86
[19] Ajitdoss L C, Smeacetto F, Bindi M, Beretta D, Salvo M and Ferraris M 2013 Mn1.5Co1.5O4 protective coating on Crofer22APU produced by thermal co-evaporation for SOFCs *Mater. Lett.* **95** 82–5
[20] Puranen J, Lagerbom J, Hyvärinen L, Mäntylä T, Levänen E, Kylmälahti M and Vuoristo P 2010 Formation and structure of plasma sprayed manganese-cobalt spinel coatings on preheated metallic interconnector plates *Surf. Coatings Technol.* **205** 1029–33
[21] Malzbender J, Batfalsky P, Vaßen R, Shemet V and Tietz F 2012 Component interactions after long-term operation of an SOFC stack with LSM cathode *J. Power Sources* **201** 196–203
[22] Wei P, Deng X, Bateni M R and Petric A 2007 Oxidation and electrical conductivity Behavior of spinel coatings for metallic interconnects of solid oxide fuel cells *Corrosion* **63** 529–36
[23] Kruk A, Stygar M and Brylewski T 2012 Mn–Co spinel protective–conductive coating on AL453 ferritic stainless steel for IT-SOFC interconnect applications *J. Solid State Electrochem.* **17** 993–1003
[24] Corni I, Ryan M P and Boccaccini A R 2008 Electrophoretic deposition: From traditional ceramics to nanotechnology *J. Eur. Ceram. Soc.* **28** 1353–67
[25] Boccaccini A R and Zhitomirsky I 2002 Application of electrophoretic and electrolytic deposition techniques in ceramics processing *Curr. Opin. Solid State Mater. Sci.* **6** 251–60
[26] Bozza F, Schafbauer W, Meulenberg W A and Bonanos N 2012 Characterization of La0.995Ca0.005NbO4/Ni anode functional layer by electrophoretic deposition in a La0.995Ca0.005NbO4 electrolyte based PCFC *Int. J. Hydrogen Energy* **37** 8027–32

Acknowledgement

Authors would like to thank for the financing: at Gdansk University of Technology by the NCN grant 2012/05/B/ST7/02153 "Functional layers for solid oxide fuel cells" and at Technical University of Denmark by a grant from the ForskEL project 2015-1-12276 "Towards solid oxide electrolysis plants in 2020" funded by Energinet.dk. D. Sz. wants to acknowledge an internship stipend by the InterPhD programme at GUT.

IOP Conf. Series: Materials Science and Engineering **104** (2015) 012018 doi:10.1088/1757-899X/104/1/012018

Thin perovskite-type ferromagnetic film (La,Sr)CoO_3

A Łoziński

Gdynia Maritime University, Poland

E-mail: lozinskiandrzej@gmail.com

Abstract. Properties of materials having a perovskite-type oxide crystal structure ABO3 can easily be modified by means of partial substitution of cations located at A or B sites. Non-stoichiometric lanthanum-strontium cobalt oxides (La,Sr)CoO_3 also referred to as LSCO exhibit electrical conductivity and ferromagnetic properties. Perovskite-type compounds $LaCoO_3$ and $SrCoO_3$ can form solid solutions in every ratios to each other. Notation (La,Sr)CoO_3, or La(1-x)SrxCoO_3 means that some lanthanum ions are substituted with the strontium ones at A sites of crystal cell. This paper describes a sol-gel manufacturing process of LSCO thin films, presents measurements of their resistivity vs. composition, shows thermal coefficient of selected composition resistivity as well as its magnetoresistive properties.

1. Introduction

Properties of materials having a perovskite-type oxide crystal structure ABO3 can easily be modified by means of particular substitution of cations located at A or B sites [1].

Perovskite-type compounds $LaCoO_3$ and $SrCoO_3$ can form solid solutions in every ratio to each other. The notation (La,Sr)CoO_3, or La(1-x)SrxCoO_3 means that some lanthanum ions are substituted with the strontium ones at A sites. Shortly we will call this system LSCO. As we have verified, substitution of some Sr ions for La ones in the LSCO system changes its electrical conductivity considerably.

2. Manufacturing process

A sol-gel, water based process was applied to obtain thin films. The starting materials were aqueous solutions of nitrates of lanthanum, strontium and cobalt. Five different solutions were prepared: 1 - $LaCoO_3$ (LCO); 2 - $La_{0.75}Sr_{0.25}CoO_3$ (75LSCO); 3 - $La_{0.5}Sr_{0.5}CoO_3$ (50LSCO); 4 - $La_{0.25}Sr_{0.75}CoO_3$ (25LSCO); 5 - $SrCoO_3$ (SCO). The flow chart shown in Figure 1 presents the preparation of initial solutions for films production. The produced solutions were used for preparing thin films by spin coating on alumina substrates in the process presented in Figure 2.

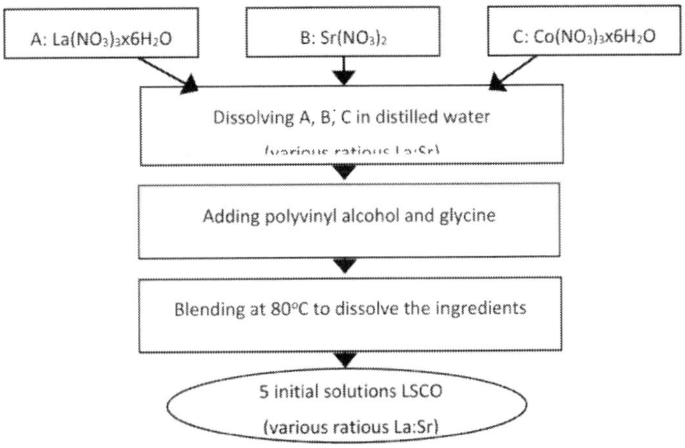

Figure 1. Flow chart of initial solutions for LSCO film preparation.

The flow chart shown in Figure 1 presents the preparation of initial solutions for films production. The produced solutions were used for preparing thin films by spin coating on alumina substrates in the process presented in Figure 2.

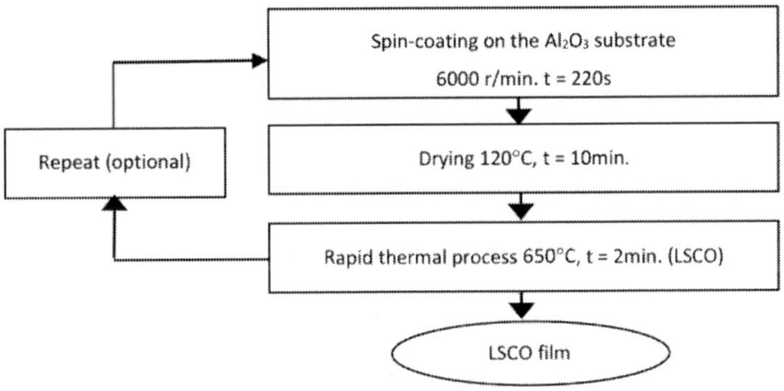

Figure2. Flow chart of LSCO film production.

Thickness of one time coated film amounts to 2.5 μm.

3. Measurements

Room temperature sheet resistivity R_S of the obtained films was measured with a four-points probe [2]. The results of these measurements are shown in Figure 3. Also for composition no. 3. the resistivity vs. temperature characteristic in 290 – 320 K range was determined, and their thermal coefficient of resistivity was calculate using the equation (1).

$$\alpha = \frac{1}{R_S}\frac{dR_S}{dT} \cdot 100\% \qquad (1)$$

Where: α – thermal coefficient of resistivity, R_S – sheet resistivity [Ω], T – temperature [K].
Figure 4. shows results of single film measurements and calculations.

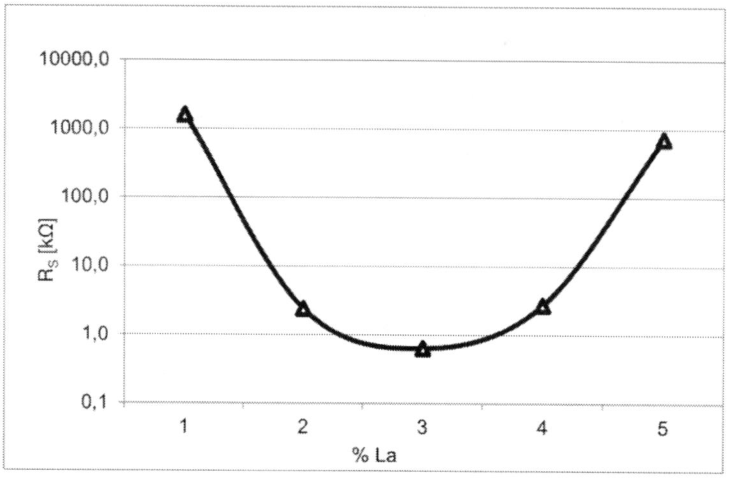

Figure 3. Room temperature resistivity of thin films v.s. the used composition.

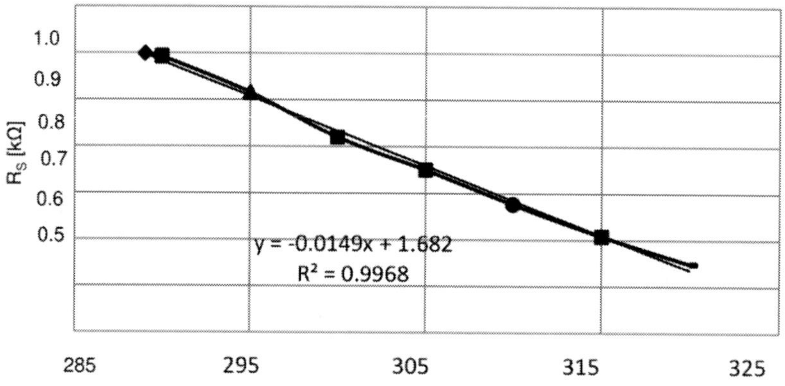

Figure 4. Resistance R_s v.s. temperature of the thin film 50 LSCO.

The thermal coefficient of resistivity was calculated from the polynomial y = -0.0149+1.682 approximating the results from Figure 4.: $\alpha = y' * 100 = -1.49$ %/K.

After processing resistance of films is unstable, yet ageing films in 200 °C for two months solves this problem. Figure 5 shows the results of this ageing. Probably it is effect of chemisorption which takes place between the gas and the solid surface of film [3].

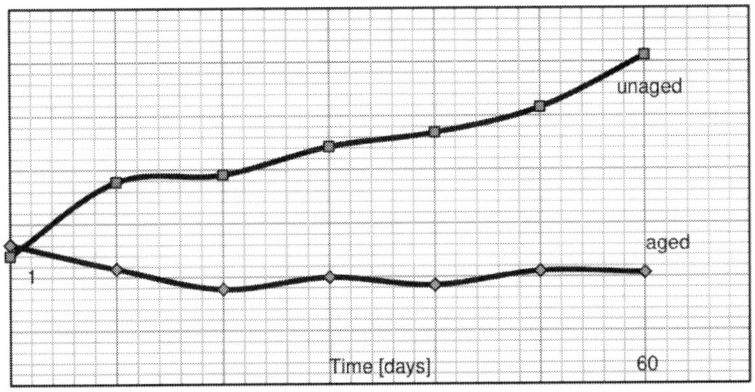

Figure 5. Results of ageing of films

Magnetoresistive properties of the thin film 50 LSCO were examined using the Helmholtz coil to obtain the uniform magnetic field. Figure 6 shows room temperature magnertoresistance of the film oriented longitudinally with respect to the magnetic field as well as the shape of the examined sample.

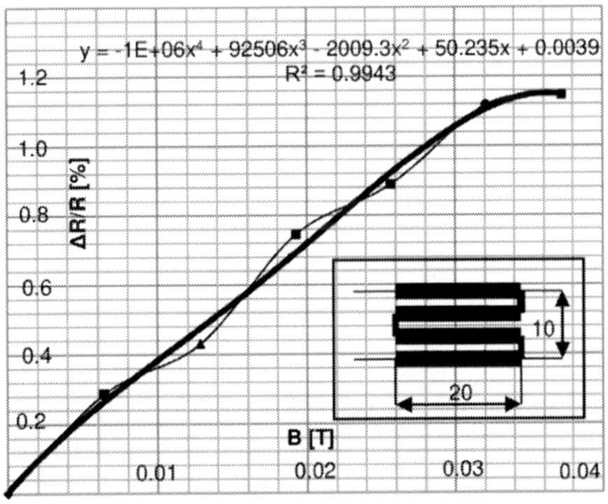

Figure 6. Magnetoresistance of the thin 50 LSCO film oriented longitudinally with respect to the magnetic field and the shape of the examined sample.

4. Results end discussion

The examined ferromagnetic composition $(La_{50},Sr_{50})CoO_3$ exhibits nearly metallic conductivity in room temperatures. Its thermal coefficient of resistivity is negative and amounts to –1.49 %/K.

The observed effect of ageing on resistance stability is typical for NTC oxidic thermistors. The longitudinal magnetoresistive effect is positive i.e. an increase in the field B causes an increase in resistance and saturation occurs at about B = 0.04 T.

The nearly metallic conductivity causes that the examined composition can be utilized as the electrode for ferroelectric films [4]. Moreover, similarity and nearing sizes of its crystal structure and perovskite ferroelectric structures, influence positively the reduction of stress usually observed between the ferroelectric film and the metallic electrode. Author has used examined composition as the electrodes for thin pyroelectric PLZT film [5].

References

[1] Glasso FS 1969 Structure, properties and preparation of perovskite-type compounds *Pergamon Press*

[2] Smits FM 1958 Measurements of Sheet Resistivity with the Four-Point Probe. *BSTJ*

[3] Kralik M. 2014 Adsorption, chemisorption and catalysis. *Springer Review, Chemical Papers, Vol **68** Issue 12 pp 1625-1638*

[4] Ohno T, at al. 2014 Preparation of lanthanum strontium cobalt oxide electrode on a Si wafer for stress engineering of ferroelectric thin films, *J. Ceram.Soc.Japan, Vol.**122** No.1421 pp.63-66*

[5] Łoziński A 2007 Thin pyroelectric PLZT film obtained with sol-gel technology. *Springer J. Electroceram 19: 303-306*

IOP Conf. Series: Materials Science and Engineering **104** (2015) 012019 doi:10.1088/1757-899X/104/1/012019

$LaNi_{1-x}Co_xO_{3-\delta}$ (x=0.4 to 0.7) cathodes for solid oxide fuel cells by infiltration

Aleksander Chrzan[1,3], Simona Ovtar[2], Ming Chen[2]

[1] Faculty of Electronics, Telecommunications and Informatics, Gdansk University of Technology, Narutowicza 11/12, 80-233 Gdansk, Poland
[2] Department of Energy Conversion and Storage, Technical University of Denmark, Risø Campus, Frederiksborgvej 399, DK-4000 Roskilde, Denmark

E-mail: aleksander.chrzan@eti.pg.gda.pl

Abstract. Performance of $LaNi_{1-x}Co_xO_{3-\delta}$ (LNC) (x=0.4 to 0.7) as a cathode in solid oxide fuel cell (SOFC) is evaluated. Symmetrical cathode/electrolyte/cathode cells for electrochemical testing are prepared by infiltration of yttria stabilized zirconia (YSZ) backbone with LNC solutions. It is showed that the cathode infiltrated with $LaNi_{0.5}Co_{0.5}O_{3-\delta}$ (LNC155) has the lowest polarization resistance and activation energy, 197 mΩ cm^2 at 600 °C and 0.91 eV, respectively. Therefore it is the most promising material of the LNC group for electrochemical applications. X-ray diffraction analysis revealed that none of the materials is single-phased after heat treatment at 800 °C as they contain residues of La_2O_3 and $La_2NiO_{4-\delta}$.

1. Introduction

Solid oxide fuel cells (SOFCs) are electrochemical devices, which when operated above 600 °C efficiently generate electricity. They can utilize hydrogen, biogas or other fuels. Relatively high operating temperature is required for a sufficient catalytic activity of electrodes and an electrolyte ionic conductivity. This causes major problems with mechanical durability, diffusion of cell components and elevates cost of fuel cell stacks fabrication. As a result, there is still ongoing research to improve the catalytic activity and the conductivity of the fuel cells that would reduce the operating temperature [1]. High research focus is on seeking for new, more efficient, more stable or low cost SOFC materials [2]. On the other hand the additional improvement of already existing materials, by varying its microstructure (e.g., producing improved cathode/electrolyte interface) is often reported [3, 4]. One of the ways to alter the structure and improve the properties of electrodes is to increase the surface area available for an electrode reaction. Nano particles have by definition high surface area and consequently high catalytic activity. The efficient way to introduce nano particles into the electrode is by infiltration process. During the infiltration porous, cathode backbone, most commonly made of electrolyte material, is partially filled with a cathode material of choice. The cathode material can be introduced both as suspension or solution [5]. In the case of solution final size and phase composition depends heavily on the heat treatment temperature. Another advantage of using infiltration process is an easy matching of thermal expansion coefficients (TEC) of different layers of SOFC. Since the backbone usually has the same TEC as the rest of the cell, the infiltrated cathode material of very different TEC does not cause an accelerated delamination. This enables usage of materials with higher TEC like $LaSr_{1-x}Co_xO_{3-\delta}$, $LaNi_{1-x}Co_xO_{3-\delta}$ (LNC) or $LaCoO_{3-\delta}$ [6, 7].

In this study we report the infiltration of yttria stabilized zirconia (YSZ) backbones with solutions of lanthanum, cobalt and nickel nitrates. LNC was chosen because of its high conductivity and stability in used temperature range [8]. The ratio of Ni to Co was varied from 4:6 to 7:3 to investigate the influence of the chemical composition on the electrochemical performance of symmetrical cells. Different reports about LCN show that the conductivity of compound varies with Ni to Co ratio, as is listed in table 1 [9-10]. The highest conductivity was reported for the compound with Ni:Co ratio of 6:4.

Table 1. Conductivity and thermal expansion coefficient of LNC in literature [9, 10].

composition	σ_{roomT} [Scm^{-1}]	α [10^{-6} K^{-1}]
LN		10
LNC191	170	
LNC182	250	
LNC173	970	
LNC164	1460	14.3
LNC155	1090	
LNC146	700	15.9
LNC137	320	
LNC128	150	
LNC119	40	
LC		22

2. Experimental

For the electrochemical impedance spectroscopy symmetrical cathode/electrolyte/cathode cells were prepared. The chosen LNC compositions were: LNC164, LNC155, LNC146 and LNC137. Cells consisted of 100 μm thick dense YSZ layer and of 47 μm thick porous YSZ backbones on both sides. Backbones porosity was 64% with the pore size between 2 to 4 μm. The cells were prepared by co-sintering of dense and porous tape casted tapes. Further detail on cell fabrication is described elsewhere [11]. Aqueous nitrate solutions of 20% gadolinium doped ceria (CGO) and LNC, containing different Ni:Co ratios were used for infiltration [12]. First, the backbones were infiltrated with CGO precursor solution to prevent reaction between YSZ and LNC. After that backbones were infiltrated with the LNC precursor solution. Each infiltration was followed by firing at 350 °C. The number of LNC infiltrations was fixed to nine. Increase of mass of the cathode material in the backbone was followed by weighing samples during the process. Platinum current collector layer was applied on the top of infiltrated cathodes and dried. Finally, the symmetrical cells with infiltrated materials were in-situ heat treated at 800 °C for 2h in an electrochemical setup.

SEM characterization of studied symmetrical cells was performed using Zeiss Supra 35 microscope. HT-XRD was carried out using Bruker D8 with MRI high temperature stage and PSD Lynx Eye detector. XRD spectra were obtained at 800 °C in air using dried LNC solutions. Electrochemical impedance spectra of the symmetrical cells were recorded using Hioki 3522 analyzer with a frequency range from 100 kHz to 1 Hz and an amplitude of 50 mV under the open-circuit conditions. The impedance spectra were measured from 800 °C to 400 °C immediately after the heat treatment. The data were corrected for the series inductance of the measurement rig which was in the range of 85–110 nH. Measurements were analyzed in ZSimpWin software to extract the series and polarization resistance.

3. Results and discussion

Figure 1 shows how the percentage of LNC mass in the total cathode mass was increasing with each infiltration step. It can be seen that the increase of infiltrated material is approximately linear, which means that pores of the backbone were not completely filled with LNC.

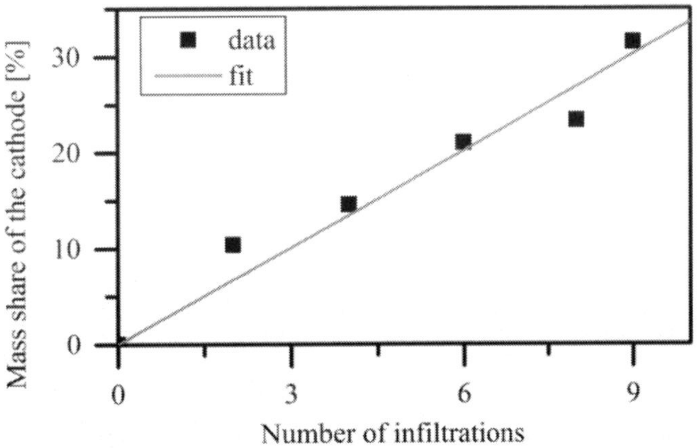

Figure 1. Average mass content of LNC in the cathode during the infiltration process.

Detailed view of the cathodes structure can be seen in figure 2. Figure 2a shows structure of LNC146 infiltrated cathode before measurement. Particles size of the infiltrated material is below the observation limit of used SEM. The infiltrated material seems to cover the whole surface of YSZ grains. Normally well defined YSZ grain boundaries are barely visible after the infiltration. Figure 2b shows the same type of structure after heat treatment at 850 °C. LNC creates porous layer over the YSZ backbone, with the particles size around 70-200 nm.

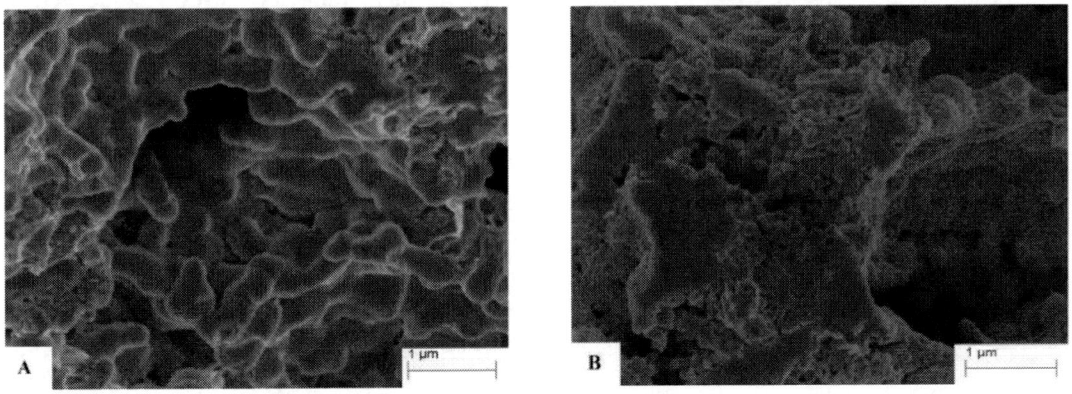

Figure 2. SEM images of the LNC infiltrated symmetrical cell before (A) and after (B) heat treatment.

Dried solutions were investigated using XRD analysis. Figure 3 shows XRD patterns of all 4 LNC materials heated in situ to 800 °C on Pt sample holder. The major phase in all samples is $LaNi_{1-x}Co_xO_{3-\delta}$. As secondary phase La_2O_3 was detected in all samples. With comparing surface area of peaks of LNC phase to La_2O_3 phase, it can be seen that the amount of La_2O_3 slightly varies with Co:Ni ratio in precursor solution. Higher amount of La_2O_3 can be found in LNC146 and LNC164 than in LNC137 and LNC155. In XRD spectra of LNC155 an additional secondary phase of $La_2NiO_{4+\delta}$ was observed. The presence of $La_2NiO_{4+\delta}$ phase in other samples cannot be excluded, due to detection limit of XRD. To conclude, none of the materials forms a pure perovskite phase at 800 °C. However, most probably due to homogeneous distribution of phases in backbone there is no evidence that it has an adverse effect on the cathodes performance.

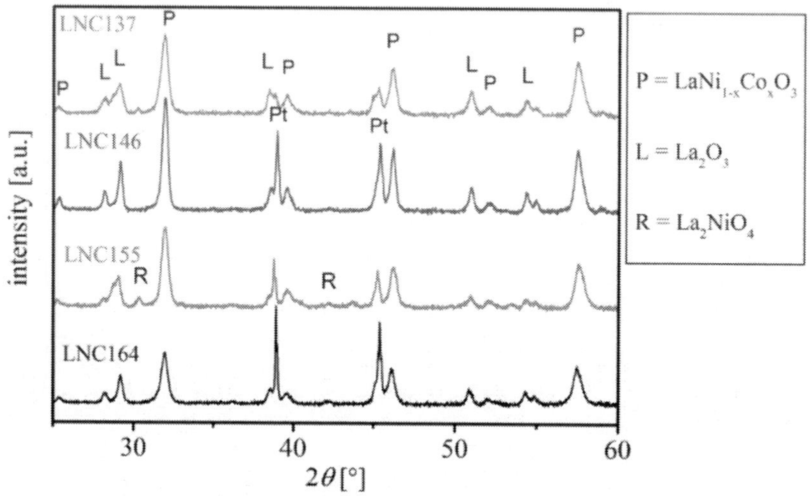

Figure 3. XRD patterns of from the top: LNC137, LNC146, LNC155 and LNC164, obtained at 800 °C.

Impedance spectroscopy performed on the symmetrical cells enabled extraction of a polarization resistance and a series resistance. Typical spectrum measured at 600 °C is shown in figure 4. The polarization resistance results from the electrochemical reactions on the cathode/electrolyte interface. After multiplying with the cathodes surface area and dividing by 2 (due to symmetrical configuration of cell) it is denoted as an area specific resistance (*ASR* [Ω cm^2]) and used to compare the cathodes electrochemical effectiveness. The series resistance almost exclusively results from electrolyte's bulk resistance and, at lower temperatures, grain boundaries resistance. It may be however influenced by the cathode's electronic conductivity and its contact area with the electrolyte. Therefore, in this paper we denote conductivity calculated from the series resistance as "apparent electrolyte conductivity" $\sigma_{el'app}$ to show how effective the cathode is in the distribution of electric current.

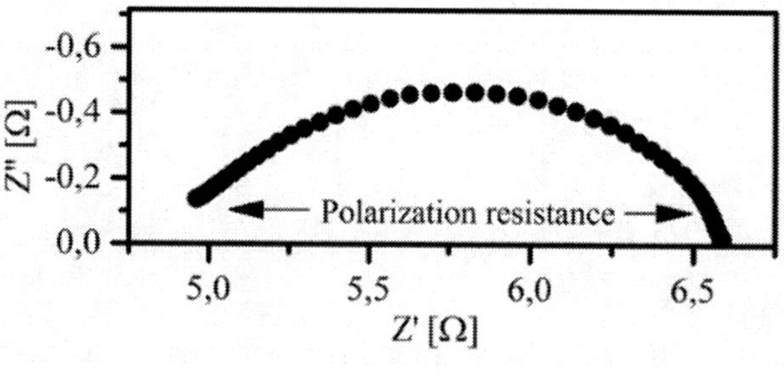

Figure 4. Typical impedance spectrum measured at 600 °C.

Arrhenius plots of ASR are shown in figure 5. Data was fitted with equation (1). It can be noticed that LNC155 is superior at all temperatures except at 700 °C. At 700 °C LNC137 seems to have the lowest *ASR*, 53 mΩ cm^2 compared to 62 mΩ cm^2 of LNC155.

$$ASR = {}^{1}/_{T}\, A\exp({}^{Ea}/_{kT}) \quad (1)$$

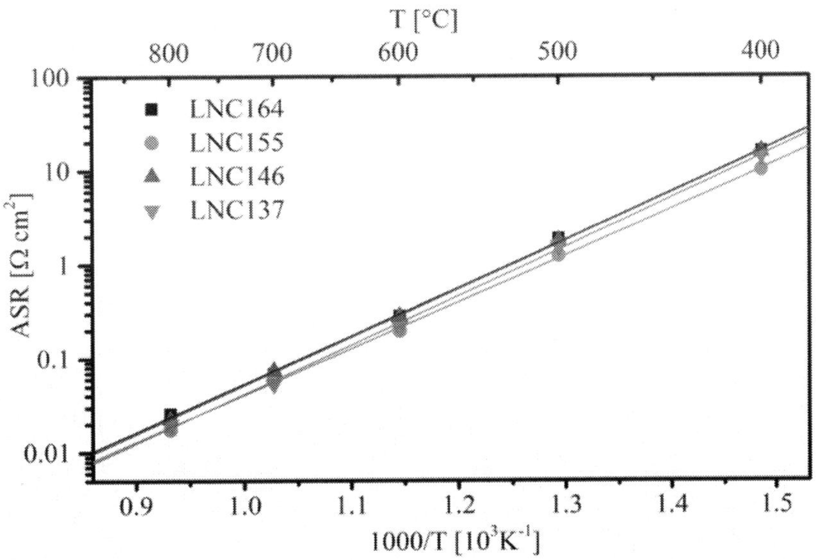

Figure 5. Arrhenius plot of the *ASR* for the symmetrical cells infiltrated with LNC164, LNC155, LNC146 and LNC137 and in situ sintered at 800 °C.

Table 2 contains parameters calculated from EIS spectra. It can be noticed that LNC137 provides the best current distribution over the electrolyte, since the sample had the highest $\sigma_{el'app}$. It may be connected to the relatively lower amount of La_2O_3 detected in the XRD analysis. However, it is LNC155 that shows the lowest *ASR* of all, which is especially visible at intermediate temperatures, i.e. 600 °C. LNC155 has also the lowest activation energy, 0.91 eV, while the rest of samples have the average activation energy of 0.95 eV. This may be due to the fact, that this composition is the only one to exhibit a detectable amount of $La_2NiO_{4+\delta}$ phase in XRD analysis, which is a popular catalyst for SOFCs [13]. The ASR for here reported cells is lower that measured for LNC146-CGO composite [14], but is higher than reported for LSC infiltrated into CGO backbone and heat treated at 600 °C [6].

Table 2. Apparent electrolyte conductivity at 800 °C, *ASR* at 600 °C and 800 °C and *ASR* activation energy calculated from EIS spectra.

	$\sigma_{el'app}$ at 800 °C [mScm⁻¹]	$ASR_{600\,°C}$ [mΩ cm²]	$ASR_{800\,°C}$ [mΩ cm²]	$E_{A,ASR}$ [eV]
LNC164	52	282	26	0.94±0.02
LNC155	51	197	17	0.91±0.02
LNC146	54	277	22	0.95±0.01
LNC137	67	219	20	0.96±0.02
LNC146-CGO [14]		440		
LSC-CGO [6]		62		0.87

4. Conclusion

Symmetrical cells infiltrated with $LaNi_{x-1}Co_xO_{3-\delta}$ (x=0.4 to 0.7) were tested. Samples infiltrated with LNC155 showed the lowest *ASR* (17 mΩ cm² at 800 °C) and the lowest *ASR* activation energy (0.91 eV). XRD analysis revealed that after sintering at 800 °C in all investigated samples secondary phases are present, with the highest amount of La_2O_3. From the electrochemical performance (low ARS and activation energy) we can conclude that LNC155 is the most promising infiltration material in LNC series for SOFC applications.

5. References

[1] Sharaf O Z and Orhan M F 2014 *Renew. Sustainable Energy Rev.* **32** 810-53

[2] Mahato N, Benerjee A, Gupta A, Omar S and Balani K 2015 *Prog. Mater. Sci.* **72** 147-337

[3] Chrzan A, Karczewski J, Gazda M, Szymczewska D and Jasinski P 2015 *J. Solid State Electrochem.* **19** 1807-15

[4] Shah M and Barnett S A 2008 *Solid State Ionics* **179** 2059-64

[5] Huang Y, Vohs J M and Gorte R J 2006 *Electrochem. Solid State Lett.* **9** A237-40

[6] Samson A J, Søgard M, Hjalmarsson P, Hjelm J, Bonanos N, Foghmoes S P V and Ramos T 2012 *Fuel Cells* **4** 511-9

[7] Kiebach R, Knöfel C, Bozza F, Klemensø T, Chatzichristodoulou C 2013 *J. Power Sources* **228** 170-7

[8] Tsipis E V and Kharton V 2008 *J. Solid State Elechtrochem.* **12** 1367-91

[9] Huang K, Lee H Y and Goodenough J B 1998 *J. Electrochem. Soc.* **145** 3220-7

[10] Hrovat M, Katsarakis N, Reichmann K, Bernik S, Kuščer D and Holc J 1996 *Solid State Ionics* **83** 99-105

[11] Samson A, Søgard M, Knibbe R and Bonanos N 2011 *J. Elechtrochem. Soc.* **158** B650-9

[12] Klemensø T, Chatzichristodoulou C, Nielsen J, Bozza F, Thydén K, Kiebach R and Ramousse S 2012 *Solid State Ionics* **224** 21-31

[13] Zhang X, Zhang H and Liu X 2015 *J. Power Sources* **269** 412-7

[14] Hjalmarsson P and Mogensen M 2011 *J. Power Sources* **196** 7237-44

Acknowledgements

The projects ForskEL 2015-1-12276 "Towards Solid Oxide Electrolysis Plants in 2020" funded by Danish Energinet.dk and OPUS DEC-2012/05/B/ST7/02153 "Functional layers for solid oxide fuel cells" funded by National Science Centre Poland are greatly acknowledged. Aleksander Chrzan acknowledges mobility support to DTU from the project POKL.04.01.01-00-368/09: "InterPhD: The development of interdisciplinary doctoral studies at the Gdansk University of Technology in modern technologies".

IOP Conf. Series: Materials Science and Engineering **104** (2015) 012020 doi:10.1088/1757-899X/104/1/012020

Commercialisation of Solid Oxide Fuel Cells – opportunities and forecasts

B Dziurdzia and Z Magonski and H Jankowski

AGH University of Science and Technology, 30-059 Krakow, Al. Mickiewicza 30, Poland

E-mail: dziurd@agh.edu.pl

Abstract. The paper presents the analysis of commercialisation possibilities of the SOFC stack designed at AGH. The paper reminds the final design of the stack, presented earlier at IMAPS-Poland conferences, its recent modifications and measurements. The stack consists of planar double-sided ceramic fuel cells which characterize by the special anode construction with embedded fuel channels. The stack features by a simple construction without metallic interconnectors and frames, lowered thermal capacity and quick start-up time. Predictions for the possible applications of the stack include portable generators for luxurious caravans, yachts, ships at berth. The SOFC stack operating as clean, quiet and efficient power source could replace on-board diesel generators. Market forecasts shows that there is also some room on a market for the SOFC stack as a standalone generator in rural areas far away from the grid. The paper presents also the survey of SOFC market in Europe USA, Australia and other countries.

1. Introduction

The promise of direct and efficient conversion of chemical to electrical energy makes fuel cell development an area of great technological interest. Fuel cells are much more efficient than heat engines, produce energy of higher density and they are environmentally friendly.

Among the types of fuel cells that are currently under development, Solid Oxide Fuel Cells (SOFCs) are gaining increased attention. SOFCs are based on ceramic electrolytes that are oxygen ion conductors but electronic insulators. They are high-temperature devices because available ceramics only achieves high ionic conductance at temperature above 850°C. The advantage of SOFCs is their high fuel flexibility compared to the other types of fuel cells.

The paper reminds the design of the SOFC stack designed at AGH, its recent modifications, measurements and prospects of commercialization on the background of the survey of SOFC market in different countries.

2. Survey of SOFC market

Fuel cell exploration started at around the same time when the era of steam engines and internal combustion engines began. Despite many spectacular applications of fuel cells, such as in the Apollo and Shuttle space missions, fuel cells have not achieved reliable operation or cost benefits comparable to internal combustion engines or gas turbines [1].

A large number of research institutes, private companies and government labs have conducted research and development on fuel cells. More than 22 billion USD has been invested in this technology in Japan, USA and Europe over the past 18 years [2] but fuel cells are still not at a widespread commercialization stage. Recently, the EU launched a new framework program (2014-20120) in which 2.8 billion EUR was allocated to the development of fuel cell and hydrogen technology [2]. It is a top priority to produce high quality fuel cells with high reliability and long-term durability at a low cost.

Table 1 summarizes the main advantages and disadvantages of fuel cells [3]

Table 1. Advantages and disadvantages of fuel cells [3]

Advantages	Disadvantages
Less or no pollution	Immature hydrogen infrastructure
High thermodynamic efficiency	Sensitivity to contaminants
Modularity and scalability	Expensive platinum catalysts (if applicable)
Fewer energy transformations	Dependance on hydrocarbon reformation
Quiet and static operation	Complex and expensive complementary components (pumps, valves etc.)
Water and cogeneration applications	Long-term durability and stability issues
Fuel flexibility	Hydrogen safety concerns
Wide range of applications	High investment cost-per-kW
	Relatively large system size and weight

Fuel cells have diverse applications ranging from micro-fuel cells with less than 1W power outputs to multi-MW power generation plants. Fuel cells can be used as stationary grid-independent (also known as stand-alone) or grid-assisted power supply. Stationary fuel cell applications cover emergency back-up power supply (EPS), remote-area power supply (RAPS), and distributed heat and power generators (CHP). Portable applications of fuel cells include power generators for outdoor personal uses (such as camping or climbing), power supply for consumer electronic devices, military equipment and transportation [3].

The subject of our interest in this paper is the application of Solid Oxide Fuel Cells (SOFCs). Significant progress has now been made in producing and field testing about 1 kW size SOFC-based CHP (combined heat and power) systems.

In Japan, over the past few years under the Ene-Farm programme about 40 000 SOFC-based heat and power (CHP) systems have been installed in homes to provide electricity and hot water. These systems are being built within the co-development agreement by Osaka Gas Co, Aisin Seiki Co., Toyota Motor Corporation using SOFCs produced by Kyocera Inc. These systems operate on natural gas and provide an electric conversion efficiency of about 45-50% [4]

Similar CHP systems called BlueGens (Figure 1) , are being produced and field tested in Australia, New Zealand, and Europe by Ceramic Fuel Cells, Ltd. (CFCL), Australia [5-7]. The company runs a dedicated BlueGen manufacturing facility in Heinsburg, Germany with a potential production capacity of 1,000 systems per year. In 2013 CFCL received an order from German energy provider EWE for 60 units of Blue Gen. BlueGens are expensive – so the Government of the North Rhine-Westphalia region began offering subsidies for the micro-CHP units. Up to 45% of the additional capital cost of a BlueGen, compared to conventional technologies, can be covered by a subsidy. CFCL plans to install up to 600 units of Blue Gen across the Germany by 2015 [6].

In summer 2013, CFCL launched also a scheme in the UK offering fully financed BlueGen units for social housing, schools and small businesses. Taking advantage of favourable feed-in tariff rates in the UK, CFCL has arranged the necessary finance to cover the capital cost of the units, whereby customers can take advantage of a free fuel cell system and reduced electricity costs by signing a ten-year contract. Financiers recoup their capital investment through feed-in tariff payments; a 6-9%

return on investment is expected by the end of the contract, after which ownership of the fuel cell passes to the customer [7].

Figure 1. BlueGen - stack of SOCFs – 1.4 kW [5]

Figure 2. Eneramic – stack of SOFCs – 100W [10]

Both the Japanese and the Australian CHP systems use SOFCs based on the conventional yttria-stabilized zirconia electrolyte and operate at about 750- 800°C. Ceres Power Ltd. of UK is developing SOFCs based on a ceria-based electrolyte for operation at 550-600°C for use in wall-mountable residential CHP systems, also of about 1 kW size [8].

The important player at the fuel cell market in Europe is the Fraunhofer Institute for Ceramic Technologies and Systems (IKTS). IKTS presented in 2014 the eneramic fuel cell system (Figure 2) based on SOFCs [9-11]. Eneramic is a light-weight electricity generator for camping and outdoor activities which can operate with commercially available liquefied petroleum gas from pressure cylinders. The system is equipped with a ceramic solid oxide fuel cell battery. The system provides output power of 100 Watts which in combination with a buffer battery is enough to cover the typical demand of electricity during camping. Figure 2 shows the eneramic energy system [10].

Eneramic system is testing as a power supply for warning traffic lights on motorways during road works [9]. The fuel cell system supplied with standard propane gas provides power for up to three weeks. Traditional batteries need to be changed twice a week.

IKTS is also involved in a German project with Vaillant to develop a wall-mounted, fuel cell micro combined heat and power system (CHP) for family homes using SOFCs manufactured by Staxera GmbH, Dresden [10].

IKTS signed a contract with Mayur Renergy Solution, India to develop SOFC generators which could operate with solar, wind and biogas installations as micro power plants installed in every house in rural areas of India and other developing countries [11].

Convion Ltd. manufactured cogeneration SOFC system C50 which can be operated with natural gas or biogas. It has nominal power output 58 kW with 53% electrical efficiency and over 85% total energy efficiency. Its operation parallel to the grid or in an island mode is possible. C50 has a modular design facilitating arrangement installations of even higher higher power output. Now, the system of 20 KW net power has been started the operation [11]. Convion used in C50 SOFC system MK351 stacks developed by Fraunhofer IKTS and Plansee. Plansee provided interconnects for the stack.

Currently, in the European project ene.field up to 1 000 micro-CHP systems will be installed in 12 EU member States and their performance will be evaluated over a 3 year period. Two fuel cell technologies will be included: high and low temperature PEM and intermediate and high temperature SOFC [12-16].

Delphi has developed two SOFC stack sizes: Delphi Gen 4 Stack up to 9 kW and Delphi Gen 3 Stack up to 1.5 kW (Figure 3). Delphi's stacks feature a modular design, ideal for integration into large power plants. They are designed for high volume manufacturing and can be implemented into a stationary or transportation application. Stacks can operate with natural gas, hydrogen, gasoline, diesel fuel, bio fuels, or other hydrocarbon fuels. They provide power density at 500 mW/cm^2. Delphi is seeking a development partner to commercialize its SOFC technology [17].

Delphi Corporation has developed also a 5 kW onboard auxiliary power units (APUs) for heavy-duty trucks [17, 6]. The SOFC system allows the drivers to turn the truck engines off at night or at a parking place and use the fuel cell APU to deliver power for the driver's cab. There is an estimated 200,000 trucks per year that would need these APUs, that provides an attractive market for SOFCs.

Bloom Energy is manufacturing and installing large scale stationary power units in the 100-200 kW range (Figure 4). Bloom Energy is the largest manufacturer of SOFC units with over 75 MW of generation capacity in the USA and it announced their intention to expand into the EU and Japanese markets [18, 6].

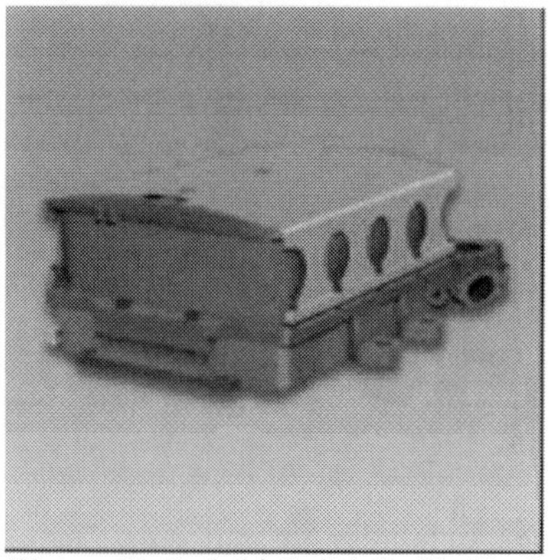

Figure 3. Delphi Solid Oxide Fuel Cell Stack [17]

Figure 4. Bloom Energy 200kW SOFC units [6]

Siemens Westinghouse had operated 100 kW SOFC generator in Europe for over 20 000 h and a 220kW SOFC/gas turbine system for over 900h in California but since then have stopped further research and development [19, 6]

To summarize, many SOFC units of varying sizes for combined heat and power generation are undergoing evaluation all over the world.

The subject of research carried out at AGH was a stack of SOFCs of output power 2 kW which characterizes by innovative design patented in many countries.

3. Stack of SOFC designed at AGH

The battery of fuel cells according to the AGH invention [20] is composed of a number of flat two-sided fuel cells, where every two-sided cell is made in the form of a ceramic plate with individual connections for supply and drainage of fluids and contacts for electric power output. Figure 5 shows such a flat two sided fuel cell, which is provided with an anode base structure, which on both sides, has a number of fine microchannels intended for distribution of fuel and transfer it to the operating anode.

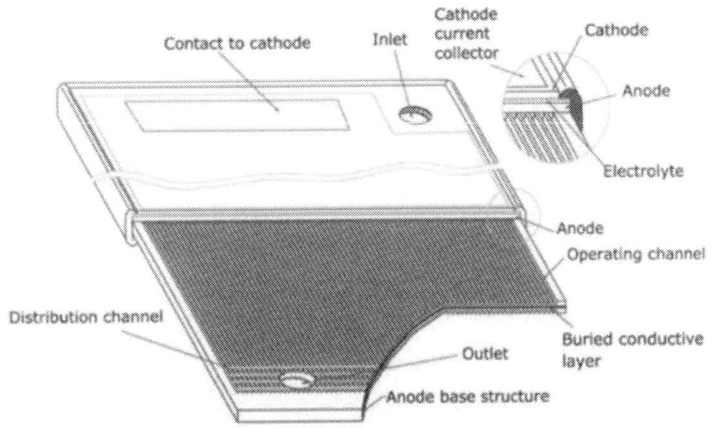

Figure 5. Flat two-sided fuel cell

The anode base structure, which serves also as an anode current collector is permanently bonded with a laminate consisting of an anode operating layer and a solid electrolyte layer. On the solid electrolyte layers, cathode layers are screen printed, on which, in turn, cathode conductive layers are applied, serving as the current collecting electrodes.

The flat two-sided cell constitutes a complete autonomous energy converter and is provided with individual input and output respectively for fuel distribution and removing combustion products, and is further equipped with electrical contacts enabling connection of the electric load.

The ceramic structure is symmetrical and includes two fuel cells on both sides. This fact is advantageous, because owing to this symmetry, tensions between the electrolyte and anode layers, which inevitably arise due to mismatch in temperature coefficients of expansion and mismatch in shrinking, are balanced.

The process of fabrication of two-sided fuel cell includes the following steps [20]:

- isostatic lamination of a central anode structure,
- micro-machining of fuel channels,
- isostatic lamination of the functional anode with the electrolyte,
- bonding of the laminated functional anode and electrolyte to the central anode structure ,
- thermal processing of the assembled structure,
- screen printing of cathodes and thermal processing,
- screen printing and thermal processing of conducting pads,
- drilling of inlet and outlet,
- sealing of the edges of ceramic structure, and finishing the rims around inlet and outlet,
- reduction of the anode in the atmosphere of hydrogen.

Two-sided fuel cells were fabricated in thick-film technology with using dedicated for fuel cells ceramic tapes and pastes manufactured by Electro-Science Laboratory (ESL), USA (Table 2).

Table 2. Ceramic materials used in a battery fabrication [20].

Part of a fuel cell	Symbol	Content	Abbr.	Firing [°C]	Thickness [mm]
Central ceramic anode structure	ESL 42421	Ni-ZrO_2(8mol%Y_2O_3) cermet	Ni/YSZ tape	1550	0.800
Functional anode	ESL 42421	Ni-ZrO_2(Y_2O_3) cermet	Ni/YSZ tape	1550	0.130
Electrolyte	ESL 42400	ZrO_2(Y_2O_3)	8YSZ tape	1550	0.010
Cathode	ESL 4421	$La_{0.8}Sr_{0.8}Fe_{0.8}Co_{0.2}O_3$	LSCF paste	1100	0.015

Two-sided fuel cell of dimensions 3" x 4" provides 6W of output power. I-V characteristics as well as results of impedance spectroscopy measurements were reported earlier [20, 21].

Flat double-sided fuel cells can be easily assembled into a stack (Figure 6). Individual two-sided cells are separated one from another with flexible ring separators which ensure the proper distance between the cells and provide easy air flow in vicinity of their cathodes. Simultaneously, the fuel can be transferred through these ring separators between the adjacent two-sided cells. Additionally, the flexible separators compensate the inevitable lack of parallelism between the planes of the neighboring two-sided cells. The flexible separators are supported with sealing pads intended to ensure gas tightness of connections. The two-sided cells in each stack are electrically connected in series with a metallic tape. All flexible separators and sealing pads are arranged concentrically and are compressed with clamping rods [21 , 22].

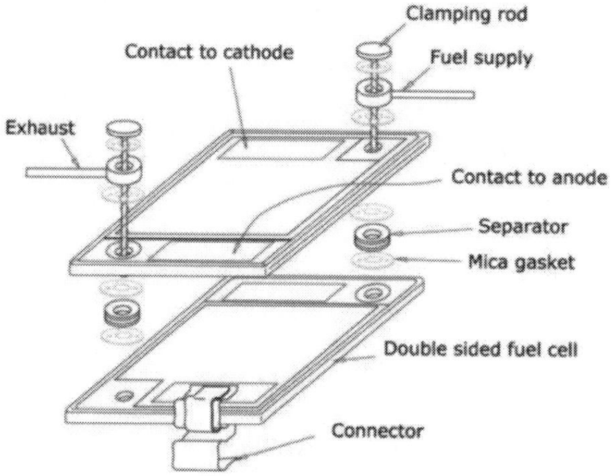

Figure 6. Arrangement of the two-sided fuel cells into a stack

The stack is located in a thermally insulated chamber. Inside the chamber, the cylindrical heaters are arranged in order to provide additional heating during the battery start up process. The application of a device forcing oxidant circulation inside the chamber allows to improve heat exchange efficiency between heaters and the battery. When the temperature inside the chamber reaches 800°C, additional heating is turned off and the operating battery delivers by itself enough heat to maintain optimum and safe working conditions. Heat excess is removed by heat absorbers located inside the thermally insulated chamber and can be used by other equipment, e.g. water heaters. Circulating stream of gases around the heat absorbers allows efficient heat exchange and maintaining uniform temperature of individual cells of the battery.

Figure 7 presents the design of a battery of output power 2 kW that consists of approximately 340 flat two-sided fuel cells. Each flat two-sided fuel cell provides about 6W of output power.

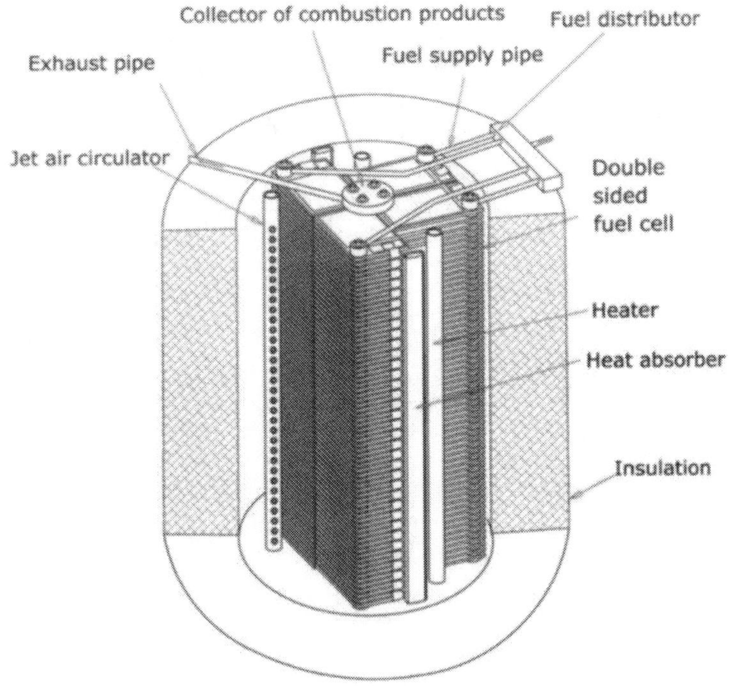

Projected parameters of the battery:

- output power 2 kW,
- number of fuel cells: 340 of 6W cells,
- start-up time below 1h,
- temperature of operation 800°C,
- power/volume ratio 1kW/liter,
- dimensions with thermal insulation: 0.45m x 0.3m x 0.3m

Figure7. Design of a battery [22]

The design of a planar double-sided ceramic fuel cell as well as a the design of a battery are protected by the following patents [23- 29].

Battery of Solid Oxide Fuel Cells (SOFCs) according to the AGH design characterizes by low thermal capacity, high resistance to rapid temperature changes and short start-up time. It is expected that for the battery of 2 kW, the start-up time will be much shorter than 1 hour. The battery doesn't contain metallic interconnectors and supporting frames that is why troubles with the application of special sealing between metal and ceramic can be avoided. The technological process of the battery is relatively simple and feasible for the mass production. The price of the device resulted from the price of the specialized ceramic foils used for fuel cell fabrication.

The possible application of such a battery is as an autonomous electricity generator used in individual households in rural areas far away from a grid, or in luxurious caravans, yachts, ships at berth. The SOFC battery operating as clean, quiet and efficient power source could replace on-board diesel generators. The battery could also operate as a complementary power supply system which can cooperate with renewable energy sources such as solar cells and wind turbines.

At this moment, a model of a battery of output 36W exists and it is shown in Figure 8. The stack consists of six double-sided fuel cells and it is located in a test chamber shown in Figure 9.

Figure 8. Battery of SOFC of the output power 36W

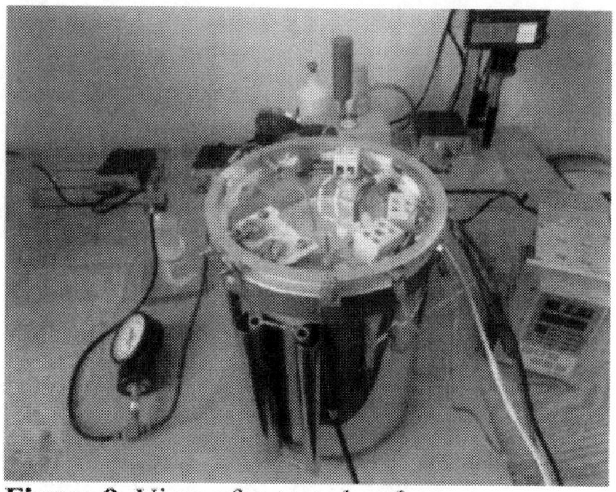

Figure 9. View of a test chamber

Figure 10 presented the I-V characteristic of this battery.

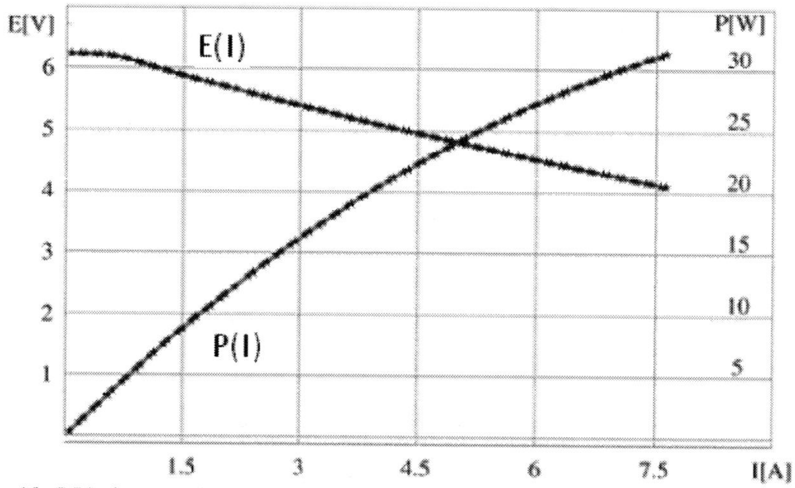

Figure 10. I-V characteristic of the battery (which consists 6 flat two-sided fuel cells).
Blue line – E(I) – output battery voltage {V] versus battery current [A]
Red line – P(I) – output battery power [W] versus battery current [A]

Figure 11 shows the measurement set-up for the current interrupt test. The principle of the current interruption method is to interrupt the current and to observe the resulting voltage transient at a cell. The ohmic loss can be measured from the difference between the voltage immediately before and after the current interruption (Figure 12).

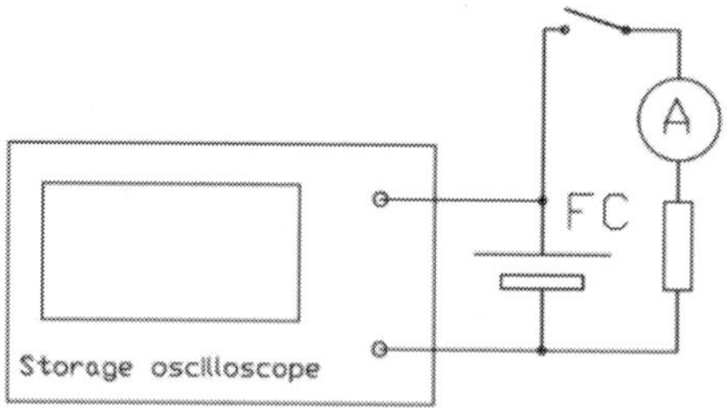

Figure 11. Measurement set-up for the current interrupt test.

Load current of 6.5A produces total output voltage drop of 280mV, which is attributed to polarization of activation, polarization of concentration, and pure resistive losses. The edge of positive voltage transition enables separation of pure ohmic resistance from other losses (150mV; 24 mΩ)

$$\Delta I = 6.5A, \Delta U = 150mV, R = 24m\Omega$$

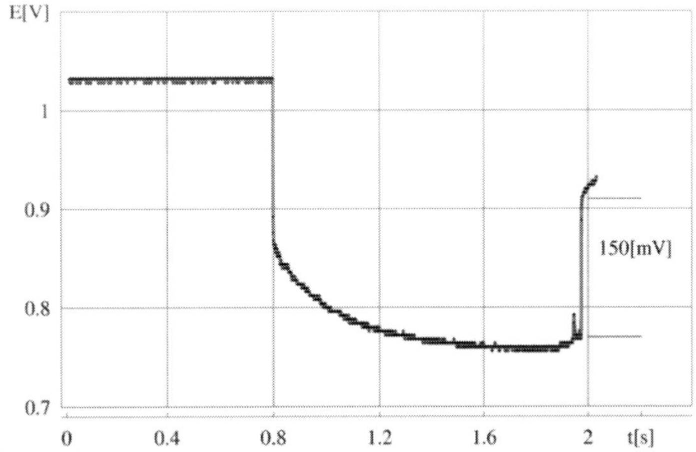

Figure 12. Output voltage measured during current interrupt test.

In the near future we are going to build in cooperation with the Institute of Electron Technology – Cracow Branch (ITE-Cracow Branch) and in support of Centre of Technology Transfer (CTT) at AGH a model of the stack of the output power 100W. We are seeking for development partners who are interested in creating a model of the stack of the output power 2 kW and implementation it into production.

The price of a battery of output power 2 kW according to the AGH invention was estimated by the professional consultant company CoWinner [30] to 37 000 PLN (about 8 800 EUR) provided that 10 units per year are being manufactured. It is assumed that the reimbursement of the expenditure for the customer will occur after 8 years of the battery exploitation.

At a European market is now available a commercial fuel cell battery BluGen [7] manufactured by the Ceramic Fuel Cell Limited Company, Australia. Some details of this solution is revealed in the U.S. Patent 7,531,053 . The battery, described there, contains multiple electrically connected planar fuel cells with massive metallic interconnectors in which channels for the passage of fuel and air were formed. The fuel cells are configured in a stack with using a metallic frame, in which ceramic plates are located. The fuel channels are arranged orthogonally in relation to the air flow channels. Glass sealant of a carefully selected transition temperature was selected to seal interfaces between metal parts and ceramics.

The BlueGen generates up to 2 KW of electricity with a peak electrical efficiency up to 60% and can export excess electricity to the grid. However, due to application of many metallic parts, the start-up time of this battery is equal to 25 hours. Various sources provide various prices of the BlueGen in Germany – from 8 000 EUR do 25 0000 EUR [1]. The pice includes government subsidy.

In Poland, the BlueGen operates in the Research&Development Centre, Institute of The Energy Systems Automation (IASE) in Wroclaw. The battery of output power 1.4kW operates there in a system with a solar cell battery and a windmill.

4. Prospects of commercialization of the SOFC battery designed at AGH

The proposal for a SOFC stack according to the AGH design, in view of the authors, makes important contribution to the construction of SOFCs. The battery design was awarded at the Invention Competitions in Taipei (2014), Paris (2015) and Cracow (2015).

This new design is advantageous, because bulky and heavy metal components are eliminated, the process of manufacturing is simplified and the technology is suitable for a large scale production. In the new design of the stack, rigid glass seals between metal and ceramic are avoided. In order to make a commercial product, the stack should be additionally provided with a number of supporting devices, like fuel reformer, fuel purifier, inverter, pumps etc. All these devices should be adapted for the battery size and specific design.

There is important issue of the total cost of the battery. So far, the components of the stack are created on the basis of green tapes manufactured by Electro-Science Laboratory (ESL) ,USA at a laboratory scale. The price of ESL green tapes is high - for example 1 square inch of the anode foil ESL 42400 costs now about 0.76 EUR. To fabricate one double sided fuel cell 60 inch square are needed. The ESL declares that the price of ceramic materials could be lowered in the case of high volume order.

At this moment, in order to lower the price of the battery two options can be considered: search for the alternative provider of green tapes abroad or producing them in Poland by a company experienced in ceramic technology. We do believe that our partners from the Institute of Electron Technology – Cracow Branch are able to prepare the required materials at a lower cost but it requires two or three years of the intensive research work, for which financing is needed.. So, we think that it is reasonable now to continue the works and verify the proposal of the battery design on the basis of ESL materials, and make the substitution with domestic materials after finishing the research works.

5. Conclusions

In the paper, the survey of the fuel cell market was presented with the focus on the application of SOFC batteries as cogeneration units of output power from one to a few kW.

AGH offers the design of the SOFC stack of output power 2 kW of innovative construction feasible for a high-volume production, The stack characterizes by a low thermal capacity and short start-up time. The design of a stack is protected with many patents and it was awarded at various innovation competitions. There is a need to build-up a full-scale model of the stack and to carry out intensive exploitation tests. At this moment the stack is relatively expensive. To solve the problem of a high price of the device, it is necessary to provide cheaper ceramic foils for the stack fabrication. If fuel cells successfully enter the mass production stage, their costs are expected to significantly drop and become consumer affordable. Further assembling of a stack with additional devices such as reformers, pumps, valves etc. are needed to make it ready for commercialization.

The stack has got the commercialization evaluation prepared by the professional consulting agency CoWinners.

The clean energy industry is expected to continue to grow rapidly in the coming years. Development of green energy resources including solar, wind, geothermal and water energy is getting inevitable in modern world. Poland generates around 90 percent of its electricity from coal and must increase renewable energy to at least 15 percent of the total energy production by 2020 to meet EU rules on carbon emissions. Poland's government approved in 2015 a draft law that lays out new long-term subsidies for renewable energy. Under this draft law, developers and owners of new renewable installations can sell their energy at auctions at a fixed price that would be guaranteed for 15 years regardless of market prices. The Act of Renewable Energy doesn't cover fuel cells, so the installations of fuel cells in Poland can't be subsidized by the government. We believe that in the future solar , wind and fuel cell installations will be operated in one complementary system to power our houses and in this way new market dor fuel cells will be opened. .

6. References

[1] Behling N H 2013 Fuel Cells Current Technology Challenges and Future Research Needs (Elsevier)

[2] Wang J 2015 Barriers of scaling-of fuel cells: Cost durability and reliability Energy **80** 509-521

[3] Sharaf O Z et al 2014 An overview of fuel cell technology: Fundamentals and applications Renewable and Sustainable Energy Reviews **32** 810-853

[4] Lewis J 2014 Stationary fuel cells-Insights into commercialization International Journal of Hydrogen Energy **39** 21896-21901

[5] Badwal S et al 2014 Review of Progress in High Temperature Solid Oxide Fuel Cells Journal of the Australian Ceramics Society **50** 23-37

[6] http://fuelcelltoday.com/media/1889744/fct_review_2013.pdf

[7] http://www.ceramicfuelcells.co.uk/technology/gennex-module/

[8] Wachsman E D and Singhal S C Solid Oxide Fuel Cell Commercialization Research and Challenges The Electrochemical Society Interface **Fall 2009** 38-43

[9] Reuber S 2014 *Eneramic – the Mobile SOFC Power Generator Well on its Way to Commercialisation* Proc. 11th European SOFC&SOE Forum (1-4 July 2014, Lucerne Switzerland) **A0602** http://publica.fraunhofer.de/eprints/urn_nbn_de_0011-n-3236764.pdf

[10] http://www.h2fc-fair.com/hm13/images/ppt/11th/1300.pdf

[11] http://www.innovations-report.com/html/report/materials-science/strong-partners-in-sofc-technology.html

[12] Gorte R J 2005 Recent Developments Towards Commercialization of Solid Oxide Fuel Cells AIChE Journal **51** 2377-2381

[13] Cottrell C A et al 2011 Strategies for stationary and portable fuel cell markets International Journal of Hydrogen Energy **36** 7969-7975

[14] Chick L et al 2015 The Case for Natural Gas Fueled Solid Oxide Fuel Cell Power Systems for Distributed Generation Fuel Cells **1** 49-60

[15] http://www.fch.europa.eu/project/european-wide-field-trials-residential-fuel-cell-micro-chp

[16] Choudhury A et al 2013 Application of solid oxide fuel cell technology for power generation – A review Renewable and Sustainable Energy Reviews **20** 430-442

[17] http://delphi.com/manufacturers/auto/fuelcells/

[18] http://energy.gov/sites/prod/files/2014/11/f19/fcto_2013_market_report.pdf

[19] http://www.hfcletter.com/issues/XIX_2/stories/91-1.html

[20] Dziurdzia B, Magonski Z 2011 Planar double-sided anode supported SOFC-novel design Journal of Microelectronics and Electronic Packaging **8** 164-170

[21] Dziurdzia B, Magonski Z, Jankowski H 2013 Battery of Solid Oxide Fuel Cells Proc Int. Conf. On Microelectronics and Packaging IMAPS-CPMT Poland (Krakow: Poland) p.1-6

[22] Dziurdzia B, Magonski Z, Jankowski H 2014 Stack of Solid Oxide Fuel Cells Microelectronics International **31** 207-211

[23] Magonski Z, Dziurdzia B 2015 *Method for fabrication of electrochemical energy converter and the electrochemical energy converter* United States Patent **US 8968959B2** http://patenty.bg.agh.edu.pl/pelneteksty/US8968959B2.pdf

[24] Magonski Z, Dziurdzia B 2013 *Method for fabrication of electrochemical energy converter and the electrochemical energy converter* Russian Patent **RU 2502158 C2** http://patenty.bg.agh.edu.pl/pelneteksty/RU2502158C2.pdf

[25] Magonski Z, Dziurdzia B 2011 *Sposób wykonania elektrochemicznego konwertera energii i elektrochemiczny konwerter energii* Polish Patent **PL 213349 B1** http://patenty.bg.agh.edu.pl/pelneteksty/PL213349B1.pdf

[26] Magonski Z, Dziurdzia B 2013 *Method for fabrication of electrochemical energy converter and the electrochemical energy converter* German Patent **DE 112010002963 B4** *http://patenty.bg.agh.edu.pl/pelneteksty/DE112010002963B4.pdf*

[27] Magonski Z, Dziurdzia B 2014 *Bateria ogniw paliwowych* Polish Patent **PL 218785 B1** http://patenty.bg.agh.edu.pl/pelneteksty/PL218785B1.pdf

[28] Magonski Z, Dziurdzia B 2014 *Battery of Fuel Cells* United States Patent **US 8778550 B2** http://patenty.bg.agh.edu.pl/pelneteksty/US8778550B2.pdf

[29] Magonski Z, Dziurdzia B 2015 *Battery of Fuel Cells* European Patent **EP 2722915 B1** http://patenty.bg.agh.edu.pl/pelneteksty/EP2722915B1.pdf

[30] Krzewinski Z et al 2015 *Wycena ścieżki komercjalizacji baterii ogniw paliwowych ze szczególnym uwzględnieniem opcji związanych z licencjonowaniem* expert assessment CoWinners for AGH 6-11

Acknowlegments
The paper is financed by the European Regional Development Fund under the Operational Programme Innovative Economy (project UDA-POIG-01-03-02-12-005/12)

IOP Conf. Series: Materials Science and Engineering **104** (2015) 012021 doi:10.1088/1757-899X/104/1/012021

The kinetic of photoreactions in zinc oxide microrods

M Fiedot, O Rac, P Suchorska-Woźniak, W Nawrot, H Teterycz

Wrocław University of Technology, Faculty of Microsystem Electronics and Photonics, Janiszewskiego 11/17, Wroclaw, Poland

E-mail: marta.fiedot@pwr.edu.pl

Abstract. Zinc oxide is the oldest sensing material used in the chemical resistive gas sensors which allow to detect many gases, such as carbon oxide, nitrogen oxides and other. This material is also widely used in medicine and daily life as antibacterial agent. For this reason this semiconductor is often synthesized on the polymer substrates such as foils and textiles. In presented results zinc oxide was deposited on the surface of poly(ethylene terephthalate) foil to obtain antibacterial material. As synthesis method chemical bath deposition was chosen. The growth of zinc oxide structures was carried out in water solution of zinc nitrate (V) and hexamethylenetetramine in 90°C during 9 h. Because antibacterial properties of ZnO are strongly depended on photocatalytic and electric properties of this semiconductor impedance spectroscopy measurements were carried out. During the measurements material was tested with and without UV light to determinate the kinetic of photoreactions in zinc oxide. Moreover the composite was analyzed by XRD diffraction and scanning electron microscope. The X-ray analysis indicated that obtained material has the structure of wurtzite which is typical of zinc oxide. SEM images showed that on the PET foil microrods of ZnO were formed. The impedance spectroscopy measurements of ZnO layer showed that in UV light significant changes in the conductivity of the material are observed.

1. Introduction

Zinc oxide is a white semiconductor that is widely used in medicine as an antifungal and antibacterial material, which is associated with its high biological activity, lack of toxicity and low price [1]. Moreover, this oxide is widely used in electronics for production of various components, such as gas sensors [2], varistors [3] or solar cells [4]. It has many useful properties such as refractive index of n = 1.95-2.10 [5], a direct band gap of 3.3 eV [6], piezoelectric coupling coefficient of 1.19 C/m^2 [7] and good photocatalytic properties [8].

The photocatalytic properties of this material are some of the causes of their good biological activity. As a result of the processes occurring on the surface of the semiconductor, the photocatalytic generation of free radicals [9-11] and the release of the Zn^{2+} ions [12-14] takes place.

In the literature, there are many studies which show that the exposure of ZnO to ultraviolet radiation leads to a significant improvement in its biological activity [15]. Therefore, it can be assumed that the mechanism associated with the photocatalysis is more important than the release of ions.

Good photocatalytic properties of ZnO are directly related to photogeneration of electric charge carriers in the material under the influence of electromagnetic radiation with an energy equal to or greater than the width of the band gap. In the case of this semiconductor, radiation is within the UV range [16].

IOP Conf. Series: Materials Science and Engineering **104** (2015) 012021 doi:10.1088/1757-899X/104/1/012021

Research conducted by Ann et al. demonstrated that exposure of zinc oxide rods to UV radiation for 20 minutes reduced the survivability of *E. coli* by 18% and *S. aureus* by 22% in comparison with the test with no additional lighting [17].

In order to better understand the antibacterial activity of ZnO and explain the effect of UV radiation on its properties, the kinetics of photoinduced electron transfer occurring in the microrods of zinc oxide deposited on a polymer film were studied in this article. Measurements were performed using impedance spectroscopy.

2. Experimental part

As a method of the surface modification of the poly(ethylene terephthalate) (PET) film, the chemical bath deposition (CBD) technique was applied. The growth of the ZnO structures was carried out in an equimolar aqueous solution of zinc nitrate (V) $Zn(NO_3)_2$ and hexamethylenetetramine (HMT) at a concentration of 100 mM. The obtained solution was stirred on a magnetic stirrer for 24 hours, after which it was filtered. Then the previously prepared polymer film sized 4 x 2 cm was placed in a beaker with the prepared solution. The mixture was heated at 90°C for 9 hours. After the deposition process, the obtained sample was rinsed to remove by-products and the excess of the modifying agent, which had not permanently bound to the surface of the test material.

The prepared sample was examined by X-ray diffraction (XRD) to determine the crystalline structure of deposited layer and define the size of crystallites. The crystallographic structure of zinc oxide was determined using Philips Materials Research Diffractometer (MRD) by means of CuKα radiation.

The microstructure of sample was examined with the use of electron scanning microscope (SEM) JSM 5800 LV of Jeol company. The microscope is equipped in ISIS 300 system for X-ray analysis of Oxford company.

The AC measurement was performed across the test structure, between the upper electrode and the lower electrode (the test material was placed between two copper electrodes). To obtain appropriate contact between sample and electrons the measuring system was pressed with 462.5 g/cm^2. The Solartron SI 1260 impedance analyser was used in the tests. The measurements of polymer composite impedance were performed in a frequency range from 10 Hz to 1 MHz, stimulating the structures by the sinusoidal voltage signal with an amplitude of 500 mV. In addition, the sample was exposed to light with a UV radiation wavelength. The radiation source was produced by MICROTHERM company.

3. Results

In order to determine the crystalline structure of deposited layer, XRD measurements were performed. The obtained diffractogram show patterns corresponding to the PET substrate and zinc oxide with the wurtzite structure (Figure 1).

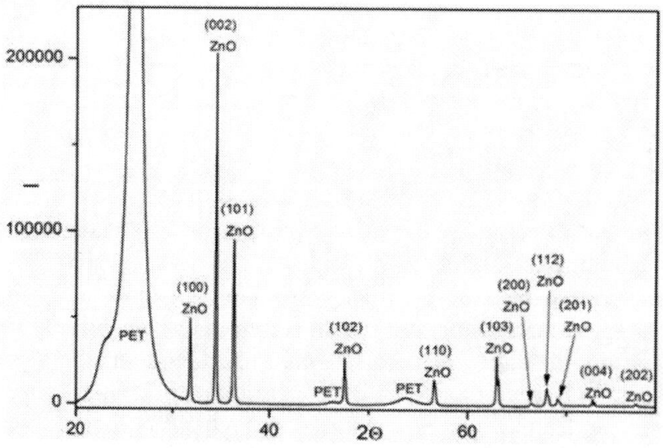

Figure 1. XRD test of the sample.

Furthermore the average size of the obtained ZnO crystallites ($L_{(002)}$) (1) was determined using the Scherrer's equation. The apparatus broadening was 0.05°. On the basis of literature data, it was assumed that the value of constant K was 0.9 [18]. By substituting relevant data to the equation, it was found that the mean size of the ZnO crystallites formed on the PET substrate was approximately 62.1 nm.

$$L_{(101)} = \frac{K \cdot \lambda}{B \cdot \cos\Theta} \tag{1}$$

where: K – Scherrer constant,
λ – length of an X-ray radiation beam [Å],
B – integral width (FWHM) of the peak (002) [rad],
θ – glancing angle a given interference band [rad];

The SEM analysis of the pure polymer substrate as well as shape and size of ZnO structures obtained in the deposition process were carried out. The PET substrate surface analysis showed it to be smooth, although there were visible spherical structures, which might be spherulitic semicrystals in polymer or post-production impurities (Figure 2a). The SEM image of the sample after the deposition process shows apparent hexagonal rods of zinc oxide with an average width of 550 nm and a height of 2 μm (Figure 2b).

Figure 2. SEM image of (a) pure polymeric films,(b) polymeric films after deposition.

Many authors associate the antibacterial properties of ZnO with its photocatalytic properties. If the oxide is exposed to radiation with energy greater than or equal to its band gap, then the transfer of electrons from the valence band to the conduction band takes place. As a result of this process, holes (h^+) migrate to the surface of material and react with water molecules (moisture) from the environment, leading to the formation of hydroxyl radicals ($OH^\bullet$). These radicals present very strong oxidizing properties, thus they are capable of degrading all kinds of organic compounds, from which the microorganisms or biological impurities are composed. As a result of these reactions, various kinds of organic substances are decomposed into water and carbon dioxide [19,20].

Moreover it is known from the literature that during zinc oxide exposure to air, the oxygen adsorption occurs on its surface. In this process oxygen molecules react with electrons present in the material (2). In result chemically adsorbed oxygen ions are formed on the surface and the superficial depletion layer with lower conductivity than in the bulk of zinc oxide emerges. During exposure of the semiconductor to UV radiation, the holes present on the ZnO surface react with the adsorbed ions, causing release of the gaseous oxygen, which desorbs from the surface (2) [21].

$$O_2 + e^- \rightarrow O_2^- \tag{2}$$

$$O_2^- + h^+ \rightarrow O_2 \uparrow \tag{3}$$

In order to analyze behavior of the tested materials under the influence of UV radiation, the transportation kinetics of carriers generated by radiation was analysed. For this purpose, the sample was placed between two electrodes, one of which had a hole for exposing the sample (Figure 3).

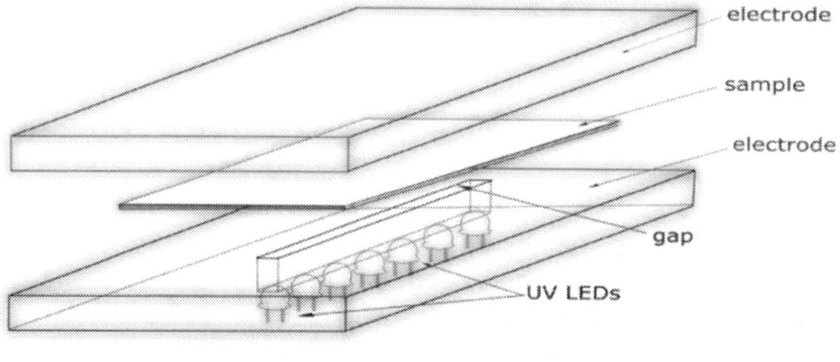

Figure 3. Diagram of the measuring system.

The electrical parameters of samples were determined by impedance spectroscopy during and after radiation. Measurement - both in the presence of UV radiation and after it was turned off - was performed every two minutes. Analysis of Nyquist plots showed that the system impedance decreases rapidly during exposure and it stabilizes after approximately 6 minutes (Figure 4). After switching off the lighting, the impedance of the sample gradually increases and approaches the baseline (Figure 5).

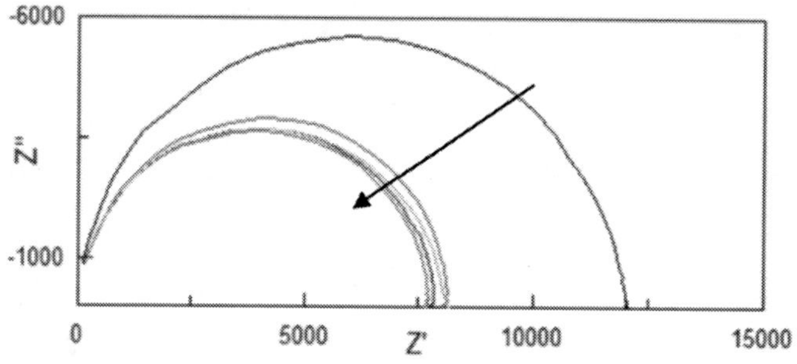

Figure 4. Nyquist plots for a sample during exposure to UV radiation.

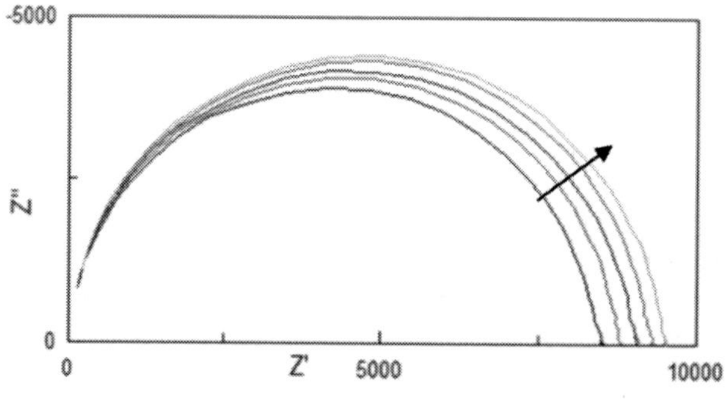

Figure 5. Nyquist plots for a sample after exposure to UV radiation.

IOP Conf. Series: Materials Science and Engineering **104** (2015) 012021 doi:10.1088/1757-899X/104/1/012021

Based on the literature data and the obtained test results, equivalent circuit model of the test object was determined. It consists of three resistors, one capacitor and two constant phase elements CPE (Figure 6) [21].

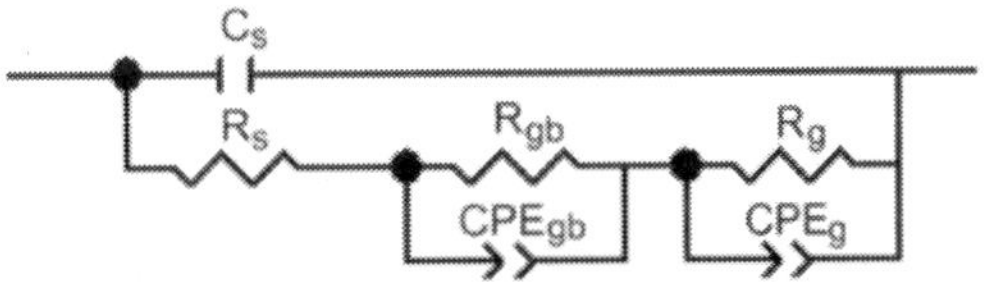

Figure 6. Equivalent circuit model of the test sample.

On the basis of the generally accepted model of zinc oxide photoconductivity, which takes into account the phenomena occurring in the bulk of material as well as on its surface, it can be concluded that C_s is substrate geometrical capacitance, R_s is contact resistance, R_{gb} and CPE_{gb} correspond to the electric parameters of the grain boundary, and R_g and CPE_g to the grain bulk.

On the basis of the Nyquist plots and using the above model, the changes in value of individual elements of the electrical equivalent circuit were determined. It was noted that the contact resistance oscillates around a constant value, which is approximately 150 Ω. In the case of surface and bulk resistance, evident changes occur under UV radiation. During radiation of the samples, both R_{gb} and R_g decrease. Decrease of bulk resistance is due to the generation of electron-hole pairs during radiation exposure. In contrast, the increase in ZnO grain boundary conductivity is due to the decrease in the width of the semiconductor depletion layer, which is formed in the surface area of ZnO in oxygen atmosphere. After switching off the radiation, as a result of the recombination of charge carriers and electric and readsorption of oxygen ions on the surface, both resistance values increase and the system returns to the state before exposure (Figs. 7, 9).

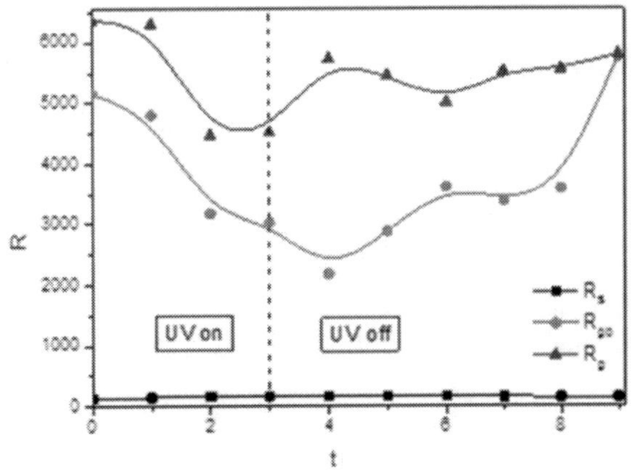

Figure 7. Changes in resistance of the system with UV and without UV.

During the radiation rapid changes in the capacitance of depletion layer along grain boundaries are also observed (Figure 8). This is due to desorption of oxygen ions from the surface of the material and decrease in thickness of the depletion layer and an increase in density of semiconductor carriers (Figure 8, 9). After turning off the UV radiation, the material gradually returns to the state before exposure and the CPEgb value increases.

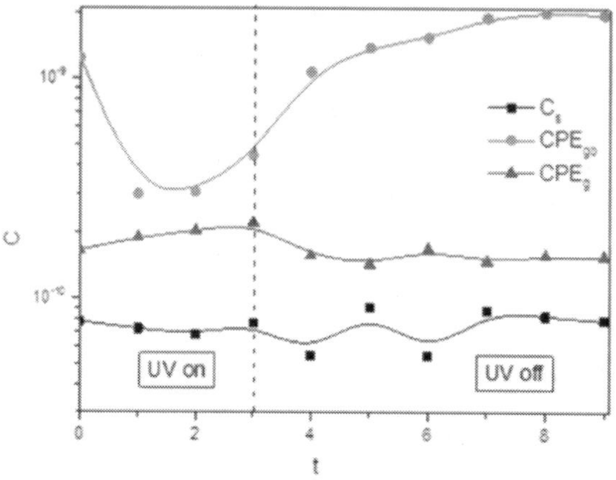

Figure 8. Changes in the system capacitance with UV and without UV.

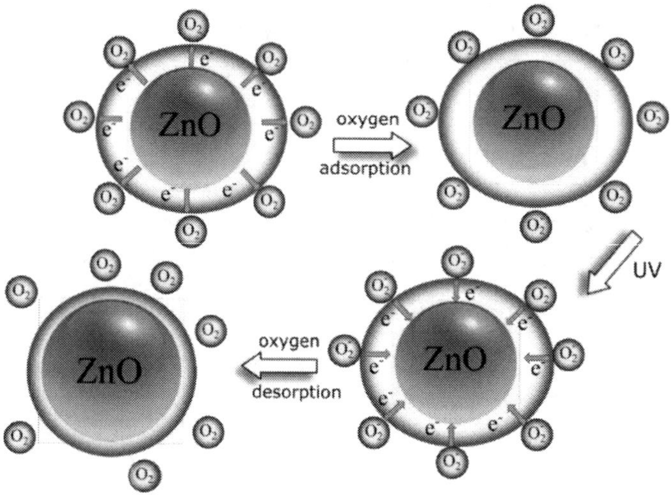

Figure 9. Changes occurring on the surface of ZnO under UV.

In conclusion, on the basis of the tests carried out using impedance spectroscopy on the rods of zinc oxide formed on the surface of the polymer film, it was found that under the influence of UV radiation, mainly resistance and capacitance of the depletion layer at the grain boundary is changed. It can therefore be assumed that the material will demonstrate high photocatalytic activity. It is further noted that separation speed of the carriers in the tested ZnO microrods is greater than their recombination rate. On this basis, authors of the publication believe that this effect can significantly contribute to increase of ZnO biological activity. Moreover, the long recombination time of the charge carriers in the system would likely result in improvement of antibacterial activity of the rods also after the radiation is stopped (up to approx. 10 minutes).

4. Conclusion

This article presents the results of research designed to analyse the phenomena occurring in microrods of zinc oxide under ultraviolet radiation.

The ZnO microrods were deposited on the polyester (PET) substrate in the hydrothermal process. The obtained structures have height of approx. 2 microns and width of 550 nm and wurtzite structure. The electrical characterization of the ZnO microrods subjected to UV irradiation, using impedance spectroscopy were carried out. A replacement model of the sample consisting of three resistors, one capacitor and two constant phase elements was proposed. The analysis of each resistance value showed that both the value of bulk resistance R_g as well as grain boundary R_{gb} decrease during irradiation and after turning off the radiation source, they return to their baseline values. It is further noted that under UV, the capacity of the grain boundary decreases and increases after the radiation is finished.

All the observed changes are related to the interaction of the ZnO surface with oxygen present in the atmosphere which absorbs electrons while adsorbing on the surface of material and turns into the ion form. In the process, the depletion layer having a higher resistance than the bulk of material is formed on the zinc oxide surface. During ZnO exposure to UV radiation electron-hole pairs are formed and oxygen is desorbed, resulting in width reduction of the depletion layer. After turning off the radiation, oxygen readsorption occurs and the system returns to its original state.

Taking into account the test results, authors are planning further work to determine direct impact of UV radiation on the antibacterial properties of polymer films and polymer fabrics coated with zinc oxide bacteria by analysing the survivability of bacteria with and without radiation. This will allow identifying the determinants of the photocatalytic, electrical properties and biological activity of the zinc oxide microrods.

5. References

[1] Moezzi A, McDonagh A M, Cortie M B 2012 *Chem. Eng. J.* **185–186** pp 1–22
[2] Ellmer K, Klein A, Rech B 2007 *Springer Science & Business Media* **104**
[3] Eda K 1989 *IEEE Elect. Insul. Mag.* **6** pp 28-30
[4] Beek W J, Wienk M M, Janssen R A 2004 *Adv. Mater.* **16** pp 1009-1013
[5] Yang Y, Sun X W, Chen B J, Xu C X, Chen T P, Sun C Q, Tay B K, Sun Z 2006 *Thin Solid Films* **510** pp 95–101
[6] Lin B, Fu Z, Jia Y 2001 *Appl. Phys. Lett.* **79** pp 943-945
[7] Wang Z L, Song J 2006 *Science* **312** pp 242-246
[8] Huang M H, Wu Y, Feick H, Tran N, Weber E, Yang P 2001 *Adv. Mater.* **13** pp 113-116
[9] Sawai J, Shoji S, Igarashi H, Hashimoto A, Kokugan T, Shimizu M, Kojima H 1998 *J. Ferment. Bioeng.* **86** pp 521-522
[10] Lipovsky A, Nitzan Y, Gedanken A, Lubart R 2011 *Nanotechnology* **22** 105101
[11] Zhang L, Ding Y, Povey M, York D 2008 *Prog. Nat. Sci.* **18** pp 939-944
[12] Kasemets K, Ivask A, Dubourguier H-C, Kahru A 2009 *Toxicol. In Vitro.* **23** pp 1116-1122
[13] Brunner T J, Wick P, Manser P, Spohn P, Grass R N, Limbach L K, Bruinink A, Stark W J 2006 *Environ. Sci. Technol.* **40** pp 4374-4381
[14] Li M, Zhu L, Lin D 2011 *Environ. Sci. Technol.* **45** pp1977-1983
[15] Sirelkhatim A, Mahmud S. Seeni A, Kaus N H M, Ann L C, Bakhori S K M, Hasan H, Mohamad D 2015 Nano Micro Lett. **7** pp 1-24
[16] Zhai T, Fang X, Liao M, Xu X, Zeng H, Yoshio B, Golberg D 2009 *Sensors* **9** pp 6504-6529
[17] Ann L C, Mahmud S, Bakhori S K M, Sirelkhatim A, Mohamad D, Hasan H, Seeni A, Rahman R A 2014 *Appl. Surf. Sci.* **292** pp 405-412
[18] Humayun Q, Kashif M, Hashim U, Qurashi 2014 *Nanoscale Res Lett.* **9** pp 1-5
[19] Konga X Y, Wang Z L 2004 *Appl. Phys. Lett.* **84** pp 975-977
[20] Zhang H, Chen B, Jiang H, Wang C, Wang H, Wang X 2011 *Biomaterials* **32** pp 1906-1914
[21] Soci C, Zhang A, Xiang B, Dayeh S A, Aplin, D P R, Park J, Bao X Y, Lo Y H, Wang D 2007 *Nano Lett.* **7** pp 1003–1009

Acknowledgements

Scientific work financed from funds for science in the years 2013 – 2017 as a research project within the program under the name „Diamond Grant" No DI2012 008542.

Electrical Characterization of Diamond/Boron Doped Diamond Nanostructures for Use in Harsh Environment Applications

Ł. Gołuński[1,3], K. Zwolski[2], P. Płotka[2],

[1] Department of Metrology and Optoelectronics; Faculty of Electronics Telecommunication and Informatics; Gdańsk University of Technology; Gdańsk; Poland.
[2] Department of Microelectronic Systems; Faculty of Electronics Telecommunication and Informatics; Gdańsk University of Technology; Gdańsk; Poland.
[3] Faculty of Physics & Astronomy Physics; University of California, Riverside; Riverside; California, USA.

E-mail: lukgolun@student.pg.gda.pl

Abstract. The polycrystalline boron doped diamond (BDD) shows stable electrical properties and high tolerance for harsh environments (e.g. high temperature or aggressive chemical compounds) comparing to other materials used in semiconductor devices. In this study authors have designed electronic devices fabricated from non-intentionally (NiD) films and highly boron doped diamond structures. Presented semiconductor devices consist of highly boron doped structures grown on NiD diamond films. Fabricated structures were analyzed by electrical measurements for use in harsh environment applications. Moreover, the boron-doping level and influence of oxygen content on chemical composition of diamond films were particularly investigated. Microwave Plasma Enhanced Chemical Vapour Deposition (MW PE CVD) has been used for thin diamond films growth. Non-intentionally doped diamond (0 ppm [B]/[C]) films have been deposited on the Si/SiO_2 wafers with different content of carbon, boron and oxygen in the gas phase. Then, the shape of the highly doped diamond structures were obtained by pyrolysis of SiO_2 on NiD film and standard lithography process. The highly doped structures were obtained for different growth time and [B]/[C] ratio (4000 - 10000 ppm). The narrowest distance between two highly doped structures was 5μm. The standard Ti/Au ohmic contacts were deposited using physical vapour deposition for electrical characterization of NiD/BDD devices. The influence of diffusion boron from highly doped diamond into non-doped/low-doped diamond film was investigated. Surface morphology of designed structures was analyzed by Scanning Electron Microscope and optical microscope. The resistivity of the NiD and film was studied using four-point probe measurements also DC studies were done.

1. Introduction

Diamond shows unique electrical, physical, chemical and mechanical properties [1–3]. Due to its properties has many applications in electronics [2–7], MEMS [8–11], bio-sensors [12], may also act as an electromechanical sensor [13–15], be used for microfluidics [16] or be applied in neuron electrodes [11]. The chemical inertness of the polycrystalline boron doped diamond (BDD) combined with its high thermal conductivity, mechanical strength and electrical conductivity, makes it an attractive material for harsh environment applications.

The characterization of the electrical properties is of importance in order to be able to control and utilize diamond as an electronic material. In order to be able to characterize diamond thin films a new electronic devices have been designed. This devices have been fabricated from non-intentionally doped (NiD) diamond thin films and highly boron doped structures.
The NiD films and BDD structures has been grown by Microwave Plasma Enhanced Chemical Vapor Deposition (MW PE CVD). NiD films have been deposited on the Si/SiO_2 wafers. For the production diamond layers on Si/SiO_2 wafers, nucleation is necessary [1,3,17–19]. The most common used nucleation method is mechanical seeding of the substrate in water solution [19,20] or DMSO [21] with the use of ultrasound bath [22,23]. There are also known methods in which a thin film with nuclei is applied on a substrate [3]. In this methods the nanodiamond powder is placed in polyvinyl alcohol (PVA) matrix [24] or photoresist [25]. Based on diamond suspension technique in polymer PVA mixture [26] has been used for nucleation. This method is easy to obtain and in the PVA are elements only appearing during diamond synthesis by microwave plasma (H, C, O).
Highly boron doped structures has been fabricated by growing BDD structures on the NiD layer by method called second grown [27–29]. The second grown was successfully used for diamond growth [16,30]. For obtain shape of highly doped diamond structures (boxes in 4 size with different distance between them) a pyrolysis of SiO_2 on NiD film and standard lithography process has been used. Deposition and etching SiO_2 were exanimated by Smirnov et al [31].
In this paper authors investigated and showed influence of diffusion boron from highly doped diamond into non-doped/low-doped diamond film. The influence of diffusion has been shown by comparison resistivity of non-doped diamond layer measured by using four-point probe and V/I characteristics of non-doped/boron doped diamond structures. Also, morphology of designed structure was analyzed by Scanning Electron Microscope (SEM) and optical microscope.

2. Experimental

2.1. Growth of structure

Before growth diamond Si/SiO_2 (400nm thick SiO_2 by wet thermal oxidation) wafers have been rinsed in acetone and 2-isopropanol for 10 minutes in ultrasonic bath. After cleaning substrates have been dried in nitrogen blow. For nucleation PVA mixture with nanodiamonds was spin-coated on SiO_2 surface. The method of obtaining PVA mixture was described elsewhere [26,32]. NiD diamond layers 400nm thick were grown by μPA CVD system (SEKI 5250). Before process movable stage with samples was heated to 700°C. Temperature during deposition where increase by plasma heating up to 1000°C. The mixture of gases CH_4, H_2 and B_2H_6 was used to growth diamond. The NiD layer was grown with flow of gases H_2 - 298.5 sccm, CH_4 - 3 sccm, O_2 - 1 sccm . Gas pressure in the chamber were around 50 torr, and plasma were activated by microwaves 2.45 GHz at power 1300 W.
The SiO_2 mask on NiD diamond layer was obtained, for that at first, a layer of SiO_2 and thickness 1.5 μm was produced by pyrolytic method. In to the system (PCVD: Oxford Plasmalab System 80 Plus) a mixture of Si_2H_4 and N_2O was inserted in the ratio 1:5 with entire flow of 1200 sccm. SiO_2 deposition process was carried out in the temperature of 350°C, pressure of 1.5 torr and microwave power of 80W. The SiO_2 growth rate with those parameters is 169nm/min. . Next, on the NiD layer the photoresist was spin coated and exposed to UV light for 25s (Karl Suss Model MA-6.). In the next step, SiO_2 was etched by buffered oxide etch (BOE) 6:1.
Etched samples go back again to μPACVD system for one hour secondary growth of highly doped diamond structures. The structures were grown with 4000 and 10000 ppm [B]/[C] ratio. After second growth of diamond the Ohmic (Ti/Au) contacts were deposited on diamond structures by thermal evaporator (Temescal BJD 1800).

2.2. Analysis of structures

The NiD thin films were analysis by spectroscopic elipsometry and 4-point probe. For determining properties of continuous diamond layer on Si/SiO_2 substrate was used Jobin-Yvon UVISEL ellipsometer (HORIBA Jobin-Yvon Inc.). The investigated wavelength range was 260–830 nm. The experiments were performed at room temperature using an angle of incidence fixed at 70°. Ellipsometric fitting was based on a four-phase optical model roughness/diamond/SiO2/Si-wafer), Elipsometry were also used also for thickness analysis of SiO_2.

The morphology of obtained highly doped diamond structures (size of crystals and continuity) was determined by optical microscope (mikroLAB MMT 120 BT) and scanning electron microscope (SEM Hitachi S-3400N). Apart from that, SEM was used to analyze the highly doped structures, to assess smoothness of the edge lines, density of diamond grains occurrence within the areas of unintentional growth. Smoothness of the diamond phase edge was determined from SEM picture after graphic processing in GIMP programme (v. 2.6.11, GIMP development team) and a curve separating diamond and non-diamond phase was determined. As the y-axis was used a line parallel to expected direction of the edge line and contacting in a specific distance the smallest value of the curve. Next, the area under the curve was integrated and divided a segment length on which it was measured. The result was averaged for 10 segments of length 30 μm.

The surface resistivity of non-doped diamond layer was measured by 4-point probe connected to Keithley 2400 and computer with dedicated software. The conductivity between structures were measured by metallic needles on micromanipulators also connected to the Keithley 2400. The 4 point probe were used to analysis resistivity of diamond layer. In case when the film thickness is much smaller than the distance between the probes, the surface resistivity can be calculated from Eq:

$$\rho = \frac{\pi}{\ln(2)} \cdot \frac{V}{I} \cdot h \cdot f_2$$

where V is voltage measured on the internal probes, I is current applied to the external probes, h is the thickness of a layere and f_2 is correction factor [33]. The size of samples (20 mm x 20 mm) and spacing between probes (1.5 mm) gives value on correction factor around 0.97 [34].

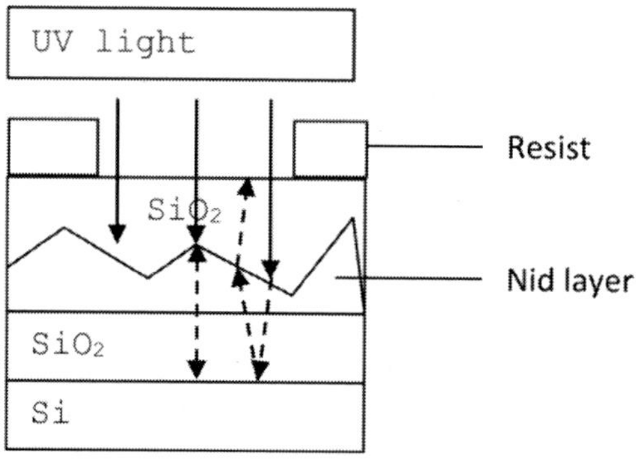

Figure 1. Imaging of contact UV light to not perpendicular surface of diamond.

3. Results and discussion

In the experiment the highly doped diamond structures in shape of pads with minimal size of 50μm were obtained. The NiD diamond layers were used as a substrate for second growth were obtained on Si/SiO_2 wafer with density of diamond nuclei 10^{10}/cm^2. The thickness of diamond film was measured by elipsometry and was 300 nm which results growth rate of 5 nm/min.

To fabricate the SiO_2 mask for creating diamond structures was necessary to create SiO_2 layer on diamond. For the next step of production structures the 1.5 μm SiO_2 has been grown. For fabrication mask typical process of photolitography has been done. In reason of changed reflection by diamond from the background (figure 1) (compared to bare Si/SiO_2) to get high quality smooth mask from SiO_2 was necessary to change times exposition to UV light. Additional layers and also not flat diamond layer provides to longer way of the light and different intensity of exposition on some areas on sample. To decrease this negative effects it`s possible to increase the time of exposition to UV light. Good effect was obtained when Si/SiO_2/diamond/ SiO_2/photoresist sample was 3 times longer exposed to UV light than normally Si/SiO_2 structure. The differences of SiO_2 edged after short can be observed on figure 2.

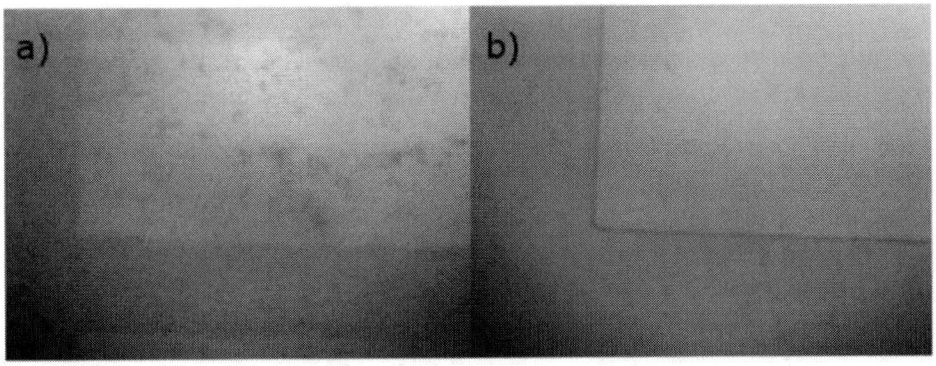

Figure 2. The shape of SiO_2 layer after etching with a) standard time of exposition to UV light (7 s) b) 3 times longer (21 s) exposition to UV light.

Figure 3. Obtained pads with cover Ti/Au picture from optical microscope.

Created and measured highly doped structures were in the shape of the pads in four sizes (50 x 200 μm, 100 x 200 μm, 150 x 200 μm and 200 x 200 μm). The pads of one size were put in the row and distance between them were increasing. The distance between next pads was starting from 5 μm from edge to edge goes by 10, 15, 20, 40, 60, 80, 100, 150 up to 200 μm (figure 3).

The SEM observations after NiD overgrowth by highly doped diamond structures show high quality edges. The amount of nuclei on the area of unintentional growth was close to 0. The minimal amount of crystals which were grown on black area in figure 4 have minor influence on the isolation between structures. Also the average smoothness of edge line was only 170 nm with maximum inclination of 500nm.

The Ohmic contacts has been used to get better contact to diamond layer in both cases to measured the surface resistivity of NiD layer and to measured contact between highly boron doped structures. The measured value of voltage at the minimum range of current were over the scale of measurement system shows that the material is an isolator.

The measured I/V characteristics have been done between every neighboring pads. The analyzed resistance between all pads of the same size shows almost linear changing resistivity to the distance

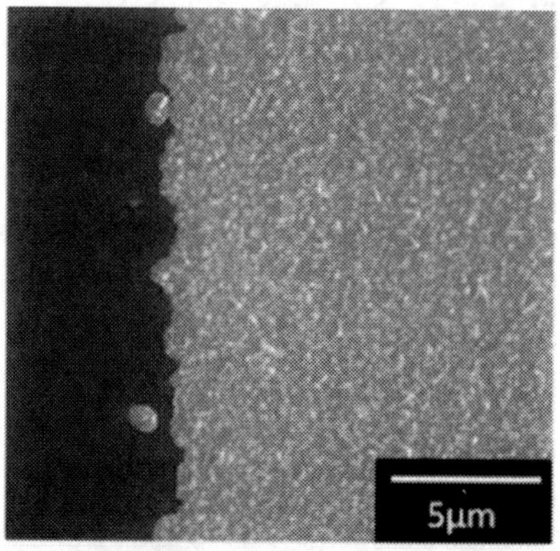

Figure 4. SEM of observed edge line with one of the crystals on non intentionaly phase growth.

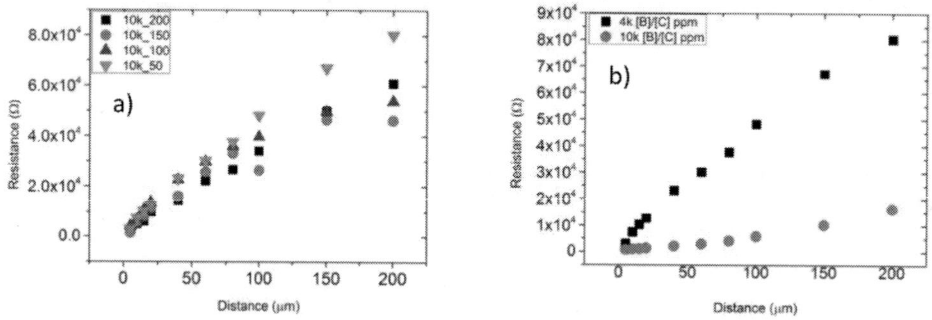

Figure. 5. (a) Comparison of resistance between different size pads (b) Resistivity comparison for two different level of dopant at pads of size 50 x 200 μm.

what was suspected. For small distances value of the resistance of all size pads it`s almost that same 6.5×10^3 Ω +/- 10^3 for 10 μm. The analysis the resistivity when distance is 150 μm the differences between pads are bigger and there are 5×10^4 Ω +/- 1.5×10^4. But it is impossible to recognize the influence of pads size on the resistivity (see figure 5). When compare all results it is possible to observe that using 10000 ppm [B]/[C] level of doping substrate you get better conductivity between the structures than when using 4000 ppm [B]/[C] in gas phase. The 4000 ppm level of doping has always 5 times bigger resistivity than 10000 ppm. It shows that some number of dopant migrate to the non-doped layer provides electrical contact between structures. This effect can be used by controlling the level of dopant migrating into NiD layer or low resistivity layer to build semiconductor devices.

4. References

[1] Butler J E and Sumant A V 2008 The CVD of Nanodiamond Materials *Chem. Vap. Depos.* **14** 145–60
[2] Geis M W, Twichell J C and Lyszczarz T M 1996 Diamond emitters fabrication and theory *J. Vac. Sci. Technol. B* **14** 2060–7
[3] Liu H and Dandy D S 1995 Studies on nucleation process in diamond CVD: an overview of recent developments *Diam. Relat. Mater.* **4** 1173–88
[4] Aleksov A, Kubovic M, Kaeb N, Spitzberg U, Bergmaier A, Dollinger G, Bauer T, Schreck M, Stritzker B and Kohn E 2003 Diamond field effect transistors—concepts and challenges *Diam. Relat. Mater.* **12** 391–8

[5] Kang W P, Davidson J L, Wisitsora-at A, Wong Y M, Takalkar R, Subramania K, Kerns D V and Hofmeister W H 2005 Diamond and carbon-derived vacuum micro- and nano-electronic devices *Diam. Relat. Mater.* **14** 685–91
[6] Gimenez S P 2010 Diamond MOSFET: An innovative layout to improve performance of ICs *Solid-State Electron.* **54** 1690–6
[7] Kang W P, Davidson J L, Wong Y M and Holmes K 2004 Diamond vacuum field emission devices *Diam. Relat. Mater.* **13** 975–81
[8] Sullivan J p., Friedmann T a. and Hjort K 2001 Diamond and Amorphous Carbon MEMS *MRS Bull.* **26** 309–11
[9] Luo J K, Fu Y Q, Le H R, Williams J A, Spearing S M and Milne W I 2007 Diamond and diamond-like carbon MEMS *J. Micromechanics Microengineering* **17** S147
[10] Hees J, Heidrich N, Pletschen W, Sah R E, Wolfer M, Williams O A, Lebedev V, Nebel C E and Ambacher O 2013 Piezoelectric actuated micro-resonators based on the growth of diamond on aluminum nitride thin films *Nanotechnology* **24** 025601
[11] Varney M W, Aslam D M, Janoudi A, Chan H-Y and Wang D H 2011 Polycrystalline-Diamond MEMS Biosensors Including Neural Microelectrode-Arrays *Biosensors* **1** 118–33
[12] Nebel C E, Rezek B, Shin D, Uetsuka H and Yang N 2007 Diamond for bio-sensor applications *J. Phys. Appl. Phys.* **40** 6443
[13] Bogdanowicz R, Czupryniak J, Gnyba M, Ryl J, Ossowski T, Sobaszek M, Siedlecka E M and Darowicki K 2013 Amperometric sensing of chemical oxygen demand at glassy carbon and silicon electrodes modified with boron-doped diamond *Sens. Actuators B Chem.* **189** 30–6
[14] Fabiańska A, Bogdanowicz R, Zięba P, Ossowski T, Gnyba M, Ryl J, Zielinski A, Janssens S D, Haenen K and Siedlecka E M 2013 Electrochemical oxidation of sulphamerazine at boron-doped diamond electrodes: Influence of boron concentration *Phys. Status Solidi A* **210** 2040–7
[15] Qureshi A, Kang W P, Davidson J L and Gurbuz Y 2009 Review on carbon-derived, solid-state, micro and nano sensors for electrochemical sensing applications *Diam. Relat. Mater.* **18** 1401–20
[16] Müller R, Schmid P, Munding A, Gronmaier R and Kohn E 2004 Elements for surface microfluidics in diamond *Diam. Relat. Mater.* **13** 780–4
[17] Lifshitz Y, Lee C H, Wu Y, Zhang W J, Bello I and Lee S T 2006 Role of nucleation in nanodiamond film growth *Appl. Phys. Lett.* **88** 243114
[18] Gruen D M 1999 Nanocrystalline Diamond Films1 *Annu. Rev. Mater. Sci.* **29** 211–59
[19] Williams O A, Douhéret O, Daenen M, Haenen K, Ōsawa E and Takahashi M 2007 Enhanced diamond nucleation on monodispersed nanocrystalline diamond *Chem. Phys. Lett.* **445** 255–8
[20] Varga M, Ižák T, Kromka A, Veselý M, Hruška K and Michalka M 2012 Study of diamond film nucleation by ultrasonic seeding in different solutions *Open Phys.* **10** 218–24
[21] Chu Y-C, Tu C-H, Jiang G, Chang C, Liu C, Ting J-M, Lee H-L, Tzeng Y and Auciello O 2012 Systematic studies of the nucleation and growth of ultrananocrystalline diamond films on silicon substrates coated with a tungsten layer *J. Appl. Phys.* **111** 124328
[22] Akhvlediani R, Lior I, Michaelson S and Hoffman A 2002 Nanometer rough, sub-micrometer-thick and continuous diamond chemical vapor deposition film promoted by a synergetic ultrasonic effect *Diam. Relat. Mater.* **11** 545–9
[23] Bogdanowicz R, Śmietana M, Gnyba M, Gołunski Ł, Ryl J and Gardas M 2014 Optical and structural properties of polycrystalline CVD diamond films grown on fused silica optical fibres pre-treated by high-power sonication seeding *Appl. Phys. A* **116** 1927–37
[24] Siuzdak K, Bogdanowicz R, Sawczak M and Sobaszek M 2014 Enhanced capacitance of composite TiO2 nanotube/boron-doped diamond electrodes studied by impedance spectroscopy *Nanoscale* **7** 551–8
[25] Wang M, Zhou Y S, Xie Z Q, Gao Y, He X N, Jiang L and Lu Y F 2013 Seed-Free Growth of Diamond Patterns on Silicon Predefined by Femtosecond Laser Direct Writing *Cryst. Growth Des.* **13** 716–22
[26] Scorsone E, Saada S, Arnault J C and Bergonzo P 2009 Enhanced control of diamond nanoparticle seeding using a polymer matrix *J. Appl. Phys.* **106** 014908
[27] Nishizawa J, Plotka P and Kurabayashi T 2002 Ballistic and tunneling GaAs static induction transistors: nano-devices for THz electronics *IEEE Trans. Electron Devices* **49** 1102–11
[28] Nishizawa J, Płotka P and Kurabayashi T 2008 GaAs area-selective regrowth with molecular layer epitaxy for integration of low noise and power transistors, and Schottky diodes *Phys. Status Solidi C* **5** 2799–801

[29] Nishizawa J, Oyama Y, Plotka P and Sakuraba H 1996 Optimization of low temperature surface treatment of GaAs crystal *Surf. Sci.* **348** 105–14
[30] Aharonovich I, Lee J C, Magyar A P, Buckley B B, Yale C G, Awschalom D D and Hu E L 2012 Homoepitaxial Growth of Single Crystal Diamond Membranes for Quantum Information Processing *Adv. Mater.* **24** OP54–9
[31] Smirnov W, Yang N, Hoffmann R, Hees J, Obloh H, Müller-Sebert W and Nebel C E 2011 Integrated All-Diamond Ultramicroelectrode Arrays: Optimization of Faradaic and Capacitive Currents *Anal. Chem.* **83** 7438–43
[32] Bogdanowicz R, Fabiańska A, Golunski L, Sobaszek M, Gnyba M, Ryl J, Darowicki K, Ossowski T, Janssens S D, Haenen K and Siedlecka E M 2013 Influence of the boron doping level on the electrochemical oxidation of the azo dyes at Si/BDD thin film electrodes *Diam. Relat. Mater.* **39** 82–8
[33] Schroder D K 2006 *Semiconductor Material and Device Characterization* (John Wiley & Sons)
[34] Smits F M 1958 Measurement of sheet resistivities with the four-point probe *Bell Syst. Tech. J.* **37** 711–8

Acknowledgments
The authors gratefully acknowledge financial support from the Polish National Science Centre (NCN) under grant no. 2011/03/D/ST7/03541 and 2014/14/M/ST5/00715. The DS funds of the Faculty of Electronics, Telecommunications and Informatics of the Gdansk University of Technology are also acknowledged

The use of thin diamond films in fiber-optic low-coherence interferometers

D. Milewska, K. Karpienko

Department of Metrology and Optoelectronics, Faculty of Electronics, Telecommunications and Informatics, Gdańsk University of Technology, Narutowicza Street 11/12, 80-233 Gdańsk, Poland

E-mail: k.karpienko@pro.wp.pl

Abstract. In this paper we present the use of thin diamond films in fiber-optic low-coherence interferometers. Two kinds of diamond surfaces were used: undoped diamond film and boron-doped diamond film. They were deposited on glass plates as well as silicon layers. A conventionally used mirror was used as a reference layer. Diamond films were deposited using Microwave Plasma Enhanced Chemical Vapour Deposition (μPE CVD) system. Measurements were performed using two superluminescent diodes (SLD) with wavelengths of 1300 mm and 1550 mm. The optimal conditions for each layers were examined: the required wavelength of the light source and the length of Fabry-Perot interferometer cavity. Metrological parameters of Fabry-Perot interferometer with different mirrors were compared. The presented thin diamond films may be an interesting alternative to the commonly used reflective surfaces.

1. Introduction

Fiber optic low-coherence interferometers are widely used as sensors of many physical [1,2] and biological quantities [3-5]. Information about the measured physical quantity is encoded in the full spectrum of light reflected from the sensing interferometer. A very important part of the sensing interferometer is a reflective layer and it has an impact on the quality of the obtained spectra. Application of diamond films in fiber-optic interferometers as a reflecting layer has the potential to improve the visibility of spectrum fringes. Moreover, they are biocompatible and provide great chemical stability. A Fabry-Perot interferometer has been successfully used as the sensing interferometer in many measurement systems [6,7]. One of the possible configurations is a fiber-optic Fabry-Perot interferometer operating in reflective mode. In this mode, the interference occurs due to multiple reflections of the light beam from two parallel surfaces: fiber tip and mirror, which is shown in Figure 1.

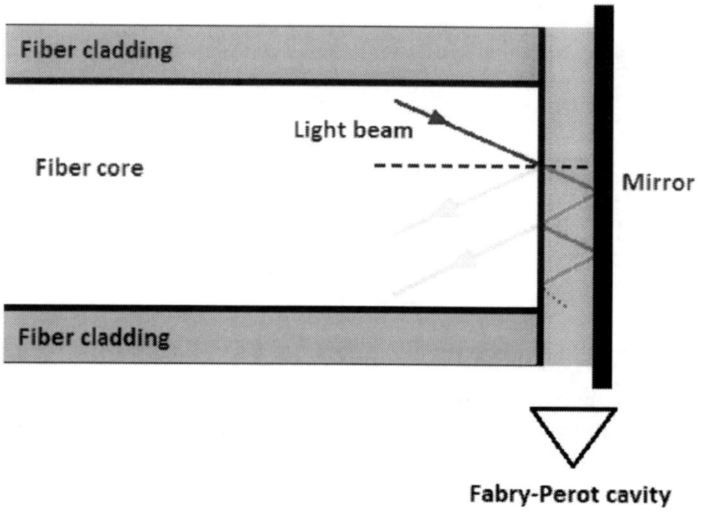

Figure 1. Fiber optic configuration of Fabry-Perot interferometer working at reflectance mode.

For the reflection mode, the optical intensity at the output of interferometer is given by [8]:

$$I_R = I_0[R + R(1-R)^2 + 2R(1-R)cos\varphi] \quad (1)$$

where: I_0 - intensity of light incident on the first boundary surface of the interferometer, R – reflectance of the reflecting surface, φ – phase shift.

The use of a broadband, low coherence light source in Fabry-Perot interferometer allows achieving high sensitivity, high resolution and dynamic measurement. However, in order to obtain the best possible metrological parameters of fiber-optic low-coherence Fabry-Perot interferometric sensors, the most important task is to maximize the visibility of the measured signal. This can be achieved by the use of the most suitable reflecting surface as a mirror in the Fabry-Perot cavity. Most commonly used as a reflective surface in the Fabry-Perot interferometers are silver mirrors. This material is cheap and sufficiently well reflecting for a light beam, however, it is sensitive to mechanical damage. Therefore, there is a need to implement other materials as the reflective surfaces so as to achieve greater resistance to harmful mechanical and chemical conditions. In addition, with a view to using a fiber-optic low-coherence Fabry-Perot interferometer as a biosensor, the reflective surface should be biocompatible with the measured sample (e.g. bodily fluids) and provide great chemical stability. Diamond thin films meet all these conditions. Moreover, application of diamond films as a reflecting layer has the potential to improve the visibility of measured signal. In the presented experiment, we investigate the possibility of using diamond thin films as reflective surfaces in Fabry-Perot interferometer.

2. Materials

Diamonds are characterized by high thermal conductivity [9] mechanical hardness, optical transparency in broad wavelength [10,11] and biocompatibility [12,13]. In addition, diamond doped with boron is a promising material to build various biocompatible and chemically stable optical biosensors.

In the experiment two kinds of diamond thin films as a reflective surfaces were investigated: undoped diamond and boron-doped diamond. Both materials were deposited on glass plates as well as silicon substrates. A silver mirror was used as a reference reflective surface.

IOP Conf. Series: Materials Science and Engineering **104** (2015) 012023 doi:10.1088/1757-899X/104/1/012023

Diamond thin films were deposited using the microwave plasma assisted chemical vapor deposition (μPE CVD). The undoped and boron-doped diamond films were synthetized with the use of MW PA CVD system (SEKI Technotron AX6200S, Japan). Substrate temperature was approximately at 500 °C and plasma excitation was achieved by the microwave radiation (2.45GHz). Base pressure inside the chamber was 10−4 Torr and process pressure 50 Torr. Mixture of hydrogen, methane (1%) and diboran (boron-doped – 5%, undoped – 0%) filled the chamber. The doping level of boron in the gas phase was 10 000 ppm. The growth time were: 60 min for silicon substrates and 270 min for glass plates. Preparation of diamond film on glass plates have been described by Sobaszek et al. [14]. Produced diamonds films have thickness of ~ 200-300 nm.

3. Experimental

The experimental setup consists of two interchangeable, broadband light sources (superluminescent diodes) produced by Superlum Ltd., Ireland. Parameters of these light sources were as follows:

— S1300-G-I-20: $\lambda = 1290$ nm, $\Delta\lambda_{FWHM} = 50$ nm,

— S-1550-G-I-20: $\lambda = 1550$ nm, $\Delta\lambda_{FWHM} = 45$ nm.

Detection of the measured signal was performed using an Ando AQ6319 optical spectrum analyzer with resolution bandwidth set to 1 nm. All devices were connected with a single-mode commercially available, telecommunications coupler (fibre SMF-28). The Fabry-Perot interferometer was formed by the fibre tip and the examined reflective surface.

4. Results

The identification of the optimal length of Fabry-Perot cavity for the reference surface – silver mirror (Figure 1a) was the first step of the experiment. Then, measurements of undoped and boron-doped diamond thin films on silicon substrates and glass plates were performed. For each reflective surface, the influence of the length of the Fabry-Perot cavity on the visibility of the measured signal was investigated. At each measurement, the cavity length was varied from 0 μm to 600 μm in increments of 50 μm. Representative measured spectra obtained with a cavity length of 200 μm are presented in Figure 2.

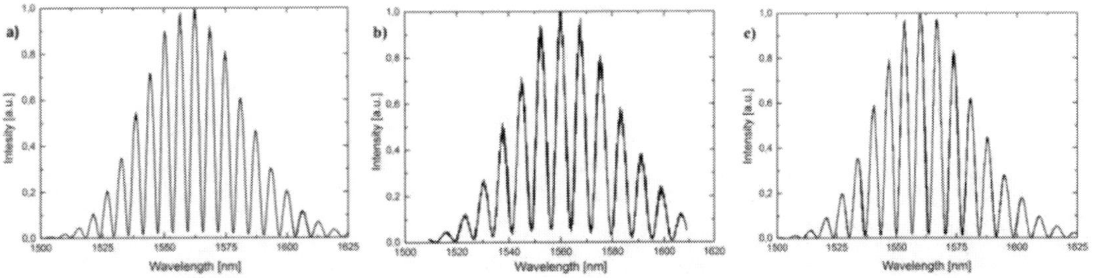

Figure 2. Representative registered spectra from the low-coherence fiber-optic Fabry-Perot interferometer, light source 1500 nm: (a) silver mirror;(b) undoped diamond on glass plate; (c) boron-doped diamond on glass plates.

It can be noted that the use of the investigated reflective surfaces did not worsen the signal visibility. However, the spectral separation was changed. This was due to the change in refractive index and does not affect the proper operation of the measurement system. Comparison of the obtained visibility for all of reflecting surfacea at a length of Fabry-Perot cavity equal to 200 μm, is shown in Table 1.

Table 1. The visibility value for different reflective surfaces of Fabry-Perot cavity and different light sources. The length of the Fabry-Perot cavity was constant – 200 μm.

Reflective surface	Light source [nm]	Signal visibility [%]
silver mirror	1300	97,2
undoped diamond on silicon substrate	1300	97
boron-doped diamond on silicon substrate	1300	51
undoped diamond on glass plates	1300	62,2
boron-doped diamond on glass plates	1300	91,1
silver layer	1500	97,3
undoped diamond on silicon substrate	1500	80,3
boron-doped diamond on silicon substrate	1500	63,4
undoped diamond on glass plates	1500	94,9
boron-doped diamond on glass plates	1500	98.1

As it can be noted from Table 1. The use of boron-doped thin films deposited on glass plates as reflective surfaces allows us to achieve a high visibility value, almost as high as when the silver mirror is used. The value of measured signal visibility when undoped diamond films deposited on silicon substrate are used is smaller, yet still acceptable and interesting for sensing applications.
The use of different mirrors influences the optimal geometrical parameters of Fabry-Perot cavity. For the use of silver mirror and silicon substrates as interferometer mirror the optimal length of cavity is 200 μm, when for thin films deposited on glass plates is 100 μm and 150 μm for undoped and boron-doped diamond on glass plates respectively.

Table 2. The value of visibility of the measured signal for thin films deposited on glass plates for different lengths of the Fabry-Perot cavity and different light sources.

Reflective surface	Light source [nm]	Length of the Fabry-Perot cavity [μm]	Signal visibility [%]
undoped diamond on glass plates,	1300	100	82,5
boron-doped diamond on glass plates	1300	150	97
undoped diamond on glass plates	1500	150	99,4
boron-doped diamond on glass plates	1500	150	99,7

The obtained results indicate that undoped and boron-doped diamond films deposited on glass plates have shorter optimal Fabry-Perot cavity length, so they can be successfully implemented in measurements of thin samples.

5. Conclusion

An investigation of the possibility of using diamond thin films as a reflective surfaces in Fabry-Perot interferometer has been presented. Measurements with undoped and boron-doped diamond thin films as reflective surfaces in fiber optic low-coherence Fabry-Perot interferometer have been performed. Analysis of results indicates that the use of diamond thin films as the reflective surface in Fabry-Perot interferometer allows achieving very good visibility of the signal for the other lengths of the Fabry-Perot cavity, than those achievable using a conventional silver mirror. This property can be successfully used to change the size of Fabry-Perot cavity, so as to adapt its size to the sample. Moreover, Fabry-Perot interferometer with diamond thin film would be more immune to mechanical damage and chemical conditions. This kind of thin films may be used in biocompatible biosensors.

6. References

[1] Jędrzejewska-Szczerska M, Wierzba P, Abou Chaaya A, Bechelany M, Miele P, Viter R, Mazikowski A, Karpienko K, Wróbel M S 2015 ALD thin ZnO layer as an active medium in a fiber-optic Fabry-Perot Interferometer *Sensors and Actuators A – Physical* **88-94** 221

[2] Karpienko K, Wróbel M, Jędrzejewska-Szczerska M 2014 Determination of Refractive Index Dispersion Using Fiber-Optic Low-Coherence Fabry-Perot Interferometer: Implementation and Validation *Optical Engineering* **7** 53

[3] Jędrzejewska-Szczerska M 2014 Response of a New Low-Coherence Fabry-Perot Sensor to Hematocrit Levels in Human Blood *SENSORS* **4** 14

[4] Jędrzejewska-Szczerska M 2013 Measurement of complex refractive index of human blood by low-coherence interferometry *The European Physical Journal Special Topics* **9** 222

[5] Wierzba P and Jędrzejewska-Szczerska M 2013 Optimization of a Fabry Perot Sensing Interferometer Design for an Optical Fiber Sensor of Hematocrit Level *Acta Physica Polonica A* **3** 124

[6] Pluciński J, Hypszer R, Wierzba P, Strakowski M, Jedrzejewska-Szczerska M, Maciejewski M, Kosmowski B B 2008 Optical low-coherence interferometry for selected technical applications *Bulletin of the Polish Academy of Sciences. Technical Sciences* **2** 56

[7] Grattan K T V and Meggitt B T 2000 *Optical fiber sensor technology* (Boston: Kluwer Academic Publisher)

[8] Egorov S, Mamaev A, Likhachiev I 1996 High reliable, self calibrated signal processing method for interferometric fiber-optic sensors *Proc. SPIE* **193-197** 2594

[9] Sukhadolau A V, Ivakin E V, Ralchenko V G, Khomich A V, Vlasov A V, Popovich A F 2005 Thermal conductivity of CVD diamond at elevated temperatures *Diamond and Related Materials* **14** 589

[10] Checoury X, Néel D, Boucaud P, Gesset C, Girard H, Saada S, Bergonzo P 2012 Nanocrystalline diamond photonics platform with high quality factor photonic crystal cavities *Applied Physics Letters* **101** 171115

[11] Stotter J, Haymond S, Zak J K, Show Y, Cvackova Z, Swain G 2013Optically transparent diamond electrodes for UV-Vis and IR spectroelectrochemistry *Electrochemical Society Interface* **12** 33-38

[12] Amaral M, Dias A G, Gomes P S, Lopes M A, Silva R F, Santos J D, Fernandes M H 2008Nanocrystalline diamond: In vitro biocompatibility assessment by MG63 and human bone marrow cells cultures *J. Biomed. Mater. Res. A* **87** 91

[13] Bajaj P, Akin D, Gupta A, Sherman D, Shi B, Auciello O, Bashir R 2007 Ultrananocrystalline diamond film as an optimal cell interface for biomedical applications *Biomedical microdevices* **6** 767 – 794

[14] Sobaszek M, Skowroński Ł, Bogdanowicz R, Siuzdak K, Cirocka A, Zięba P, Gnyba M, Naparty M, Gołuński Ł, Płotka P 2015 Optical and electrical properties of ultrathin transparent nanocrystalline boron-doped diamond electrodes *Optical Materials* **42** 24-34

Acknowledgement

This study was partially supported by the National Science Centre under the grant No. 2011/03/D/ST7/03540, as well as DS Programs of the Faculty of Electronics, Telecommunications and Informatics of the Gdańsk University of Technology.

Optically transparent boron-doped nanocrystalline diamond films for spectroelectrochemical measurements on different substrates

M. Sobaszek[1], K. Siuzdak[2], Ł. Skowroński[3], R. Bogdanowicz[1] and J. Pluciński[1]

[1]Department of Metrology and Optoelectronics, Faculty of Electronics, Telecommunications and Informatics, Gdansk University of Technology, 11/12 G. Narutowicza St., 80-233 Gdansk, Poland
[2]Centre for Plasma and Laser Engineering, The Szewalski Institute of Fluid-Flow Machinery, Polish Academy of Sciences, 14 Fiszera St., 80-231 Gdansk, Poland
[3]University of Technology and Life Sciences, Institute of Mathematics and Physics, 7 Kaliskiego St., 85-789 Bydgoszcz, Poland

E-mail: michal.sobaszek@pg.gda.pl

Abstract. Fabrication process of optically transparent boron nanocrystalline diamond (B-NCD) electrode on silicon and quartz substrate was shown. The B-NCD films were deposited on the substrates using Microwave Plasma Assisted Chemical Vapor Deposition (MWPACVD) at glass substrate temperature of 475 °C. A homogenous, continuous and polycrystalline surface morphology with high sp^3 content in B-NCD films and film thickness depending from substrate in the range of 60-300 nm was obtained. The high refraction index and transparency in visible (VIS) wavelength range was achieved. Moreover, cyclic voltammograms (CV) were recorded to determine reaction reversibility at the B-NCD electrode. CV measurements in aqueous media consisting of 1 mM $K_3[Fe(CN)_6]$ in 0.5 M Na_2SO_4 demonstrated relatively fast kinetics expressed by a redox peak splitting below 503 mV for B-NCD/silicon and 110 mv for B-NCD/quartz.

1. Introduction

Nanocrystalline diamond (NCD) film, which exhibit outstanding properties such as: hardness, chemical resistance, biocompatibility [1,2] and so on, also have optical transparency in a broad wavelength range from ultraviolet (UV) to far infrared (IR). Natural diamond is a wide-band semiconductor $E_g = 5.45$ eV that can be easily doped by boron atoms using the *in situ* Microwave Plasma Assisted Chemical Vapour Deposition (MWPACVD) process. Due to this properties, boron-doped diamond films are promising materials for optically transparent electrodes that have various applications, *e.g.* spectroelectrochemical measurements. Fused silica substrates allow working in wavelength range from 160 to 4000 nm and are resistant to hazardous media. Silicon substrates not only allow working in the IR wavelength range but also are optimal material for planar waveguides, resonators etc. The optical transparency of the MWPACVD polycrystalline diamond are strongly influenced by surface roughness, boron doping film thickness or nondiamond phases and other impurities.

Boron-doped diamond films are commonly used as materials for electrodes for sensing various organic substances or as electrodes to utilization hazarodous organic coumpouds like phenolons, etc. . Hence, they are promising materials for developing sensors based on optically transparent electrodes or as protective coatings for biomedical applications [3], due to optical transparency and biocompatibility. Zak *et al.* [4] proposed boron-doped diamond as an optically transparent electrode material. The authors reported a 0.38 mm thick freestanding electrode for spectroeletrochemical measurements. Stotter *et al.* [5,6] applied boron-doped diamond electrodes to spectroeletrochemical studies in the UV-VIS wavelength region showing good transparency at 50 – 60%.

The present paper reports thin and continuous optically transparent and conducting boron-doped nanocrystalline diamond (B-NCD) films on quartz and silicon substrates using a MWPACVD. To achieve good quality of optical and conducting properties, optimization of growth parameters is needed. The usage of a truncated cone-shaped substrate holder [7] enables an increase of microwave power density, yielding a high sp^3/sp^2 ratio, small roughness, lower growth rate, and more effective boron incorporation to diamond lattice [8].

The thin B-NCD film were performed by MWPACVD. The film morphology was investigated by numerical analysis of scanning electron microscopy (SEM). The chemical composition of the films was investigated by means of Raman spectroscopy. The thickness and optical properties in the UV-VIS wavelength rage were estimated using spectroscopic ellipsometry (SE). Electrochemical measurements e.g. cyclic voltammetry and chronoamperometry were used to investigated redox reaction reversibility at the electrode and to carry out spectroelectrochemical measurement.

2. Experimental

2.1. A CVD nanocrystalline diamond film deposition

B-NCD films were synthetized in an MWPACVD system (SEKI Technotron AX5400S, Japan) on optically polished quartz and single-crystal p-type Si (100) substrates. The substrates were seeded by means of spin-coating, spun three times by 60 seconds with 4000 rpm, using nanodiamond DMSO/PVA base slurry with crystallites size 4-7 nm. The substrate temperature was maintained at 475 °C during the deposition process. A special truncated cone-shaped shim was used during the growth of diamond films. Excited plasma was ignited by microwave radiation (2.45 GHz) [9]. The plasma microwave power, optimized for diamond synthesis, was kept at 1300 W. In this study, the molar ratio of CH_4-H_2 mixture was kept at 1% of gas volume, whose total flow rate was set at 300 sccm. The base pressure was about 10^{-6} Torr (10^{-4} Pa) and the process pressure was kept at 50 Torr (6.6 kPa). The boron level expressed as the [B]/[C] ratio in the gas phase was 0, 2000, 5000, 7500 or 10,000 ppm. Diborane (B_2H_6) was used as a dopant precursor. The mirror polished substrates were covered by B-NCD film, essentially providing full surface encapsulation. The growth time was 1 hour for Si substrates and 3 hours for quartz.

2.2. Electrochemical and spectroelectrochemical setup

Cyclic voltammetry (CV) measurements were carried out in aqueous media consisting of 5 mM $K_3[Fe(CN)_6]$ in 0.5 M Na_2SO_4 at a scan rate of 100 mV s^{-1}. Electrochemical cell consisted of a three electrode system. B-NCD, platinum wire and Ag/AgCl/0.1 KCl were used as working, counter and reference electrodes, respectively. The area of the working electrode exposed to electrolyte was 0.5 cm^2. The electrochemical experiment was performed using an Autolab potentiostat/galvanostat (PGSTAT30 and GPES 4.9 software/Nova 1.10.2, Netherlands). The spectroelectrochemical measurements were carried out in thin-layer electrochemical cell consisting 30 µL of solution.

2.3. Surface and structure analytical techniques

A scanning electron microscope (S-3400 N, Hitachi, Japan) with a tungsten source and variable chamber pressure (VP-SEM) was utilized to inspect the surface of synthesized thin films. The molecular composition of films was studied by means of Raman spectroscopy using a Raman confocal microscope (Horiba LabRAM ARAMIS, Japan). Spectra were recorded in a range of diode pumped solid state (DPSS) laser in combination with a 100x objective magnification (NA = 0.95) and 50 lm confocal aperture.

The ellipsometric parameters Ψ, Δ and depolarization factor (%Depol) were measured for three angles of incidence (65°, 70°, 75°) in range from (300 - 1000 nm) with a step of 0.02 eV at the room temperature of 20 °C by means of a rotating analyser device (V-VASE, J.A. Woollam Co. Inc, USA (196 – 2200 nm).

A five-layer structural model of the samples (ambient / SRL- surface roughness layer / B-NCD / intermix / Si – substrate) was applied to determine the effective thickness (d_r) of the SRL and the thickness of the B-NCD film (dl) and its effective complex dielectric function , where ε_1, ε_2 and are the real part, the imaginary part of $\tilde{\varepsilon}$ and the effective complex refraction index, respectively. The thickness of the intermix layer [10] had been derived in a previous investigation and was set as di~2 nm for all samples in this study. Optical responses of roughness and intermix layers were estimated using Bruggeman effective medium approximation (EMA). The optical response of a B-NCD film was predicted using the sum of Drude and Lorentzian oscillators. Generally, a semi-classical form of the complex effective dielectric function can be written as [11]:

$$\langle\tilde{\varepsilon}\rangle = \varepsilon_\infty - \frac{(\hbar\omega_p)^2}{E^2 + i\Gamma E} + \sum_k \frac{A_k E_{0k}^2}{E_{0k}^2 - E^2 - i\gamma_k E} \tag{1}$$

In Eq. (1) ε_∞, ω_p and Γ define the high-frequency dielectric constant, the unscreened plasma frequency and the free-carrier damping, respectively. The inter-band transitions are represented by the Lorentz oscillator located at photon energy E_{0k} with amplitude A_k and broadening parameter γ_k. In this study, a single Lorentzian oscillator was used to characterize the UV-VIS absorption.

3. Result and discussion

3.1. Surface morphology and chemical composition of B-NCD films

Fully investigation of B-NCD/silicon samples is described in our previous paper [9]. SEM surface investigation shows fully encapsulation on both substrates without any cracks, defects or impurities. The average grain size achieved for silicon were below 100 nm due to 1 hour of deposition. Figure 1 shows high resolution SEM micrographs for B-NCD/quartz sample with boron doping at 5000 ppm level.

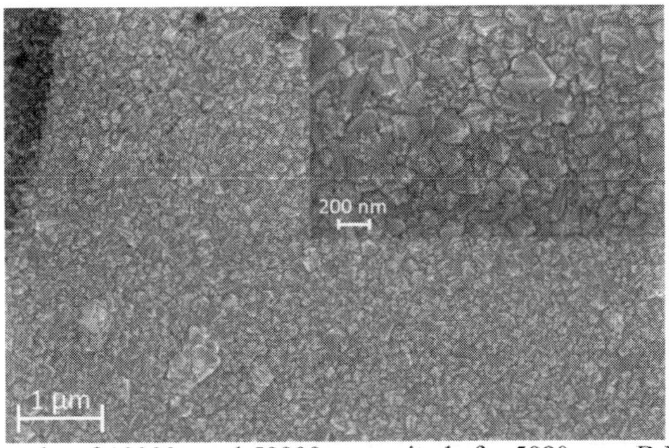

Figure 1. SEM micrographs of 10000x and 50000x magnitude for 5000 ppm B-NCD film on quartz substrate.

It can been see that diamond film is continuous and uniform with grain size in range from 50 up to 250 nm and should be notice that time of deposition is 3 hours. In case of quartz substrates shorter time of diamond film deposition occurs with cracks that are connected with internal stress deriving from difference of thermal expansion coefficient between quartz and diamond. Moreover film cracking have minor impact on optical properties but significant on electrical parameters.

Raman spectrum for B-NCD/quartz samples shown in Figure 2 proves presence of strong bands assigned to the diamond lattice, and confirms high content of sp^3 carbon in the film structure. Its shift from typical for diamond 1332 to about 1305 cm^{-1} with boron doping increasing and is caused by a stress introduced by boron atoms [12]. Moreover, the boron presence in sample B-NCD-10 (10000 ppm [B]/[C]) can be confirmed by a band at 1211 cm^{-1} and strong signal intensity drop above 1650 cm^{-1}. The peaks at band 1131 to 1148 can be assigned as nanocrystalline diamond or tetrahedral amorphous carbon (ta-C) with about 90% sp^3 bonding [13–15]. The D band at 1474 to 1478 cm^{-1} can be assigned to the mixture of amorphous carbon phase sp^2 and sp^3, while the G band at 1537 to 1555 cm^{-1}, to sp^2-bonded amorphous carbon.

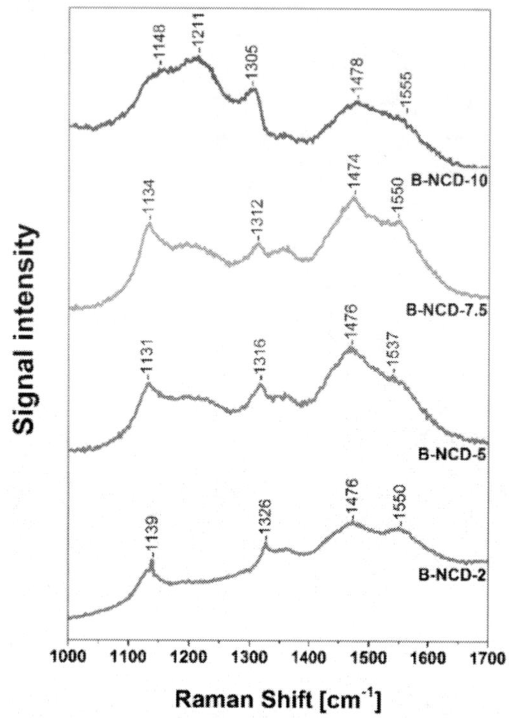

Figure 2. Raman spectra of the ultrathin B-NCD films deposited by MWCVD on quartz substrates, depicting the typical signatures of diamond with different [B]/[C] ratios.

3.2. Optical properties of boron nanocrystalline diamond

The refraction index) *n* is a crucial parameter in design of various kinds of waveguides, sensors, coatings etc. It is also important in coupling different materials. Since *n* is correlated with the materials density [16,17], a high refractive index determines a high optical density. Figure 3(a) and 3 (b) shows the *n* for B-NCD/quartz and B-NCD/silicon samples, respectively.

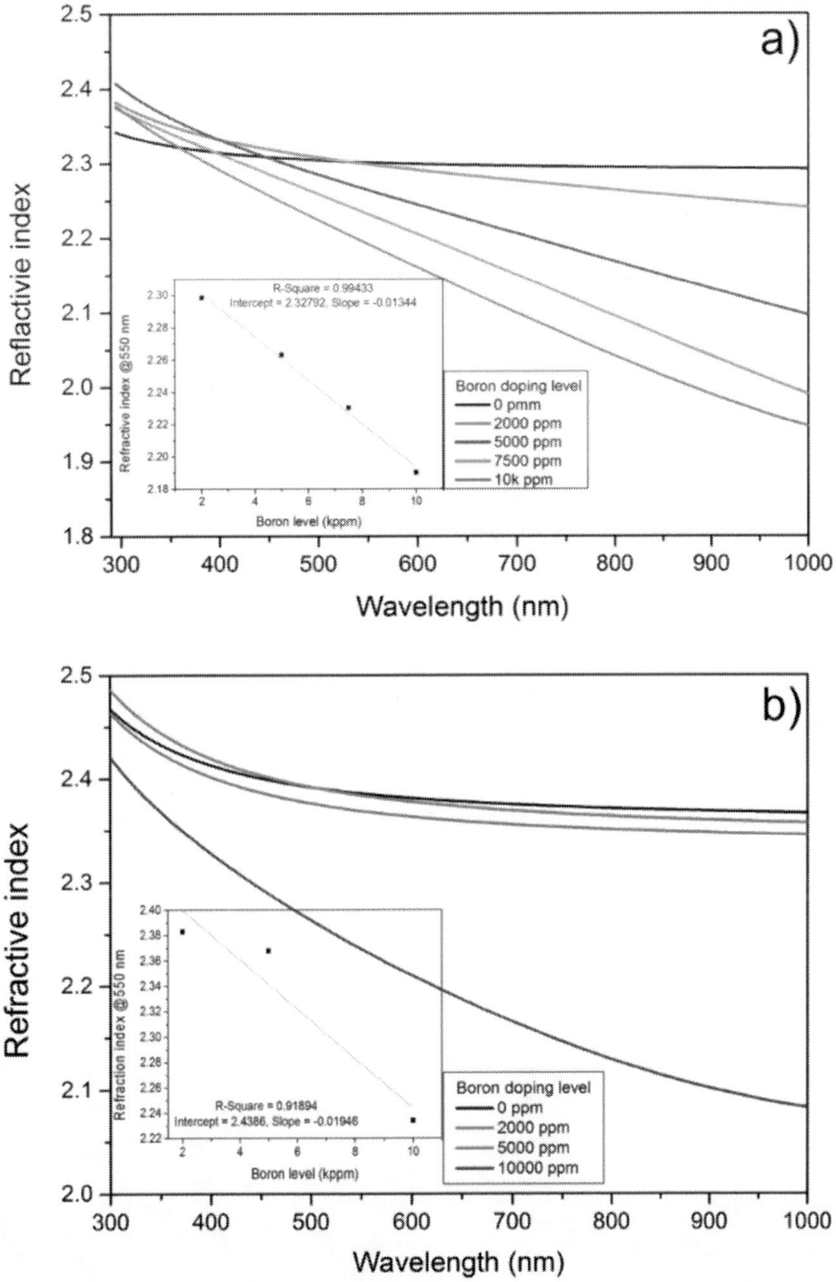

Figure 3. Refractive index of B-NCD samples (a) on quartz and (b) on silicon substrates, insets shows linearity of changing RI with increasing doping for 550 nm.

The obtained values for films on quartz ranged from 2.4 to 1.96 with increasing wavelength. In the case of the non-doped sample, the change of n is minor with increasing λ and remains slight for the low doped sample (2000 ppm). For the higher doping level, the impact of changing λ on RI is significant. Change of the n at 550 nm with increasing boron doping is linear (Figure 3a inset) with R-square equal to 0.994.

Table 1. The effective thickness and surface roughness of B-NCD/quartz samples.

[B]/[C] ppm	Thickness nm	Surface roughness nm
0	291.79	17.04
2000	303.10	15.31
5000	272.48	10.76
7500	271.72	9.12
10000	247.10	8.50

The boron doping level influences film thickness, which constantly decreases from 303 nm to 247 nm going from 2000 to 10000 ppm samples. Moreover, this is accompanied by a decrease of the surface roughness, yielding half the roughness for 10000 ppm, compared to the undoped sample (see Table 1).

The RI of natural diamond is 2.42 at 550 nm wavelength [18] and B-NCD/silicon film are close to this value for 2000 and 5000 ppm doping. Furthermore, the change of n with increasing wavelength is nearly constant. The significant change of n for 10000 ppm can be explained by the transition of conductivity from semiconducting to semi-metallic/metallic [9]. The refractive index of the 10000 ppm doped sample was similar to that characteristic for metals especially in IR wavelength range [18]. This phenomena can be explained by the high boron content in the diamond film. Boron doping induces an acceptor level at 0.37 eV leading to light absorption for higher energies [19].

Also the change of refractive index, with R square 0.918, is not that linear as in the case of B-NCD/quartz and exhibits a high decrease of n for most doped sample. Hence, the higher values of n for diamond films on silicon substrates indicate a higher density material. The achieved values of RI agree with those found by others groups [20–22].

Figure 4 shows intrinsic transmittance for B-NCD calculated using the determined values of optical constants and thickness and measured transmittance for B-NCD/quartz. All diamond films (B-NCD/silicon) has good transmittance, above 65% for low doped films and 50% for 10000 ppm.

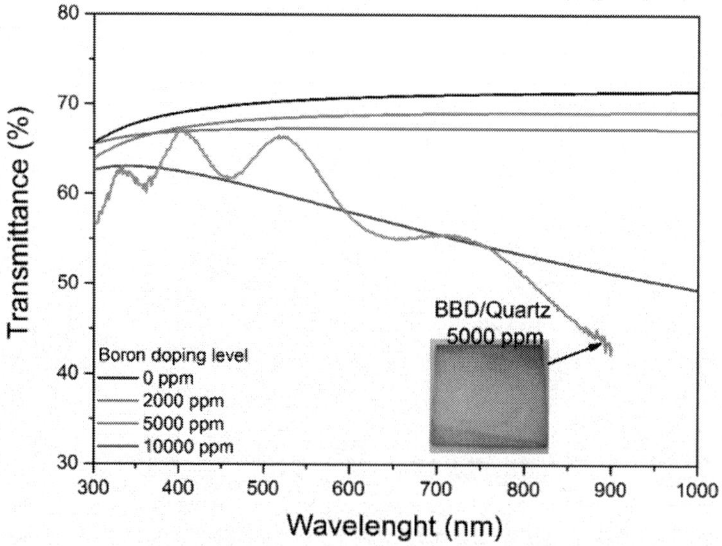

Figure 4. Calculated transmittance form spectroscopy ellipsometry of B-NCD samples on silicon and measured transmittance for B-NCD/quartz.

3.3. Electrochemical and spectroelectrochemical testing

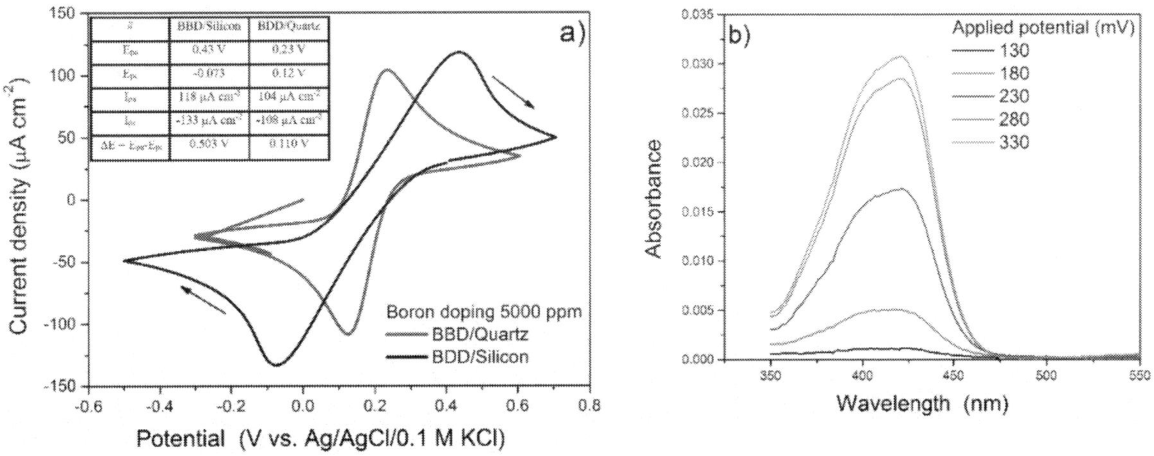

Figure 5. (a) Comparison of CV i-E curves for B-NCD/quartz and B-NCD/silicon for reversibility of redox reaction of 1 mM $Fe(CN)_6^{3-/4-}$ in 0.5 M Na_2SO_4 with a scan rate of 100 mV s^{-1}, inset table shows potential and current peaks and ΔE for each i-E curve; (b) recorded spectra of absorbance of 1 mM ferrocyanide ($Fe(CN)_6^{4-}$) with different applied potentials.

Cyclic voltammetry (CV) is a very useful and easy method for investigation electrochemical properties of the material interface being with contact with an electrolyte. Fig 5a shows the CV i-E curves for redox reaction of 1mM $Fe(CN)_6^{3-/4-}$ in 0.5 M Na_2SO_4 solution on B-NCD films doped at 5000 ppm level deposited on silicon and quartz substrates, with a scan rate of 100 mV/s. The $Fe(CN)_6^{3-/4-}$ redox system is sensitive to surface properties such as sp^2 bonded carbon or density of electronic states near the formal potential [23–25]. The redox system undergoes a reversible 1-electron transfer according to the following reaction:

$$\begin{array}{ccc} Fe(CN)_6^{3-} + e^- & \Leftrightarrow & Fe(CN)_6^{4-} \\ yellow & & colorless \end{array} \tag{2}$$

The recorded value of ΔE for B-NCD on silicon is wide and reaches 503 mV and is almost five times higher than those on quartz ΔE = 110 mV, at scan rate 100 mV/s. However, it should be noted that the film thickness strongly influences on electrochemical properties [26]. The B-NCD/quartz shows outstanding electrochemical features combined with good optical transparency and therefore is a promising diamond layer to develop new class of sensors. Moreover, Figure 5b shows a series of spectra recorded during electrochemical oxidation of 1 mM $Fe(CN)_6^{4-}$ in 0.5 M Na_2SO_4 at potential between 130 and 330 mV that cover the states of solution from the transparent form (130 mV) to fully oxidized form (yellow), which is completely achieved at 330 mV. The process is completely reversible and reproducible showing usefulness of the B-NCD film as optical diamond electrodes in spectroeletrochemistry.

4. Conclusion

In this work we have instigated optical and electrochemical response of optically transparent boron nanocrystalline diamond films on different substrates. The usage of a special sample holder allow achieving good nanocrystalline diamond film with remarkable electrochemical and optical properties. However, it must be emphasized that the working area, grain size and thickness of the B-NCD electrode have a strong influence on carrier transport in the system and optical properties. Application of electrically conductive B-NCD films allows for developing both optical and electrochemical sensing devices, and, what is more, allows for combining both capabilities in one sensing structure.

5. References

[1] Bajaj P, Akin D, Gupta A, Sherman D, Shi B, Auciello O and Bashir R 2007 Ultrananocrystalline diamond film as an optimal cell interface for biomedical applications *Biomed. Microdevices* **9** 787–94.

[2] Amaral M, Dias A G, Gomes P S, Lopes M A, Silva R F, Santos J D and Fernandes M H 2008 Nanocrystalline diamond: In vitro biocompatibility assessment by MG63 and human bone marrow cells cultures *J. Biomed. Mater. Res. A* **87A** 91–9.

[3] Mitura S, Mitura A, Niedzielski P and Couvrat P 1999 Nanocrystalline Diamond Coatings *Chaos Solitons Fractals* **10** 2165–76.

[4] Zak J K, Butler J E and Swain G M 2001 Diamond Optically Transparent Electrodes: Demonstration of Concept with Ferri/Ferrocyanide and Methyl Viologen *Anal. Chem.* **73** 908–14.

[5] Stotter J, Haymond S, Zak J K, Show Y, Cvackova Z and Swain G M 2003 Optically Transparent Diamond Electrodes for UV-Vis and IR spectroelectrochemistry *Interface* **12** 33–8.

[6] Stotter J, Zak J, Behler Z, Show Y and Swain G M 2002 Optical and electrochemical properties of optically transparent, boron-doped diamond thin films deposited on quartz *Anal. Chem.* **74** 5924–30.

[7] Muchnikov A B, Vikharev A L, Gorbachev A M and Radishev D B 2011 Comparative study of homoepitaxial single crystal diamond growth at continuous and pulsed mode of MPACVD reactor operation *Diam. Relat. Mater.* **20** 1225–8.

[8] Vikharev A L, Gorbachev A M, Muchnikov A B, Radishev D B, Kopelovich E A and Troitskiy M M 2012 Investigation of the optimized parameters of microwave plasma-assisted chemical vapour deposition reactor operation in a pulsed mode *J. Phys. Appl. Phys.* **45** 395202.

[9] Sobaszek M, Skowroński Ł, Bogdanowicz R, Siuzdak K, Cirocka A, Zięba P, Gnyba M, Naparty M, Gołuński Ł and Płotka P 2015 Optical and electrical properties of ultrathin transparent nanocrystalline boron-doped diamond electrodes *Opt. Mater.* **42** 24–34.

[10] Woollam J A 2010 *Guide to Using WVASE32®* (310 Madison Avenue, Suite 905, New York, NY 10017: Wextech Systems Inc.).

[11] Fujiwara H 2007 *Spectroscopic Ellipsometry: Principles and Applications* (Chichester, England ; Hoboken, NJ: Wiley-Blackwell).

[12] Ager III J W, Walukiewicz W, McCluskey M, Plano M A and Landstrass M I 1995 Fano interference of the Raman phonon in heavily boron-doped diamond films grown by chemical vapor deposition *Appl. Phys. Lett.* **66** 616–8.

[13] Ferrari A C and Robertson J 2001 Origin of the $1150 cm^-1 Raman mode in nanocrystalline diamond *Phys. Rev. B* **63** 121405.

[14] Kuzmany H, Pfeiffer R, Salk N and Günther B 2004 The mystery of the 1140 cm−1 Raman line in nanocrystalline diamond films *Carbon* **42** 911–7.

[15] Prawer S, Nugent K W, Jamieson D N, Orwa J O, Bursill L A and Peng J L 2000 The Raman spectrum of nanocrystalline diamond *Chem. Phys. Lett.* **332** 93–7.

[16] Robertson J 2002 Diamond-like amorphous carbon *Mater. Sci. Eng. R Rep.* **37** 129–281.

[17] Davis R F 1993 *Diamond films and coatings development, properties, and applications* (Park Ridge, N.J.: Noyes Pub.).

[18] Palik E D 1991 *Handbook of Optical Constants of Solids, Vol. 2* (Boston: Academic Press)

[19] Williams O A and Nesládek M 2006 Growth and properties of nanocrystalline diamond films *Phys. Status Solidi A* **203** 3375–86.

[20] Zimmer A, Williams O A, Haenen K and Terryn H 2008 Optical properties of heavily boron-doped nanocrystalline diamond films studied by spectroscopic ellipsometry *Appl. Phys. Lett.* **93** 131910–131910 – 3.
[21] Hu Z G, Prunici P, Hess P and Chen K H 2007 Optical properties of nanocrystalline diamond films from mid-infrared to ultraviolet using reflectometry and ellipsometry *J. Mater. Sci. Mater. Electron.* **18** 37–41.
[22] Gupta S, Weiner B R and Morell G 2001 Spectroscopic ellipsometry studies of nanocrystalline carbon thin films deposited by HFCVD *Diam. Relat. Mater.* **10** 1968–72.
[23] Granger M C and Swain G M 1999 The Influence of Surface Interactions on the Reversibility of Ferri/Ferrocyanide at Boron-Doped Diamond Thin-Film Electrodes *J. Electrochem. Soc.* **146** 4551–8.
[24] Kneten K R and McCreery R L 1992 Effects of redox system structure on electron-transfer kinetics at ordered graphite and glassy carbon electrodes *Anal. Chem.* **64** 2518–24.
[25] Chen P, Fryling M A and McCreery R L 1995 Electron Transfer Kinetics at Modified Carbon Electrode Surfaces: The Role of Specific Surface Sites *Anal. Chem.* **67** 3115–22.
[26] Wang S, Swope V M, Butler J E, Feygelson T and Swain G M 2009 The structural and electrochemical properties of boron-doped nanocrystalline diamond thin-film electrodes grown from Ar-rich and H2-rich source gases *Diam. Relat. Mater.* **18** 669–77.

Acknowledgment

This study was supported by the Polish National Science Centre (NCN) under Grant No. 2011/03/D/ST7/03541. DS funds of the Faculty of Electronics, Telecommunications and Informatics of the Gdansk University of Technology under No. MNiD/2014/2015/D are also acknowledged.
Financial support for the instrumentation was obtained from The European Regional Development Fund(Stage 2 of the Regional Centre of Innovativeness) and the Polish Ministry of Science and Higher Education.
The authors gratefully acknowledge financial support from the Polish National Science Centre (NCN) under Grant No. 2012/07/D/ST5/02269.

39th International Microelectronics and Packaging IMAPS Poland 2015 Conference IOP Publishing
IOP Conf. Series: Materials Science and Engineering **104** (2015) 012025 doi:10.1088/1757-899X/104/1/012025

Low temperature growth of diamond films on optical fibers using Linear Antenna CVD system

M. Ficek[1], S. Drijkoningen[2], J. Karczewski[3], R. Bogdanowicz[1], K. Haenen[2,4]

[1]Department of Metrology and Optoelectronics, Faculty of Electronics, Telecommunications and Informatics, Gdansk University of Technology, 11/12 G. Narutowicza St., 80-233 Gdansk, Poland

[2]Institute for Materials Research (IMO), Hasselt University, Wetenschapspark 1, B-3590 Diepenbeek, Belgium

[3]Faculty of Applied Physics and Mathematics, Gdansk University of Technology, 11/12 G. Narutowicza St., 80-233 Gdansk, Poland

[4]IMOMEC, IMEC vzw, Wetenschapspark 1, B-3590 Diepenbeek, Belgium

E-mail: mateuszficek@gmail.com

Abstract. It is not trivial to achieve a good quality diamond-coated fibre interface due to a large difference in the properties and composition of the diamond films (or use coating even) and the optical fibre material, i.e. fused silica. One of the biggest problems is the high temperature during the deposition which influences the optical fibre or optical fibre sensor structure (e.g. long-period gratings (LPG)). The greatest advantage of a linear antenna microwave plasma enhanced chemical vapor deposition system (LA MW CVD) is the fact that it allows to grow the diamond layers at low temperature (below 300°C) [1].

High quality nanocrystalline diamond (NCD) thin films with thicknesses ranging from 70 nm to 150 nm, were deposited on silicon, glass and optical fibre substrates [2]. Substrates pretreatment by dip-coating and spin coating process with a dispersion consisting of detonation nanodiamond (DND) in dimethyl sulfoxide (DMSO) with polyvinyl alcohol (PVA) has been applied. During the deposition process the continuous mode of operation of the LA MW CVD system was used, which produces a continuous wave at a maximum power of 1.9 kW (in each antenna). Diamond films on optical fibres were obtained at temperatures below 350°C, providing a clear improvement of results compared to our earlier work [3].

The samples were characterized by scanning electron microscopy (SEM) imaging to investigate the morphology of the nanocrystalline diamond films. The film growth rate, film thickness, and optical properties in the VIS-NIR range, i.e. refractive index and extinction coefficient will be discussed based on measurements on reference quartz plates by using spectroscopic ellipsometry (SE).

1. Introduction

Optical fibres have been extensively used for development at number of sensing devices [4]. The advantages of fibre-based sensors include the resistance to high temperature, immunity to high electromagnetic field and electrical inertness [5]. Optical fibres designed and fabricated mainly for the applications in telecommunication systems are typically covered with polymer coatings which protect the fibre against mechanical damages and harsh environmental conditions. However, the coating does not protect the fibre well against some acids. Diamond is a material with wide range of extraordinary properties. Transparent over a broad spectrum region, highly chemically and abrasively resistant, nontoxic and biocompatible diamond films are often used as protection coatings in various fields [6–8]. On other hand, low diamond adhesion to substrates (in particular glass or metals) or damage of substrates during chemical vapor deposition (CVD) are the limiting factors for its wider commercial use [9–11]. The use of LA MW CVD has been recently developed [1,12,13] for growth of NCD layers. In such systems, deposition of high quality diamond at lower temperatures and also on larger areas are more feasible than in ordinary systems [1]. Preparing an optical fibres for sensing system is very necessary. Substrate temperature have the crucial impact influence on diamond synthesis process and kinetics of pyrolysis reaction on the growth surface [14,15]. In our study, we proposed to use large areas and low temperature LA MW CVD system. Diamond films on optical fibres were obtained at temperatures below 350°C and optical fibre cover diamond film length more than 20 cm, providing a clear improvement of results compared to our earlier work [3].

2. Experimental

For the growth of NCD films a linear antenna microwave plasma enhanced chemical vapor deposition system (LA MW CVD) system was used. The growth chamber allows the deposition on substrates up to the size of 300 mm×300 mm. Microwaves are delivered into the growth chamber by four pairs of coaxial antennas enclosed in quartz envelopes and powered from both ends. During the deposition process the continuous mode of operation of the LA MW CVD system was used, which produces a continuous wave at a maximum power of 1.9 kW (in each antenna). The base pressure in the chamber was 10^{-4} mbar. Then the chamber was filled with a mixture of hydrogen, methane and carbon dioxide. Pressure in the chamber was kept at 0.1 mbar with the total flow rate of gases reaching 150 sccm at methane molar ratio of 2.5% and carbon dioxide molar ratio of 6%. During the process the stage was heated up to 350°C by induction heater and controlled by thermocouple. The seeding procedure took 7 hour. In order to investigate we prepared a single mode optical fibre (cleaved Corning SMF28, cladding diameter of 125 mm, approx. 30 cm in length) whose polymer coating has been mechanically stripped, and a set of reference samples, i.e., fused silica glass and p-100 Si wafers. The growth process included a double immersion of the optical fibre in the suspension, each time for 1 min. In the second step, the fibre was turned around and its other end was dipped. Moreover, the reference quartz slides were seeded in the same seeding media by means of spin-coating. The suspension was prepared in a two-step procedure. Firstly, 1 g of solid polyvinyl alcohol (average molar mass of 18,000 g mol^{-1}) was suspended in 99 g of dimethyl sulfoxide at a temperature of 80°C, resulting in a 1% w/w solution. Secondly, after the suspension had reached room temperature, 100 g of nanodiamond suspension (dimethyl sulfoxide–nanodiamond 0.5% w/w) was carefully added [16].

3. Results and discussion

3.1. Surface morphology

Scanning Electron Microscope FEI Quanta FEG 250 Scanning Electron Microscope (SEM) using 10kV beam accelerating voltage with SE-ETD detector (secondary electron - Everhart-Thornley detector) working in high vacuum mode (pressure 10^{-4} Pa) was used to observed the structure of the surface of optical fibre.

The morphology studies were performed with optical microscopy using combination with a 20x objective magnification and numerical aperture 0.4.

IOP Conf. Series: Materials Science and Engineering **104** (2015) 012025 doi:10.1088/1757-899X/104/1/012025

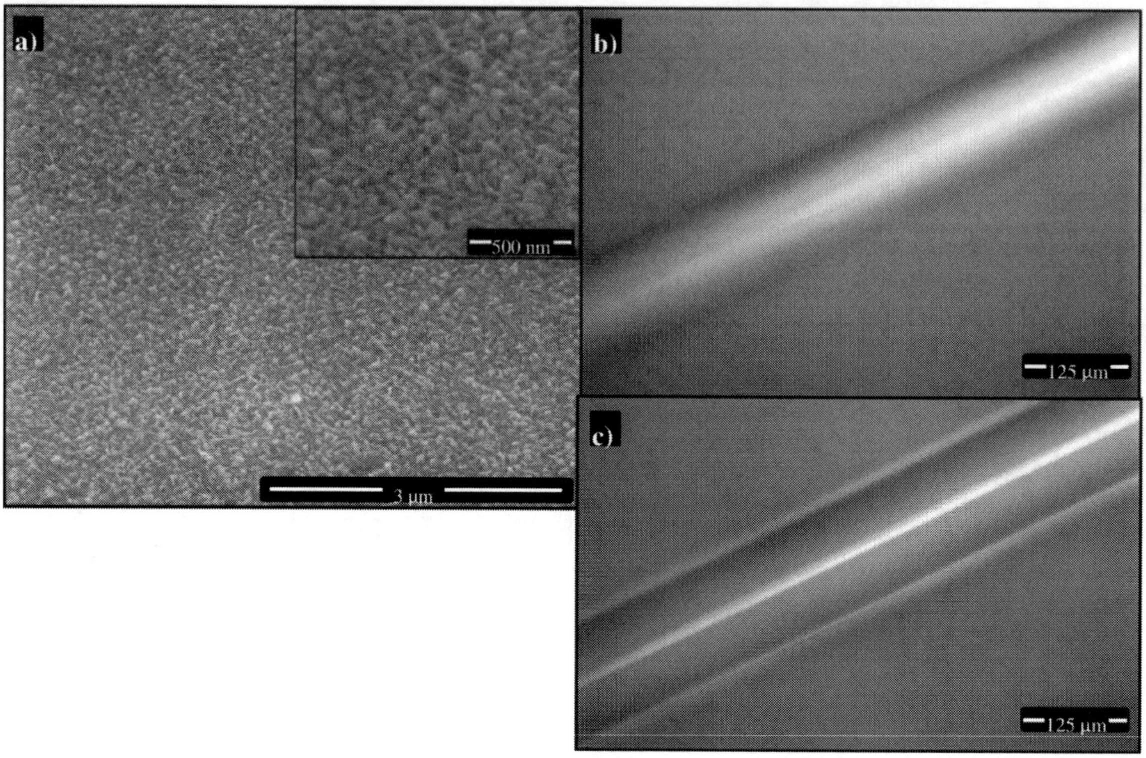

Figure 1. SEM image (a) and optical microscope image (b), (c) of diamond films on fused silica optical fibre.

Figure 1 shows the optical microscope images and SEM images of optical fibre. Use the LA MW CVD system to cover optical fibre results in continuous, homogeneous thin diamond films. Moreover, in sample it can be observed that the size as well as structure of optical fibre was not defected. The values of mean grain diameter ranged from 50 to 100 nm. Figure 1c shows the optical fibre with light guided in core. It could be concluded the core is not damaged during the process. The thin diamond films look similar in all 30cm length. The cladding diameter of optical fibre is 125 nm with core diameter of 8.5 nm. In our study, the edges of crystallites are clearly visible and the film is uniform.

3.2. Optical properties

Spectroscopic Ellipsometry (SE) investigations were carried out with a phase-modulated ellipsometer Jobin-Yvon UVISEL (HORIBA Jobin-Yvon Inc., Edison, USA). The investigated wavelength range was 260 to 830 nm. The experiments were performed at room temperature using an angle of incidence fixed at 70°. Ellipsometric fitting was based on a four-phase optical model (air/surface roughness film (SRL)/ diamond/ silica glass). The dispersion of silica glass was taken from the database [17]. The diamond film has been here assumed to be an isotropic, homogeneous material and its dispersion was fitted to the Tauc-Lorentz oscillator (TL) model. This model has been used recently for amorphous semiconductors by Logothetidis et al. [18,19]. Such materials exhibit a peculiarity due to the presence of two separated contributions of inter-band electronic transition related to sp^2 and sp^3 bonded carbon [20]. The parameters of the TL model were fitted for each of the analysed films. Finally, the assumed optical model was fitted to the experimental data using the nonlinear Levenberg-Marquardt regression method for mean-square error minimization (MSE) [21]. As a result of SE analysis, the thickness and optical constants, i.e. refractive index n(λ) and extinction coefficient k(λ) were obtained.

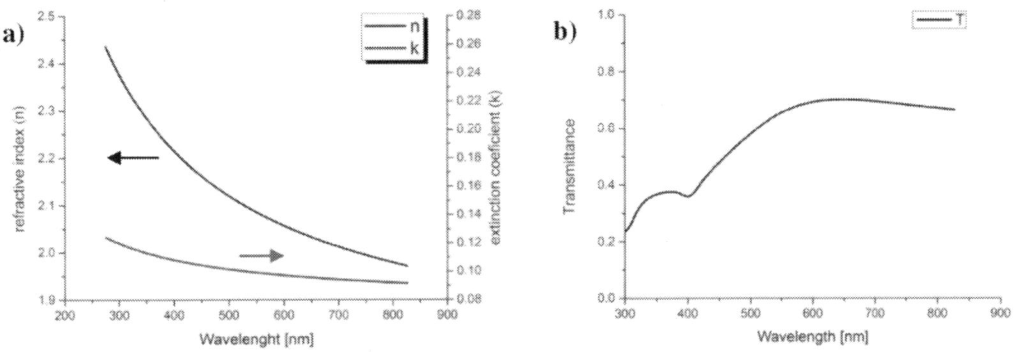

Figure 2. Dispersion of *n* and *k* (a) and transmittance (b) of diamond film measured at quartz slide.

Due to the difficulty of taking ellipsometric measurements directly on the curved surface of optical fibres, the measurements were performed on the reference diamond films deposited on quartz slides during the same deposition process. The applied fitting procedure resulted in the accurate values of film thickness, film roughness as well as dispersion of refractive index and extinction coefficient. Changes of the refractive index and extinction coefficient of the NCD films are presented in figure 2a. The obtained values of refractive index were high, ranging from 2.45 to 2.0 for the NCD films. Hu et al. [22] reported the values of n between 2.16 and 2.08 for the NCD films. The lower values of n (compared to single crystalline diamond) indicate the lower physical density of the films [23]. The k values, which correspond to optical absorption in the deposited films, were below 0.1 at 550 nm, indicating low light absorption over the entire length of the optical fibre. The transmittance in NCD films grown on quartz glass substrates were shown in figure 2b.It could be observed that the transmittance values are below 40% in the range of 300-400 nm. They increase to values above 60% over the 550 nm. The transmittance values below 60% at 550 nm. The length was optimized taking into consideration the application of NCD in the optical fibre grating devices [24]. It should be mentioned that the observed effects can be partially explained by the denser microcrystalline structure of NCD films deposited directly on quartz slides compared to that of the NCD films deposited directly on the fibres. The results obtained in our study correspond well with the SEM images shown in figure 1, which has been discussed above.

4. Conclusion

We have demonstrated that the conformal NCD films deposited on the fused silica fibres. Moreover, the growth parameters of the proposed process have been optimized, resulting in a low-temperature growth of the NCD films on 30 cm-long sections of the optical fibres. The samples exhibited relatively low deviations of refractive index (2.1 ± 0.05) and extinction coefficient (0.1 ± 0.02) along the length of 30 cm, as estimated at a wavelength of 550 nm. Diamond films on optical fibres were obtained at temperatures below 350°C.

5. References

[1] Kromka A, Babchenko O, Izak T, Hruska K and Rezek B 2012 Linear antenna microwave plasma CVD deposition of diamond films over large areas *Vacuum* **86** 776–9

[2] Drijkoningen S, Nesladek M, Pobedinskas P and Haenen K 2015 Ultralow temperature diamond growth with high frequency pulsed microwave linear antenna technology *Nanodiamond* New Diamond and Nano Carbons 2015 (Shizuoka GRANSHIP, Japan)

[3] Bogdanowicz R, Sobaszek M, Ryl J, Gnyba M, Ficek M, Gołuński Ł, Bock W J, Śmietana M and Darowicki K 2015 Improved surface coverage of an optical fibre with nanocrystalline diamond by the application of dip-coating seeding *Diam. Relat. Mater.* **55** 52–63

[4] Malgorzata Jedrzejewska-Szczerska P W 2014 ALD thin ZnO layer as an active medium in a fiber-optic Fabry-Perot interferometer *Sens. Actuators Phys.*

[5] Culshaw B and Kersey A 2008 Fiber-Optic Sensing: A Historical Perspective *J. Light. Technol.* **26** 1064–78

[6] Bajaj P, Akin D, Gupta A, Sherman D, Shi B, Auciello O and Bashir R 2007 Ultrananocrystalline diamond film as an optimal cell interface for biomedical applications *Biomed. Microdevices* **9** 787–94

[7] A. V. Sukhadolau E V I 2005 Thermal conductivity of CVD diamond at elevated temperatures *Diam. Relat. Mater.* **14** 589–93

[8] Panizza M and Cerisola G 2005 Application of diamond electrodes to electrochemical processes *Electrochimica Acta* **51** 191–9

[9] Prelas M A, Popovici G and Bigelow L K 1997 *Handbook of Industrial Diamonds and Diamond Films* (CRC Press)

[10] Carbon D-L 2001 Tribology of Diamond, Diamond-Like Carbon, and Related Films

[11] Amaral M, Gomes P S, Lopes M A, Santos J D, Silva R F, Fernandes M H 2008, Nanocrystalline Diamond as a Coating for Joint Implants: Cytotoxicity and Biocompatibility Assessment *J. Nanomater*, **2008** e894352

[12] Potocký Š, Babchenko O, Hruška K and Kromka A 2012 Linear antenna microwave plasma CVD diamond deposition at the edge of no-growth region of C–H–O ternary diagram *Phys. Status Solidi B* **249** 2612–5

[13] Varga M, Remes Z, Babchenko O and Kromka A 2012 Optical study of defects in nano-diamond films grown in linear antenna microwave plasma CVD from H2/CH4/CO2 gas mixture *Phys. Status Solidi B* **249** 2635–9

[14] Maciej Kraszewski 2013 Laser Reflectance Interferometry System with a 405 Nm Laser Diode for in Situ Measurements of CVD Diamond Thickness *Metrol. Meas. Syst.* **20**

[15] Davis R F 1993 *Diamond films and coatings: development, properties, and applications* (Park Ridge, N.J.: Noyes Pub.)

[16] Mateusz Ficek 2015 Nanocrystalline CVD Diamond Coatings on Fused Silica Optical Fibres: Optical Properties Study *Acta Phys. Pol. Ser. A* **127**

[17] Palik E D 1998 *Handbook of Optical Constants of Solids* (Academic Press)

[18] Logothetidis S, Charitidis C, Gioti M, Panayiotatos Y, Handrea M and Kautek W 2000 Comprehensive study on the properties of multilayered amorphous carbon films *Diam. Relat. Mater.* **9** 756–60

[19] Gioti M, Papadimitriou D and Logothetidis S 2000 Optical properties and new vibrational modes in carbon films *Diam. Relat. Mater.* **9** 741–5

[20] G. E. Jellison Jr V I M 2000 Characterization of thin-film amorphous semiconductors using spectroscopic ellipsometry *Thin Solid Films* **377** 68–73

[21] Tompkins H G and Irene E A 2005 *Handbook of ellipsometry* (Norwich, NY; Heidelberg, Germany: William Andrew Pub. ; Springer)

[22] Hu Z G, Prunici P, Hess P and Chen K H 2007 Optical properties of nanocrystalline diamond films from mid-infrared to ultraviolet using reflectometry and ellipsometry *J. Mater. Sci. Mater. Electron.* **18** 37–41

[23] Jr G E J and Modine F A 1996 Parameterization of the optical functions of amorphous materials in the interband region *Appl. Phys. Lett.* **69** 371–3

[24] Robert Bogdanowicz 2015 Improved surface coverage of an optical fibre with nanocrystalline diamond by the application of dip-coating seeding *Diam. Relat. Mater.* **55** 52–63

Acknowledgements

This work was supported by the Polish National Science Centre (NCN) under the Grant No. 2011/03/D/ST7/0354. The DS funds of Faculty of the Electronics, Telecommunications and Informatics of the Gdansk University of Technology are also acknowledge.

IOP Conf. Series: Materials Science and Engineering **104** (2015) 012026 doi:10.1088/1757-899X/104/1/012026

In-situ and ex-situ resistance measurements of polypyrrole film using double-band electrode

K Cysewska, P Jasiński

Gdańsk University of Technology, Faculty of Electronics, Telecommunications and Informatics, ul. Narutowicza 11/12, 80-233 Gdańsk, Poland

E-mail: pijas@eti.pg.gda.pl

Abstract. Many in-situ techniques are performed in order to determine the resistance of conducting polymer film. However, the resistance measured in solution can be the combination of polymer resistance and that of other components, such as resistance of supporting electrolyte. Therefore, in this work, the influence of the solution on the resistance of polypyrrole (PPy) film has been studied. PPy film was electrchemically synthesized onto the iron double-band electrode in a one step process from an aqueous solution of 0.1 M pyrrole and 0.1 M sodium salicylate. Resistance determination of PPy film was based on impedance spectroscopy measurements and was performed in-situ, ex-situ and in a function of electrode potential. Based on these measurements electrical equivalent circuit of PPy coated Fe electrode in contact with solution or with the air have been studied. It was noticed that the concentration of electrolyte in measuring solution did not influence on the polymer resistance measured by in-situ technique. It was also noticed that measurements of polymer resistance conducted in-situ differ from that obtained in ex-situ experiments. The differences were related with the diffusion reactions occurring between the polymer film and air/solution.

1. Introduction

Nowadays, conducting polymers have attracted attention of many researchers from different field of science [1-2]. Because of the presence in their backbone conjugated double bond they possess electronic properties similar to metals [3]. One of the most interesting among conducting polymers is polypyrrole (PPy) [3-5]. Polypyrrole exhibits very good environmental stability [4], good adhesion to the substrate [3], excellent mechanical and thermal properties and higher conductivity compared to many other conducting polymers [4]. Because of these properties PPy can be used in various applications including anti-corrosive coatings [6], biosensors [7], electrochromic windows [4], organic electronics [8] and gas sensor [9]. Polypyrrole behavior depends strongly on its oxidations state. Reduced PPy films behave as electronic insulators, however oxidation of the polymer backbone in a presence of some dopants results in formation of charge carriers, which confer electronic conductivity to the polymer [10]. Therefore, polymer may be switched between electrically insulating and conductive state in response to changes in its oxidation state [11]. Many methods have been performed in order to determine the resistance (or conductivity) of conducting polymer films, also during their doping and dedoping processes [12-18]. Two-probe in-situ measurement was performed in order to determine the conductivity of polypyrrole film [18]. It was noticed that the lower was the current density applied during electropolymerization, the higher was the conductivity of polymer film. In-situ

IOP Conf. Series: Materials Science and Engineering **104** (2015) 012026 doi:10.1088/1757-899X/104/1/012026

resistance measurement in a function of electrode potential has been performed for PPy in 0.5 M H_2SO_4 [12]. Here, resistance measurement was controlled by potentiostat and insulated Pt wire probe which was positioned close to the surface of the Pt foil where the polymer was deposited. The determination of the resistance was based on applied current and voltage drop occurring across the polymer film [12]. Based on these measurements, it was noticed that studied polymer was highly resistive at -0.2 V, however anodic polymerization results in decrease of its resistance [12]. Resistance in a function of doping potential has been also determined by using the resistometer in 1 M NaCl supporting electrolyte [13]. Here, the resistivity of the film decreased with increasing electrode potential. The thicker was the synthesized polymer, the more conductive it was at positive potentials [13]. X. Du et al. [14] used the platinum double-band electrode in order to determine resistance of differently synthesized PEDOT film. The conductivity of the polymer film was determined by both: the rates of EDOT polymerization and PEDOT overoxidation [14]. Also standard ex-situ methods were performed in order to determine the resistance/conductivity of polymeric film [15]. For example, four-probe measurement has been conducted in order to determine the conductivity of nanotubular polyaniline (PANI) [15]. The results showed that the resistivity of PANI depended on the concentration of phosphoric acid used during the polymer synthesis [15].

It was noticed that the in-situ resistance measurement of polymer usually differs from a standard ex-situ methods [16]. It was noted that the polymer measured in-situ can be the combination of polymer resistance and other components, such as resistance of solution, contact resistance, double-layer capacitance or charge transfer resistance between the polymer film and solution [13, 16, 18]. Xue et al. [16] performed four-probe in-situ resistance measurements in order to determine the influence of the contact resistance on electrical properties of polyaniline film. It was noticed that appropriate construction of working electrode used in this measurement, allowed avoiding the influence of this component. In the work of the Kankare and Kupila [18], solution resistance, charge transfer resistance and double layer capacitance were taken into account during the determination of the polymer conductivity. In order to decrease the influence of the solution resistance, the concentration of the supporting electrolyte has been lowered. Also, to minimize the influence of the capacitances, the measuring frequency had to be as low as possible. In some reports [17], the minimization of the capacitance was obtained by applying dc excitation. However, it resulted in the presence of the omnipresent 1/f noise which prevented the accurate measurement of small response signal [18].

Many authors [12-17] determine the resistance or conductivity of polymer film by performed different in-situ techniques. However, the resistance measured in-situ is the combination of polymer resistance and that of the supporting electrolyte [13]. In order to avoid this influence, frequently the conductivity (concentration) of electrolyte is lowered [18]. However, it is not clear, how much the resistance of electrolyte should be lowered in order to get the "true" value of polymer film.

Therefore, the aim of the work is to examine the influence of the solution on the resistance of polypyrrole film. In-situ and ex-situ two-probe methods for resistance measurement have been performed. Presented here ex-situ method does not require detachment of the polymer film from the electrode surface, which avoids the possible damages of the polymer. Determination of the resistance was based on the impedance spectroscopy measurements and was also determined in a function of electrode potential.

2. Experimental

Polypyrrole films were electrochemically synthesized on iron (Chempur Feinchemikalien und Forschungsbedarf GmbH) in a one step process from aqueous solution of pyrrole monomer ($\geq$ 98 % Sigma Aldrich) and sodium salicylate (EMSURE).

Different setup configurations were used depending on the type of the measurement. Electropolymerization processes and linear scan voltammetry were performed in a one-compartment cell with a three electrode system controlled by potentiostat VersaSTAT 4. An Ag/AgCl (R-10 Hydromet) electrode was used as the reference electrode and a platinum sheet (10 mm x 10 mm) as a counter electrode. Impedance spectroscopy and resistance measurements were carried out in a 2-

electrode system controlled by VersaSTAT 4 and/or Gamry Instrument (Interface 1000). The working electrode for all measurements was an iron double-band which was prepared based on description provided by Kankare and Kupila [18]. Electrode consists of two separated iron sheets (each with dimension of 18 mm x 1.5 mm) with a gap between the sheets of 80 μm. All surrounded by epoxy resin with leads attached to each of the sheet. The construction of working electrode allowed for determination of polymer resistance deposited between the iron sheets. It also allowed obtaining reproducible electrode surface after grinding and polishing processes.

The Fe double-band electrode was mechanically polished with abrasive papers: 600 and 1200 grade, respectively and rinsed with acetone before each experiment. After each measurement, the electrode was rinsed with deionized water and dried.

The coating process was conducted potentiostatically at potential of 1.2 V from aqueous solution containing 0.1 M pyrrole and 0.1 M sodium salicylate with deposition time limited by a electropolymerization charge of 3.1 $C \cdot cm^{-2}$. Selected electropolymerization charge provided the connection between the iron sheets which allowed for determination of polymer resistance.

Impedance spectroscopy measurements were performed in the air or in aqueous solution of 0.1, 0.01 or 0.001 M sodium salicylate between the frequencies of 10 kHz to 0.01 Hz against open circuit potential at an amplitude of 5 mV.

Two different methods (in-situ and ex-situ two-probe) were performed in order to determine the changes of polymer resistance as a function of electrode potential. Determination of resistance in both methods was based on impedance spectroscopy (IS) measurement, however the re-oxidation (doping) of polymer was performed by linear sweep voltammetry (-1.2 V-1.2 V vs. Ag/AgCl) in aqueous solution of 0.1 M or 0.001 M sodium salicylate with the scan rate of 10 $mV \cdot s^{-1}$. The impedance measurements were conducted at frequency of 0.01 Hz against the open circuit potential and at an amplitude of 5 mV. At this frequency the phase of impedance of 0° was obtained, and therefore, the modulus of impedance was assumed to be resistance of the polymer film. The in-situ resistance measurements were conducted simultaneously with the doping process and were controlled by two potentiostats. The first (VersaSTAT 4) was used for linear scan voltammetry, while the second (Gamry Interface 1000) was used as impedance recorder. Ex-situ two probe measurements were performed in the air and were controlled by potentiostat (VersaSTAT 4). Here, re-oxidation processes and resistance determination were conducted alternately. Polymer film after each doping stage and before resistance measurement was rinsed with distilled water and dried (for approximately 5 minutes in the warm air stream). The resistance changes of PPy in a function of time measured in the air were also studied. Here, resistance determination was based on the IS measurements performed at frequency of 1 Hz. Applied here frequency allowed determining the resistance of PPy film from impedance spectra (measured in the air). Each experiment has been performed three times and the averages of these are presented.

3. Results and discussion

Impedance spectroscopy measurements have been done for Fe and PPy coated Fe electrode. The measurements were performed in the air (ex-situ) or in aqueous solution containing 0.1, 0.01 or 0.001 M sodium salicylate (in-situ). For all measurements polypyrrole was synthesized at the same condition.

IOP Conf. Series: Materials Science and Engineering **104** (2015) 012026 doi:10.1088/1757-899X/104/1/012026

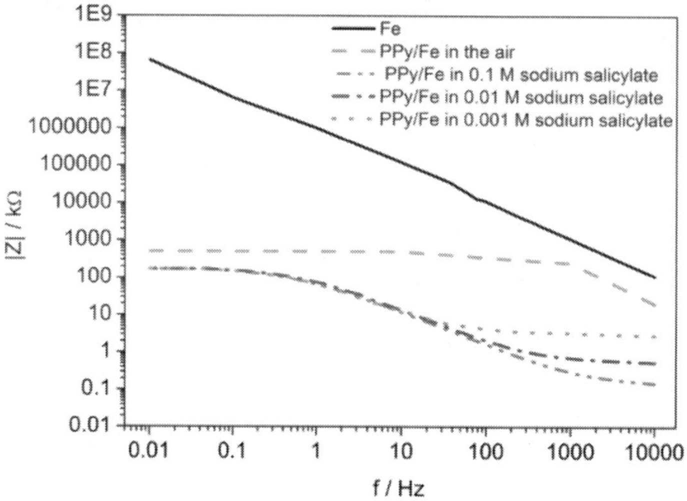

Figure 1. Impedance spectrum of Fe and PPy coated Fe measured in the air or in aqueous solution of 0.1, 0.01 and 0.001 M sodium salicylate.

The results from these measurements are presented in figure 1. First of all, it can be seen that coating the metal with the conducting polymer (PPy) resulted in decrease in impedance. The drop of impedance occurred sudden at the time when the Fe electrodes became connected. In order to understand this process, stages of polymer growth during the electropolymerization process should be studied (figure 2). During the electropolymerization polypyrrole gradually is deposited onto Fe electrodes.

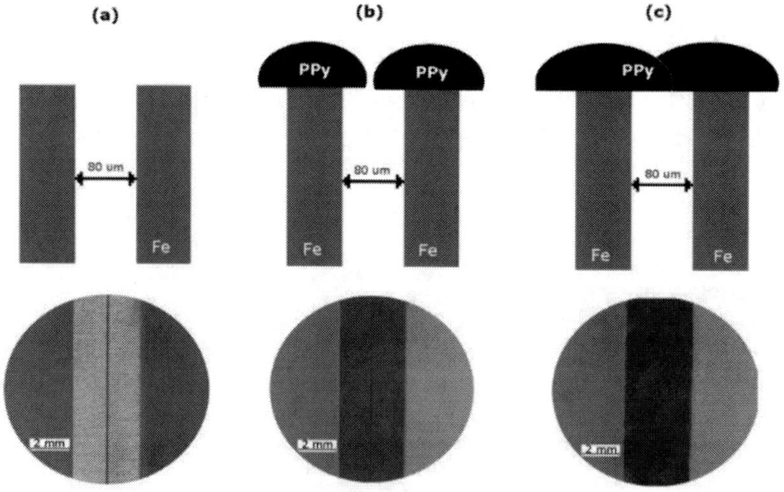

Figure 2. Schematic and macroscopic picture of PPy growth during (b, c) electropolymerization on Fe double-band electrode (a).

At the beginning of the electropolymerization PPy is deposited on both iron electrodes (figure 2a) in the both longitudinal and transverse directions (figure 2b). Impedance spectrum of sample at this

IOP Conf. Series: Materials Science and Engineering **104** (2015) 012026 doi:10.1088/1757-899X/104/1/012026

polymerization stage exhibits capacitance behaviour [6]. However, progressive polymerization results in thicker polymer film which at some time leads to connection of PPy films synthesized at each of iron electrodes (figure 2c). The moment at which two Fe electrodes became connected is seen as rapid decrease of modulus of impedance. Therefore, when the Fe electrodes became connected with the polymer, it is possible to measure its electrical properties.

Different impedance spectrum of PPy coated iron electrode was obtained (figure 1) depending on conditions at which experiments were performed. The bare iron electrode measured ex-situ exhibited pure capacitance behaviour. Polypyrrole coated Fe electrode measured ex-situ exhibited almost pure resistance behaviour which was seen at almost all applied frequency range. Different behaviour was observed when the same sample was measured in different concentrated sodium salicylate aqueous solution. Here, three regions, represented by three components in impedance spectrum can be observed (see table 1).

Table 1. Different impedance behaviour and corresponding to it component presence during in-situ impedance spectroscopy measurements of PPy coated Fe.

Frequency range (Hz)	Impedance behaviour/components
10k - 100	Resistance/Solution resistance
100 – 0.1	Capacitance/Double-layer capacity
0.1 – 0.01	Resistance/Resistance of polymer

Hence, two-probe impedance measurement performed in the air allowed determining resistance of deposited polymer film almost in all frequency range (1 kHz - 0.01 Hz). However, in-situ resistance measurement is a combination of polymer resistance and other components, such as double-layer capacitance and resistance of solution. Figure 3 and 4 present the approximate equivalent electrical circuit of PPy coated Fe-double band electrode in contact with air or with solution, respectively. When the resistance measurements are performed in the air, only the "true" value of polymer resistance (R_f) is measured (see figure 3). In this case, at almost all frequency range the impedance spectrum exhibited pure resistance behaviour (figure 1).

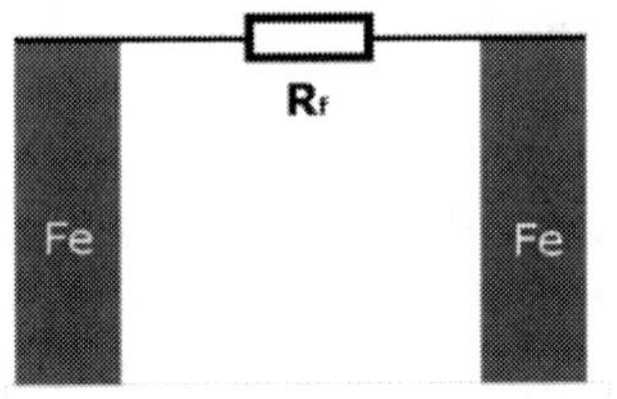

Figure 3. Electrical equivalent circuit of PPy coated Fe-double band electrode in contact with air. R_f – the resistance of polymer film.

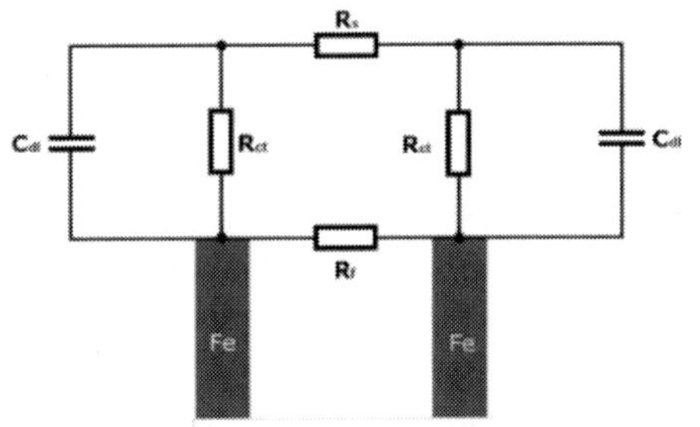

Figure 4. Electrical equivalent circuit of PPy coated Fe-double band electrode in contact with solution. R_f – the resistance of polymer film, R_s – the solution resistance, R_{ct} – the charge transfer resistance between the polymer film and solution, C_{dl} – the double layer capacitance between the polymer film and solution.

The influence of other components such as, double layer capacitance, solution or charge transfer resistance had not impact on this measurement. It is different, when the measurement is performed in solution (see figure 4). Therefore, in order to get the "true" value of polymer resistance, there is a need to eliminate some components, such as: double-layer capacitance (C_{dl}) between the polymer film and solution, resistance of solution (R_s) and charge transfer resistance (R_{ct}) between the polymer film and solution [18]. Based on the impedance measurement of PPy/Fe electrode performed in solution (figure 1) it can be noticed that double layer capacitance has an influence in a frequency range between approximately 1k – 0.1 Hz. However, lower frequencies allow eliminating this component [18]. Other important component which influence on the measurement of polymer resistance is solution resistance (R_s). In order to avoid this influence, frequently the concentration of electrolyte is lowered [18]. However, based on conducted here IS measurements (figure 1) it can be noted that polymer resistance measured in-situ does not depend on the concentration of electrolyte. It can be seen that the resistance of polymer film mainly depends on the frequency at which IS measurements are performed. It also depends on the surroundings conditions in which the experiments are conducted (in the air or in solution). Preliminary studies showed that the resistance of PPy measured in the air decreased with time which can be seen in figure 5. Therefore, the differences in polymer resistance measured in-situ and ex-situ probably can be related with the diffusion reactions, e. g. with the adsorption and desorption of oxygen molecules from the polymer film [23]. The further studies are conducted in this field.

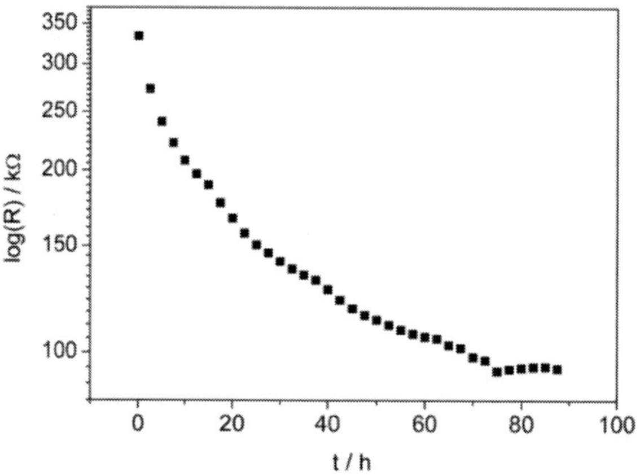

Figure 5. Resistance of PPy measured in the air. The results are shown for every 50 data point.

Other component, which can be presence during the in-situ resistance measurement, is a charge transfer resistance. This component is related with the process of electron transfer from one phase (e. g. polymer) to another (e. g. solution). Therefore, it depends on electrochemical reaction which occurred between two phases during the measurement. For in-situ impedance spectroscopy measurement of PPy coated Fe electrode this component can be neglected [18]. However, when the polymer during the in-situ resistance measurement is additionally subjected to different potentials, the charge transfer resistance can influence on the results. It is related with different reactions (e. g. oxidation/reduction of deposited polymer film) which are induced by applied potentials [20, 21].
Therefore, the resistance behavior of PPy depending on the applied electrode potential has also been studied and the preliminary results are presented in figure 6 and 7. In the first measurement (figure 6) PPy was gradually re-oxidized (doped) in 0.1 sodium salicylate aqueous solution and the resistance measurements was conducted in the air or in aqueous solution of 0.1 M sodium salicylate. In the second experiment (figure 7) re-oxidation of the polymer was performed in aqueous solution of 0.001

M sodium salicylate, while the resistance measurements were conducted in the air or in aqueous solution of 0.001 M sodium salicylate.

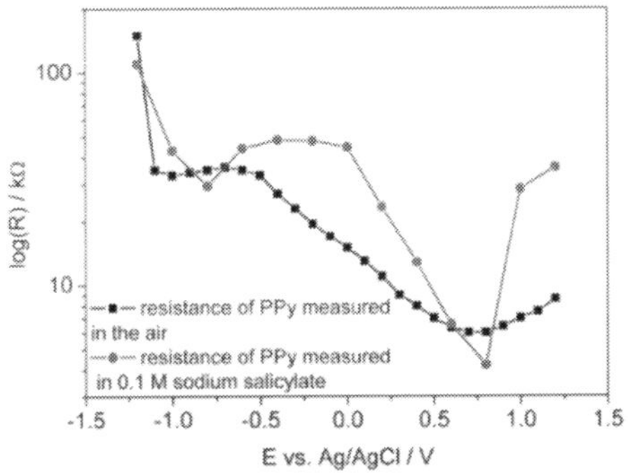

Figure 6. Resistance of PPy as a function of electrode potential measured in the air (■) or in aqueous solution of 0.1 M sodium salicylate (●). PPy was re-oxidized in 0.1 M sodium salicylate aqueous solution by sweeping the potential from -1.2 V to 1.2 V with the scan rate of 10 $mV\cdot s^{-1}$.

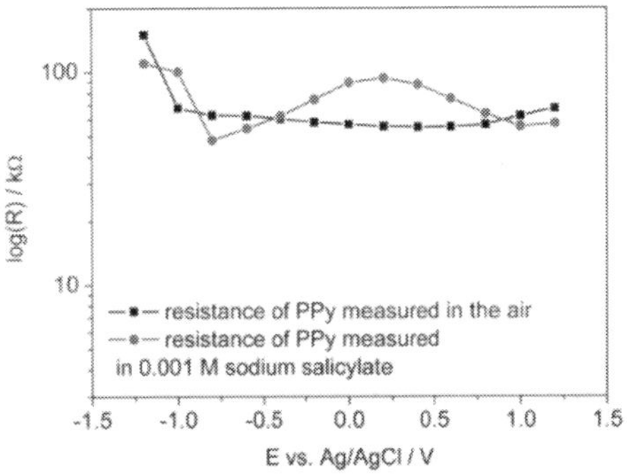

Figure 7. Resistance of PPy as a function of electrode potential measured in the air (■) or in aqueous solution of 0.001 M sodium salicylate (●). PPy was re-oxidized in 0.001 M sodium salicylate aqueous solution by sweeping the potential from -1.2 V to 1.2 V with the scan rate of 10 $mV\cdot s^{-1}$.

In both cases potential was swept between -1.2 and 1.2 V. It can be seen that almost in every case, the resistance decreased with increasing value of electrode potential which is in agreement with the literature [12, 16]. The exception was the resistance behaviour of PPy obtained in-situ in 0.001 M aqueous solution of sodium salicylate. Probably, the electroactivity of polymer measured in these conditions induces reactions that increase the resistance at these potential ranges. The sudden increase of the resistance seen at approximately 0.75 V can be related with the overoxidation process of PPy film [22]. It can be also noticed, that lowering the concentration of electrolyte resulted in a smaller changes of polymer resistance during its re-oxidation. Probably, salicylate ions in small concentration inhibit the incidence of the reactions which are responsible for the changes of polymer resistance. The same was observed by others [13]. Based on these measurements, generally, it can be seen that the changes of resistance in a function of electrode potential which were performed in solution differ from that conducted in the air. The differences can be related here with a charge transfer resistance. However, as it is known from previous results, it can be also related with the diffusion processes presence between the polymer layer and the solution or air. Xue et al. [16] also noticed that in-situ resistance measurements of polyaniline in a function of electrode potential can also be distorted by contact resistance between the polymer and electrode band which depended on the applied doping potential.

4. Conclusions

In this work, influence of the solution on resistance of polypyrrole film during in-situ measurements has been studied. For these measurements polymer was electrochemically polymerized onto the iron double-band electrode from aqueous solution of 0.1 M pyrrole and 0.1 M sodium salicylate. In-situ and ex-situ resistance measurements have been performed based on impedance spectroscopy measurements. Different impedance spectrum of PPy coated iron electrode was obtained depending on conditions at which experiments were performed. Polypyrrole coated Fe electrode measured ex-situ exhibited almost pure resistance behaviour at almost all applied frequency range. This resistance was related with the resistance of PPy film deposited between the iron electrodes. It was noticed that in-situ resistance measurement was a combination of polymer resistance and other components, such as resistance of solution and double-layer capacitance between the polymer and solution. In order to

check the influence of the solution on the in-situ resistance measurements, impedance spectroscopy of PPy/Fe electrode were performed in differently concentrated sodium salicylate aqueous solution. It was noted that polymer resistance measured in-situ does not depend on the concentration of electrolyte. However, it mainly depended on the frequency at which impedance measurements were performed and also on the surroundings conditions in which the experiments are conducted (in the air or in solution). It was noticed that the resistance of PPy film exposed to air decreased with time. Therefore, the differences in resistance measured in-situ and ex-situ was related with the diffusion processes which probably occurred between the polymer film and air/solution.
Resistance measurement during the re-oxidation of the polymer has been also studied. It was noticed that lowering the concentration of electrolyte in doping solution resulted in a smaller changes of polymer resistance during its re-oxidation. The results also showed that the changes of resistance in a function of electrode potential which were performed in solution differ from that conducted in the air. The differences were related with a charge transfer capacitance and with the diffusion processes occurring between the polymer film and air/solution.
Further research will include the study of diffusion processes occurring between the polymer film and the solution/air during the resistance measurements. It will also include the study of the degree of influence of the charge transfer on the resistance of polymer film during its doping/dedoping processes.

5. References

[1] Das T K and Prusty S 2012 *Polym.-Plast. Technol.* **51** 1487
[2] Gerard M, Chaubey A and Malhotra B D 2002 *Biosens. Bioelectron.* **17** 345
[3] Vernitskaya T V and Efimov O N 1997 *Russ. Chem. Rev.* **66** 443
[4] Wang L, Li X and Yang Y 2001 *React. Funct. Polym.* **47** 125
[5] Sadki S, Schottland P, Brodie N and Sabouraud G 2000 *Chem. Soc. Rev.* **29** 283
[6] Włodarczyk K, Singer F, Jasiński P and Virtanen S 2014 *Int. J. Electrochem. Sci.* **9** 7997
[7] Kumar A, Rajesh, Chaubey A, Grover S K and Malhotra B D 2000 *J. Appl. Polym. Sci.* **82** 3486
[8] Bof Bufon C C and Heinzel T 2006 *Appl. Phys. Lett.* **89** 012104
[9] Patois T, Sanchez J B, Berger F, Rauch J Y, Fievet P and Lakard B 2012 *Sensor. Actuat. B.* **171-172** 431
[10] Mostany J and Scharifker B R 1997 *Synthetic Met.* **87** 179
[11] Rodriguez I, Scharifker B R and Mostany J 2000 *J. Electroanal. Chem.* **491** 117
[12] Gholamian M, Suresh Kumar T N and Contractor A Q 1986 *Proc. Indian Acad. Sci. (Chem. Sci.)* **97** 457
[13] Talaie A 1995 *Iran. J. Polym. Sci. Technol.* **4** 26
[14] Du X and Wang Z 2003 *Electrochim. Acta* **48** 1713
[15] Long Y, Chen Z, Zheng P, Wang N, Zhang Z and Wan M 2003 *J. Appl. Phys.* **93** 2962
[16] Xue W, Jiang X and Harima Y 2009 *Anal. Chem.* **81** 2364
[17] Łapkowski M and Proń A 2000 *Synthetic Met.* **110** 79
[18] Kankare J and Kupila E L 1992 *J. Electroanal. Chem.* **322** 167
[19] Cysewska (nee Włodarczyk) K, Karczewski J and Jasiński P 2015 *Electrochim. Acta* **176** 156
[20] Liu Y C, Hwang B J, Jian W J and Santhanam R 2000 *Thin Solid Films* **374** 85
[21] Sung H, Lee T and Paik W 1995 *Synthetic Met.* **69** 485
[22] Bazzaoui M, Martins J I, Costa S C, Bazzaoui E A, Reis T C and Martins L 2006 *Electrochim. Acta* **51** 4516
[23] Adachi A and Yamauchi J 1996 *Bull. Chem. Soc. Jpn.* **69** 811

Acknowledgments
This work was supported by a statutory grant for research from the Gdansk University of Technology.

IOP Conf. Series: Materials Science and Engineering **104** (2015) 012027 doi:10.1088/1757-899X/104/1/012027

Modification of conductive polymer PEDOT:PSS layer by SWCNT

Aneta Araźna[1], Kamil Janeczek[1], Konrad Futera[1], Andrzej Kozioł[2]

[1] Tele- and Radio research Institute, Warsaw, Poland
[2] Warsaw University of Technology, Faculty of Chemistry, Warsaw, Poland

E-mail: aneta.arazna@itr.org.pl

Abstract. In this study, the effect of addition of single-walled carbon nanotubes (SWCNTs) on the morphological properties of poly(3,4-ethylenedioxythiophene/poly(4-styrenesulfonate) (PEDOT:PSS) has been investigated using scanning electron microscope (SEM) as well as the surface roughness evaluation by atomic force microscopy (AFM). The spin-coated films of PEDOT:PSS and PEDOT:PSS with addition of SWNTs (0.5 % by wt.) on glass were compared in our study. Experimental results show that surface roughness of PEDOT:PSS layers increases after addition SWNTs to the solution - from 3.50 nm for the samples without SWNTs to 4.65 nm for the samples with SWNTs. The analysis of SEM images showed that there are some agglomerations of SWNTs on the surface of PEDOT:PSS/SWCNT composite film.

1. Introduction

Water soluble conducting polymer poly(3,4-ethylenedioxy-thiophene)–poly-(styrene sulfonate) (PEDOT:PSS), has been widely used as electrode materials for capacitors, and in printed circuit boards and microactuators. PEDOT:PSS offers many advantages over other conducting polymers because of its ease of processability, commercial availability, high optical transparency in visible light, good mechanical strength, superb thermal stability and excellent atmospheric stability. High quality films can be easily formed on various substrates by conventional solution processing techniques such as spin-coating techniques [1,2,3,4].

PEDOT:PSS is generally used as antistatic coatings and as a hole injection layer in organic devices due to high work function, high hole affinity, and good transparency [2,3,5,6,7]. However, pristine PEDOT:PSS has a low conductivity (generally below 10 S/cm) [6]. Because of this recently, efforts have been focused on improvement of electrical conductivity of PEDOT:PSS films. Electrical conductivity of PEDOT:PSS can be enhanced by adding small amount of solvents such as dimethyl formamide (DMF), dimethyl sulfoxide (DMSO), tetrahydrofuran (THF), glycerol, and sorbitol to the PEDOT:PSS aqueous solution [1,3,8].

Another way to improve conductivity of polymer system is adding conductive fillers into this system. Carbon nanotubes (CNTs) have many advantages, above all excellent electrical and mechanical properties and exceptional flexibility. Additionally, CNTs can be dispersed in PEDOT:PSS solutions to obtain highly conductive polymer–CNT hybrid composite films. They are promising anode materials for flexible FOLED applications [9,10,11,12,13]. However, CNTs can change morphology of conductive layers. For OLED application, where layers have nanometers thickness, a morphological aspect of polymer layers is very important. It has significant effect on performance and lifetime of OLEDs devices.

In the present work, scanning electron microscope (SEM) as well as atomic force microscopy (AFM) are combined to study morphology of spin-coated PEDOT:PSS and PEDOT:PSS/SWCNT films.

2. Experiment details

Commercially available SWCNTs (outer diameter: 0.7-1.1 nm; length: 0.3-2.3 μm) from Sigma Aldrich Company were used in this work. The conducting polymer used was poly(3,4-ethylenedioxythiophene) poly(styrenesulfonate) (PEDOT:PSS) Orgacon Screenprint Ink Transparent EL-P3040 purchased from Agfa (electrical resistance - 680 Ohm/square). The preparation of the PEDOT:PSS/SWCNT solution included two steps. Firstly, the as-received SWCNTs were dispersed in distilled water (concentration of SWCNTs was 1 mg/l) by magnetic stir and sonication. In the next step, the above prepared SWCNT solution was mixed with the PEDOT:PSS solution (in ratio 0.05:1) and mixed for 30 min using an ultrasonic horn.

Two kinds of samples were prepared– the first one with pristine PEDOT:PSS and the second one with a mixture of PEDOT:PSS and SWCNTs (0.5 % by wt.). The analyzed layers were deposited on glass surface (2.5 cm × 2.5 cm squares size) using spin-coating technique (at 7500 rpm). Before layer deposition the glass surfaces were ultrasonically cleaned - washed in an ultrasonic bath of acetone and ethyl alcohol each for 5 min, then dried in a flow of air. After deposition the tested layers were dried on a hot plate for 10 min at 80 °C.

The PEDOT:PSS and PEDOT:PSS/SWCNT surfaces were characterized by AFM with a Digital Instruments Dimension 3100 in the tapping mode. Measurements were conducted at room temperature in the air, using the pyramidal Si_3N_4 tip. Surface roughness of the tested samples was calculated with AFM software.

Morphology of the tested coatings surface was characterized by scanning electron microscopy (SEM) using a JEOL JSM-7600F microscope.

The resistance of tested layers was measured with the four-point probe method using the digital multimeter Agilent 34401A.

3. Results of investigations

The morphology of the spin-coated pristine PEDOT:PSS and mixture of PEDOT:PSS/SWCNT films were investigated using SEM and AFM images.

The SEM micrographs – figures 1 and 2 show that both type of samples have porous and not uniform surface. Additionally, some aggregation was observed in the PEDOT:PSS/SWCNT composite film from the entanglement of SWCNT chains. The results are similar to that received in the works [14,15] where morphology of the spin-coated MWNT/PEDOT:PSS film was investigated. The authors of those work stated that PEDOT:PSS film shows smooth surface morphology without any agglomeration, but the MWNT aggregation was observed in the MWNT/PEDOT:PSS composite film. However, authors of [13] work stated that the PEDOT–CNT composite exhibits a porous morphology which is due to the embedding of CNTs into the polymer.

When CNTs are deposited on a substrate by solution methods, they are randomly tangled together on the surface and normally form an isotropic network. Due to the large aspect ratios of the CNTs, the solution deposition methods tend to form a large amount of voids in the film. This caused that it is very difficult to achieve intimate contact between CNT tube to CNT tube due to physical shape restriction and stiffness of the material. The polymer has filled most voids existed in the CNT network, aligning the surface of CNT film. Conductive polymer helps forming electrical contacts between CNT tubes and increasing the effective electric conductive tube–tube junctions in the CNT network. The composition of the CNT suspension is critical in the fabrication of uniform and reproducible PEDOT–CNT thin films [13,16].

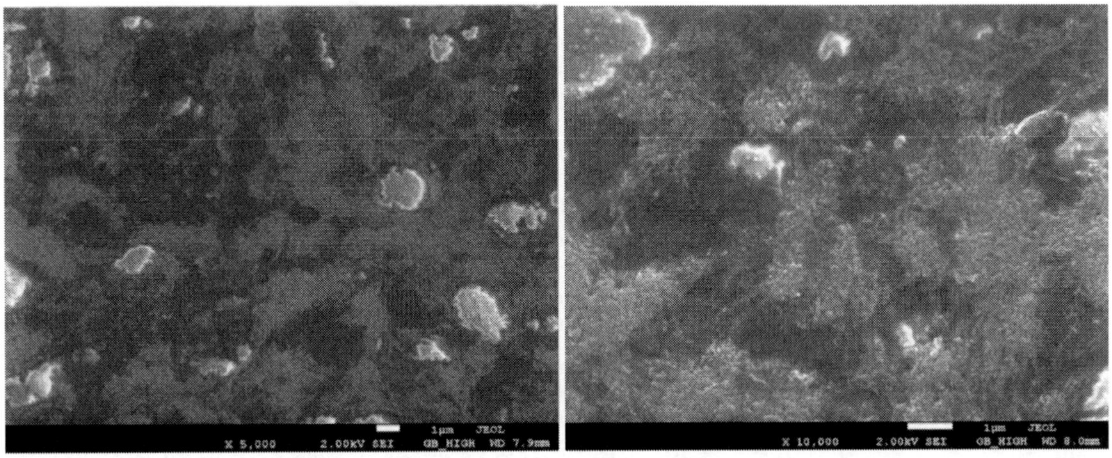

Figure 1. Example of SEM images of pristine PEDOT:PSS film.

Figure 2. Example of SEM images of PEDOT:PSS/SWCNT composite film.

We found that addition of SWCNT to the PEDOT:PSS solution led to the increase of the surface roughness of tested samples. Figures 3 and 4 show the examples of AFM micrographs of the PEDOT:PSS (figure 3) and PEDOT:PSS/SWCNT (figure 4) layers. Additionally, the values of root-mean-square roughness R_{rms}, average roughness R_a and peak-to-valley roughness R_{pv} are listed in Table 1.

Table 1. The values of surface roughness of pristine PEDOT:PSS and composite of PEDOTPSS/SWCNT samples.

	R_a [nm]	R_{rms} [nm]	R_{pv} [nm]
PEDOT:PSS	3.500	11.274	15.472
PEDOT:PSS/SWCNT	4.654	15.031	22.467

IOP Conf. Series: Materials Science and Engineering **104** (2015) 012027 doi:10.1088/1757-899X/104/1/012027

The value of R_a roughness increased from 3.5 nm to 4.654 nm and the value of R_{rms} roughness increased from 11.274 nm to the 15.031 nm. The samples with PEDOT:PSS layer had more smooth and homogeneous surface than the samples with PEDOT:PSS/SWCNT layers. That was confirmed during analysis of profiles of the fluctuation of surface roughness for all samples. The higher value of R_{pv} for PEDOT:PSS/SWCNT layer suggests that this layer has more spikes on the surface. This fact was confirmed by analysis of 3D images of the testes samples as well as by view of profiles of the fluctuation of surface roughness.

The results are similar to that received in the works [14,15]. In [14] the effect of CNT oxidation on the film properties of CNT/PEDOT:PSS composite was investigated. The authors of this work stated among other things that surface roughness of PEDOT:PSS film increase from 0.815 nm to 25.4 nm after addition of as-purchased MWCNT (0.980 % by wt.). In addition, in this work it has been concluded that oxidation of CNTs using acid solution leads to decrease in the value of CNT/PEDOT:PSS film roughness (to 10.7 nm). This is connected with formation of functional group and the ring-opening or chain break in the CNT during the oxidation process. This leads to fabrication of CNT suspension with small amount of CNT aggregation and consequently to preparation more uniform CNT/PEDOT:PSS film [14]. However, in [15] the properties of MWCNT/PEDOT:PSS films with various concentration of MWCNT were investigated. There was stated that roughness of MWCNT/PEDOT:PSS film (0.248 % by wt.) is almost ten times higher than for pristine PEDOT:PSS film. In this work it has been also found that roughness of PEDOT:PSS film increases with growth of MWCNT concentration in the solution.

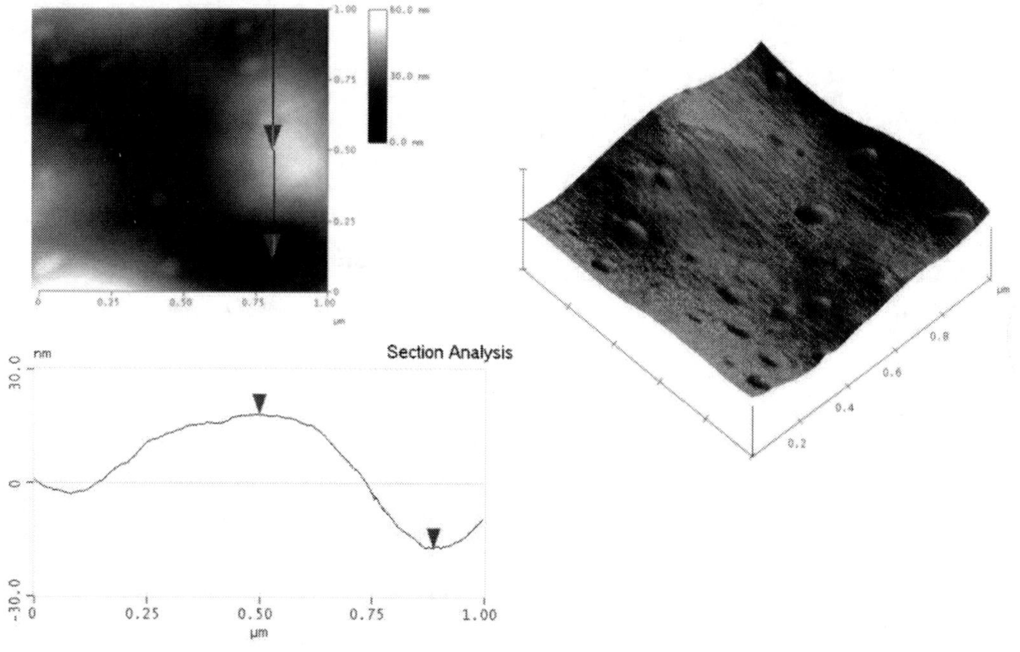

Figure 3. The example of AFM 2D and 3D images as well as profile of the fluctuations of the surface roughness of pristine PEDOT:PSS film.

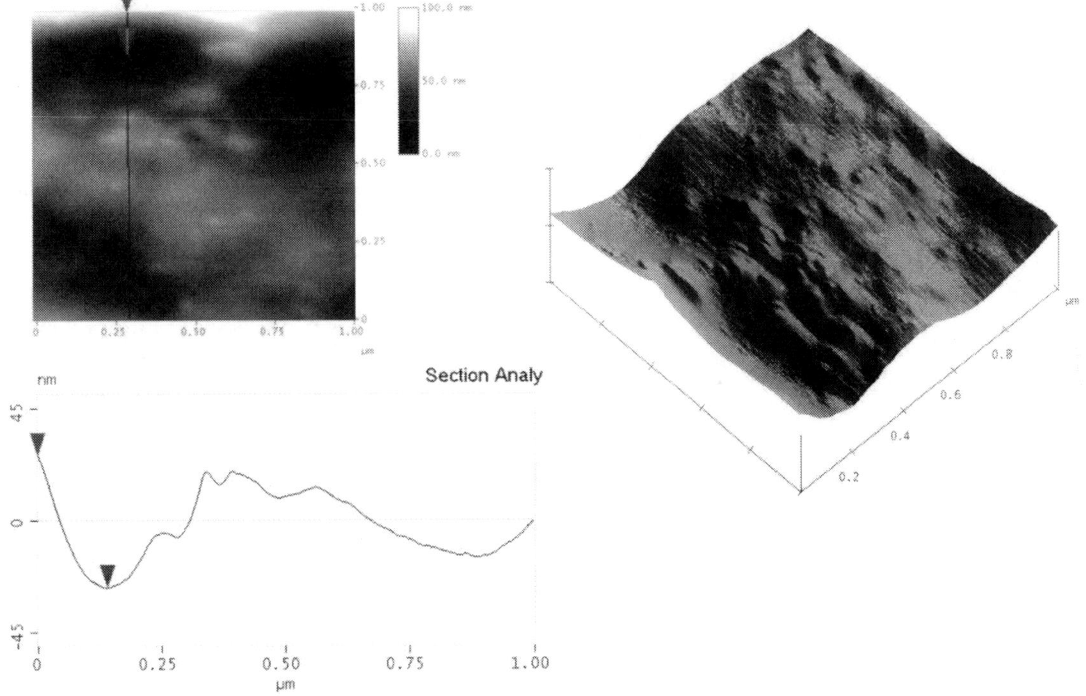

Figure 4. The example of AFM 2D and 3D images as well as profile of the fluctuations of the surface roughness of PEDOT:PSS with addition of SWCNT (0.5 % by wt.) film.

During analysis of SEM and AFM test results it could be expected that PEDOT:PSS/SWCNT layers will have higher values of sheet resistance than PEDOT:PSS films. The dispersion of SWCNTs in polymer carrier was poor. There were observed some agglomeration of SWCNTs. They caused increase of surface roughness, but they can also contribute to increase of electrical resistivity of the tested layers [11,14]. However, we observed that 0.5 % (by wt.) addition of single walled carbon nanotubes to the PEDOT:PSS solution caused the values of sheet resistance of the tested films to decrease. The sheet resistance of pristine PEDOT:PSS and PEDOT:PSS/SWCNT layers of the same thickness (about 800 nm) were 398 ± 23 Ohm/square and 254 ± 24 Ohm/ square, respectively. The results are similar to that obtained in the works [9] and [15]. There was found in the work [9] that sheet resistance of PEDOT:PSS films with 0.5 % (by wt.) SWCNTs is about 70 Ohm/square lower than sheet resistance of pristine PEDOT:PSS films of the same thickness. Also in the work [15] authors found that the resistivity of PEDOT:PSS films decreased with the addition of MWCNTs and for layers of PEDOT: PSS with additive of 0,495 % (by wt.) was twenty times lower than for pristine PEDOT:PSS layers. Our results are also similar to that presented in the work [17] where authors reported that resistivity of PEDOT:PSS/CNTs pastes layers decreases with increasing surface roughness of that layers.

4. Conclusion

The PEDOT:PSS and PEDOT:PSS/SWCNT films on glass substrate by spin-coating method were prepared and morphological properties of those films were studied using AFM and SEM methods. Some aggregation from the entanglement of SWCNT chains in the PEDOT:PSS/SWCNT composite film on the SEM micrographs was observed. Both type of samples had porous and not uniform surface. There was stated from the analysis of AFM micrographs that the PEDOT:PSS/SWCNT composite film had higher surface roughness than the pristine PEDOT:PSS film. The value of R_a

roughness increased from 3.5 nm to 4.654 nm and the value of R_{rms} roughness increased from 11.274 nm to the 15.031 nm after addition of SWCNTs to the solution of PEDOT:PSS. There was also observed that 0.5 % (by wt.) addition of single walled carbon nanotubes to the PEDOT:PSS solution caused the values of sheet resistance of the tested films to decrease. The sheet resistance of pristine PEDOT:PSS and PEDOT:PSS/SWCNT layers were 398 ± 23 Ohm/square and 254 ± 24 Ohm/ square, respectively.

The small surface roughness (R_{rms} < 1 nm [10]) is required for the functional layers in the organic electronic applications. Therefore, high roughness PEDOT:PSS/SWCNT composite film may be a kind of disadvantage in this applications. From the other hand, despite the increase in the surface roughness the sheet resistance of tested layers decrease, after adding CNTs to the polymer carrier. The major advantage of CNT networks is their mechanical flexibility in terms of bending. They are promising anode materials for flexible FOLED applications. There is demonstrated in literature a few ways to improve transparent electrodes based on CNTs such as improving material purity, chemical doping of CNT network or using separated CNTs. The attempts to modify the PEDOT:PSS/SWCNT composite layers in order to reduce their roughness will be a subject of the next article.

5. References

[1] Wang G-F *et al* 2009 *Nanoscale res Lett* **4** 613
[2] Kim W H *et al* 2003 *Journal of Polymer Science: Part B: Polymer Physics* **41** 2522
[3] Ouyang J *et al* 2004 *Polymer* **45** 8443
[4] Yan H and Okuzaki H 2009 *Synthetic Metals* **159** 2225
[5] Kim W H *et al* 2002 *Applied Physics Letters* **80** 3844
[6] Kim Y H *et al* 2011 *Adv. Funct. Mater.* **21** 1076
[7] Vitoratos E *et al* 2009 *Organic Electronic* **10** 61
[8] Dimitriev O P *et al* 2009 *Synthetic Metals* **159** 2237
[9] Wang G-F *et al* 2008 *Nanotechnology* **19** 145201
[10] Hu L *et al* 2010 *Nanotechnology* **21** 155202
[11] Najeeb *et al* C K 2010 *Nanotechnology* **21** 385302
[12] Ham H T *et al* 2008 *Polymer Engineering and Science* **48** 1
[13] Gerwing R *et al* 2012 *Frontiers in Neuroengineering* **5** 1
[14] Yun D-J and Rhee S-W 2012 *Applied Materials & Interface* **4** 982
[15] Yun D-J *et al* 2011 *Applied Materials & Interface* **3** 43
[16] Xiao G *et al* 2010 *Thin Solid Films* **518** 2822
[17] Janeczek K *et al* 2013 *Materials Science and Engineering B* **178** 511

IOP Conf. Series: Materials Science and Engineering **104** (2015) 012028 doi:10.1088/1757-899X/104/1/012028

Polypyrrole based gas sensor for ammonia detection

K J Dunst, K Cysewska, P Kalinowski, P Jasiński

Gdańsk University of Technology, Faculty of Electronics, Telecommunications and Informatics, ul. Narutowicza 11/12, 80-233 Gdańsk, Poland

E-mail:pijas@eti.pg.gda.pl

Abstract. The nature of polypyrrole response to toxic gases does not allow using the sensor in a conventional way. The main aim of this study is to acquire the information about the concentration using different approaches: a linear approximation, a non-linear approximation and a tangent method. In this paper a two-steps procedure for sensor response measurements has been utilized. Polypyrrole films were electrochemically synthesized on the interdigitated electrodes. Gas sensing measurements of polypyrrole based sensor were carried out at room temperature. The influence of the flow rate on the sensing performance to NH_3 were investigated. The preliminary studies of aging of the sensor were also explored.

1. Introduction

Today, protection of the environment and persons is more important than before. For that reason there is a need for improved or new sensors for measuring both physical and chemical changes, such as humidity, temperature, pressure and chemical species [1,2]. For this purpose there were many attempts in order to fabricate the suitable sensing materials. Nowadays there is a great interest in making conducting polymer sensors [3]. Conducting polymers, especially polypyrrole (PPy), are emerging as intelligent materials and they have a wide range of applications in the field of optical, electronic, electrochromic devices and sensors. The reasons for such widespread interest of polypyrrole (PPy) are its easy deposition from aqueous and non-aqueous media, adherence to many types of substrates, stability in air and aqueous media and high electrical conductivity [4]. Controlling the materials properties of conducting polymers such as morphology, electroactivity and conductivity, which are strongly dependent on how conducting polymer are synthesized, is required for the successful implementation of conducting polymers in devices [5]. Polymer gas sensing films can generally be prepared by several techniques including a solution casting method [6], a thermal evaporation [7], a drop-coating [8] and a Langmuir-Blodgett technique [9]. The simplicity of preparation and the possibility to control experimental conditions by electrochemical techniques make it the most employed way. In this case, the electrochemical synthesis of conducting polymer is the most convenient [4]. Moreover, it is now well-known that many chemical or physical properties of electrodeposited polypyrrole can be strongly varied by modifying the electropolymerization parameters including: current density [10], temperature [11], solvent [12,13], pH [14], counter-ion [15], counter-ion concentration [16], pyrrole concentration [17] and electrodeposition potential [18].The most important advantages of polypyrrole based gas sensors is their easy and cheap fabrication [19], temperature operation and high sensitivity [19]. Therefore, polypyrrole has been widely used as an effective material for the detection of toxic, hazardous and flammable gases [20].

The nature of PPy response to toxic gases does not allow using the sensor in a conventional way. Therefore, an alternative approach is sought [4,21]. The main aim of this study is to acquire the information about the concentration using different approaches: a linear approximation, a non-linear approximation and a tangent method. In this paper a two-steps procedure for sensor response measurements has been utilized. In the first step, the gas is turned on for 5 minutes (gas-on) and in the second step is turned off for 30 minutes (air-on) forming a 35 minutes cycle. Such times were chosen to compare the results with the literature. Gas sensing measurements of PPy/$LiClO_4$ were carried out at room temperatures. The influence of the flow rate has also been investigated. The preliminary studies of aging of the sensor were also explored.

2. Experimental

2.1. Electrochemical polymerization

Polypyrrole films were electrochemically synthesized on the interdigitated gold electrodes (BVT Technologies) in a one step process from aqueous solution of 0.1 M pyrrole monomer (97 %) and 0.1 M lithium perchlorate (≥ 98 %) (all from Sigma Aldrich).

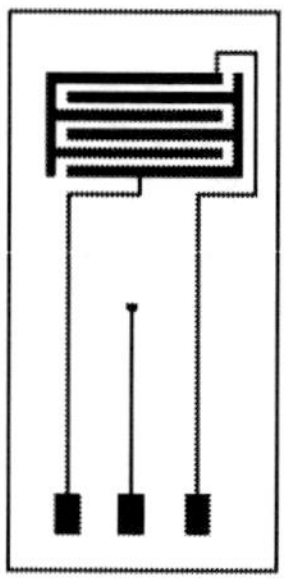

Gap between fingers: 50 μm
Finger width: 50 μm
Finger length: 7 mm

Figure 1. Schematic picture of the interdigitated gold electrode.

Electropolymerization processes were performed potentiostatically at potential of 1 V vs. reference electrode (80 C·cm^{-2}) in a one-compartment cell with a three electrode system. An Ag/AgCl (R-10 Hydromet) electrode was used as the reference electrode and a platinum gauze electrode (80 mesh 25 x 35 mm) as the counter electrode. The working electrode was the interdigitated gold electrodes with a finger width of 50 μm and a gap size of 50 μm (see figure 1). Before and after each experiment working electrode was rinsed with deionized water and isopropanol.

2.2. Gas sensing measurements

The sensor was placed inside a glass chamber of 123 mL volume, which was always kept under continuous flowing of testing gas mixture or air at a constant flow rate of 50 mL min^{-1} (sccm – Standard Cubic Centimeters per Minute) or 100 mL min^{-1}. NH_3 with a concentration of 10-100 ppm was utilized to evaluate the sensing performance. To alleviate the baseline drift, the sensing films were exposed to synthetic air overnight before data collection. Then, the sensors were cyclically exposed for 5 min to ammonia (gas-on step). Between each exposition to ammonia, the sensors were exposed to the synthetic air during 30 minutes to allow desorption of ammonia molecules from the sensors (air-on step). The resistance was monitored continuously over time (a measurement was made every 5 s). The testing gas was generated by mixing premixed gas (200 ppm of NH_3) with synthetic air controlled by the mass flow controllers (Tylan GmbH). The resistance changes of based sensors were monitored with a computer controlled multimeter (RIGOL DM3068). All the gas sensing measurements were performed at room temperatures in dry gases.

3. Results

3.1. Sensing characteristic

To take advantage of the polypyrrole polymer as a sensing layer, a two-step procedure has been conducted. The first step begun when the gas was turned on for the 5 minutes (gas-on step), while the second when the gas was turned off for 30 minutes (air-on step). These two steps formed a 35 minutes cycle (see Figure 2a). The literature does not show a repetition of the cycles for this type of sensor. Patois et al. [4] and Carquigny et al. [21] show only the first cycle of PPy sensor response, allowing them to establish that this polymeric material can be used as a gas sensor. The repeatability of the PPy sensor response to 80 ppm of NH_3 was presented for the first time in Figure 2a. It can be noted that this sensor does not work as a typical sensor, where the level of resistance determines the concentration. As the NH_3 gas is introduced into the chamber, the resistance of the sensors always increases. Ammonia reacts with the PPy and induces a modification in electrical properties of the sensor's sensitive layer. The decrease of the sensor's resistance observed under exposition to the synthetic air, corresponds to desorption of ammonia molecules from the polypyrrole film. The mechanism of interaction between a polymer film and an ammonia has been already studied by Carquigny et al. [21]. Since polypyrrole can be considered as a p-type semi-conducting material composed of both neutral and oxidized monomer units, the ammonia adsorption onto the PPy films was shown to begin with the loss of an electron by the nitrogen's doublet of some nitrogen atoms of the polymer backbone, leading to the formation of NH^+ radical groups. This electron transfer between ammonia molecule and the polymer's positive hole induces a diminution of the positive charge density which leads to an increase in the resistance. Consequently, after adsorption of NH_3, the polymer becomes less conducting, and the measured resistance increases. On the contrary, during desorption of the ammonia molecules, the sensor's resistance decreases [4,21]. Figure 2b displays the comparison of the gas sensor responses to 10-100 ppm of NH_3 depending on the flow rate of NH_3/air cycles with each cycle lasting for 35 minutes. The behavior of the gas sensors was estimated using the normalized resistance changes (R/R_0), where R_0 is an initial resistance of the first cycle. Figure 2b shows that the curves have different nature depending on the concentration of the pollutant gas and the flow rate in the gas chamber. It is expected that one can find a specific parameter in the cycle so specific gas concentrations can be determined.

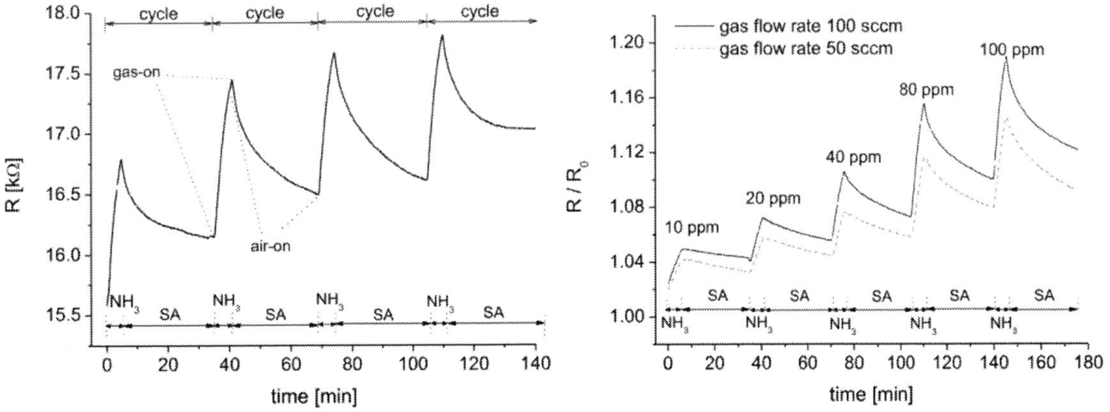

Figure 2. PPy sensor response to 80 ppm of NH_3 (a), and to different concentrations of NH_3 (b).

Therefore, the different parameters of the cycles were calculated and plotted as a function of the concentration. Namely, the data was approximated to the exponential function (parameters t_{AN1}, t_{AN2}, NR_1 and NR_2), the linear function (parameters a_1, a_2, t_{AL1}, t_{AL2}) and the tangent function (parameters t_{S1}, t_{S2}). The calculations were carried out for both the gas-on step (t_{AN1}, NR_1, a_1, t_{AL1} and t_{AS1}) as well

IOP Conf. Series: Materials Science and Engineering **104** (2015) 012028 doi:10.1088/1757-899X/104/1/012028

as air-on step (t_{AN2}, NR_2, a_2, t_{AL2} and t_{AS2}) which is presented in figure 3, 4, 5, 6 and 7 respectively. The t_{AN1}, t_{AN2}, NR_1 and NR_2 parameters were calculated based on equation:

$$R = R_c + A \cdot (1\text{-}\exp(-t / t_{ANx}), \quad (1)$$

where R_c and A are the parameters of exponential function (where x = 1 or 2 for gas-on and air-on step respectively). The values of t_{AN1} were approximately linear with concentration for the gas-on step (Figure 3a). As can be seen t_{AN1} depends on the concentration. The resistance variations marked as normalized response (NR_1 and NR_2) were calculated according to the following equation:

$$NR\,[\%] = A / R_c \cdot 100\%, \quad .. (2)$$

The parameter NR (originally called sensitivity in [22]) was also calculated. The values of NR were approximately linear with concentration for the gas-on step, as well as for air-on step (Figure 4a and 4b). However, NR_1 values depend on the flow rate. In the case of air-on step, NR_2 values are very similar, regardless of the flow rate. To determine t_{AL1} and t_{AL2} parameters and the values corresponding to the slope (a_1 and a_2) the first 12 measurement points (obtained during the first minute of gas-on step) for each acquisition were used for the approximation. This values were presented in Figure 5 a,b and 6 a,b. The a_1, a_2, t_{AL1} and t_{AL2} parameters were calculated based on equation:

$$R = a_x \cdot t + b_x, \quad (3)$$

where a_x, b_x are the parameters of linear function (where x = 1 or 2 for gas-on and air-on step respectively).

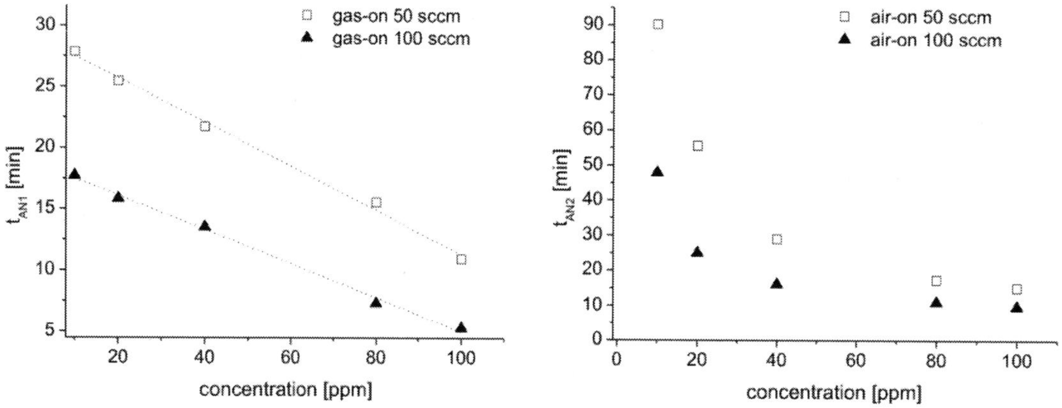

Figure 3. The t_{AN1} (a) and t_{AN2} (b) parameters calculated using exponential function.

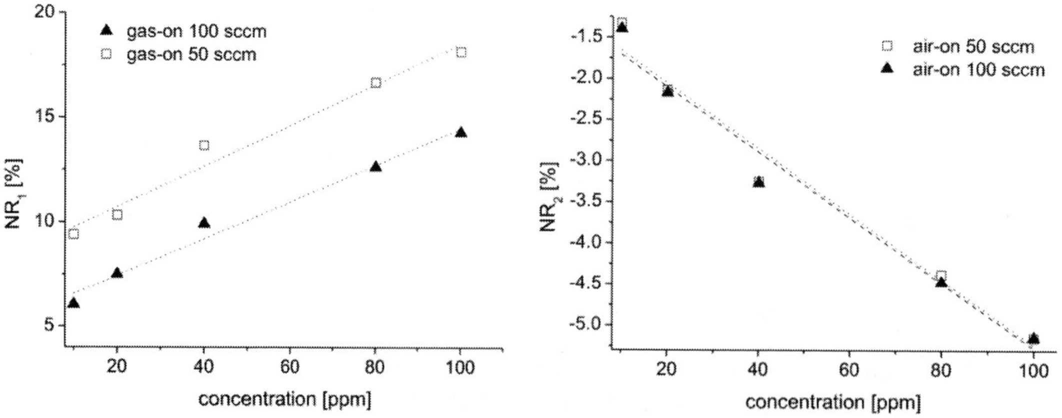

Figure 4. The NR_1 (a) and NR_2 (b) parameters calculated using exponential function.

The value of t_{AL2} for the concentration of 10 ppm and a flow rate of 50 sccm could not be calculated. It can be noticed, that both a t_{AL1} parameter and the slope a_1 are linear with concentration for the gas-on step for both the flow rate of 50 sccm and 100 sccm. According to Figure 3b, 5b and 6b it can be seen that the air-on step brings less information about the concentration, because of the non-linearity of mentioned characteristics.

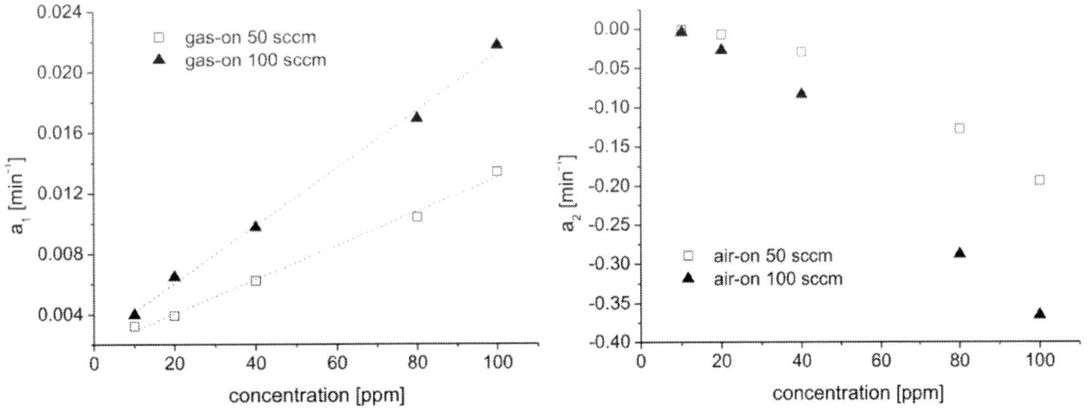

Figure 5. The a_1 (a) and a_2 (b) parameters (slope) calculated using linear function.

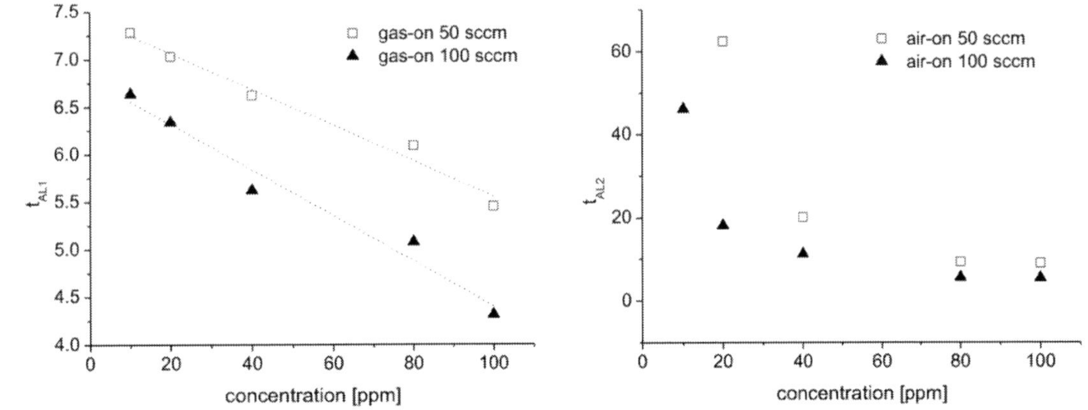

Figure 6. The t_{AL1} (a) and t_{AL2} (b) parameters calculated using linear function.

The procedure of obtaining t_{S1} and t_{S2} parameters was based on the calculation of the tangent function to the sensor response curve after one minute from the beginning of the gas-on or air-on stages. In order to do this, the derivative of the sensor response in point corresponding to one minute exposure to the target gas or synthetic air in the cleaning phase was estimated. Then, the formula of the linear tangent function was obtained and the value corresponding to the maximum resistance (during gas-on step) or minimum resistance (during air-on step) was selected for the calculations of t_{S1} and t_{S2} parameters, respectively. The formula for calculation the tangent function is presented below:

$$f_{tangent} = R_S'(t_{60}) \cdot (t - t_{60}) + R_{60}, \quad (4)$$

where t_{60} represents the time of one minute to the target gas or air in the cleaning stage, R_{60} is the value of sensor resistance corresponding to that time and $R_S'(t_{60})$ is derivative of the sensor response curve after one minute exposure. The obtained results are shown in Figure 7 a and b. The results shown, that despite the fact that values of the t_{S1} and t_{S2} parameters are decreasing for the increasing values of gas concentration, it is impossible to estimate useful trend for gas concentration estimation. This was possibly caused by the fact, that consecutive measured points in raw sensor response didn't create smooth curve. Calculated tangent functions were very sensitive to the disturbances in sensor responses and due to this fact, parameters t_{S1} and t_{S2} cannot be utilized for further analysis.

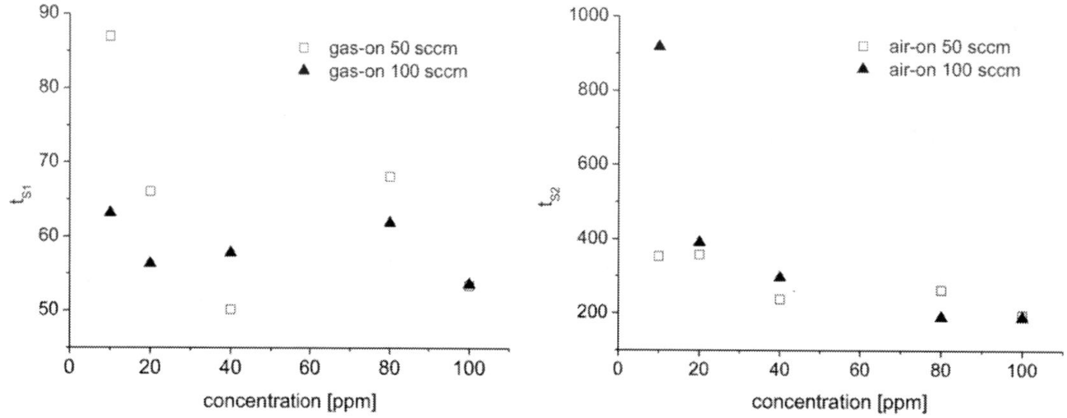

Figure 7. The t_{S1} (a) and t_{S2} (b) parameters calculated using tangents method.

Among investigated parameters in this preliminary study, the best indicator of the gas concentration was based on linear approximation of the measured points during the first minute of exposure to target gas. The method based on the exponential approximation was unpractical due to its complexity. However, method based on the t_{S1} and t_{S2} parameters calculated from the tangent function was too sensitive for the small disturbances in sensor response. Further investigations are conducted to find more robust and discriminative parameters, e.g. by utilizing multivariate methods.

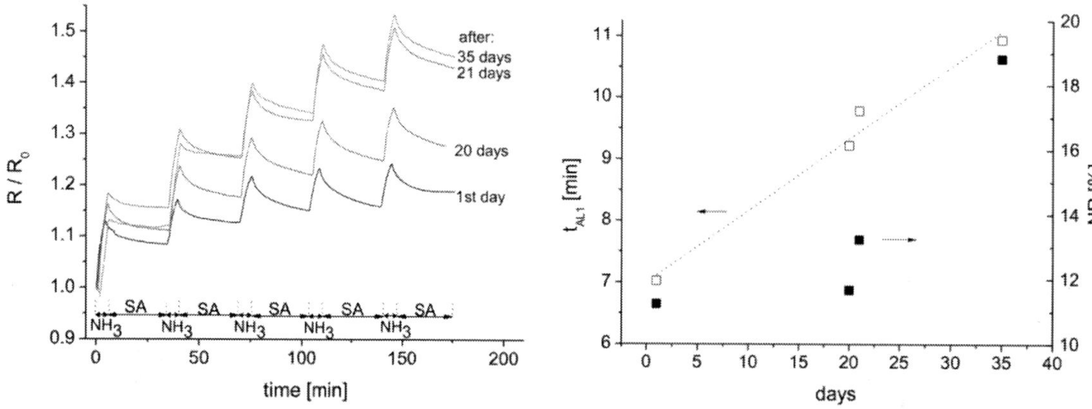

Figure 8. The aging process of PPy sensor (80 ppm of NH_3, flow rate of 100 cm^3/min).

3.2. Stability of PPy sensor

The stability of the sensor is very important parameter to be considered. Stability performance of PPy sensor to 80 ppm of NH_3 gas after 1, 20, 21 and 35 days is shown in Figure 8. It can be noted, that the drift in sensor response is noticeable (Figure 8a) but for this data it seems to have linear trend. The values of t_{AL1} parameter is approximately linear with the number of days for the gas-on step (Figure.

8b). Polymer based sensor usually possess a common drawback of decrease in normalized response due to ageing-induced effects [22]. In the report [22], the PPy sensor exhibited maximum normalized response for 100 ppm of 36%; however, it dropped from 36% to 20% after 35 days. However, in this study the normalized response increase from 11% to 19% after 35 days (see Figure 8b). This result indicate that the PPy sensor prepared using presented procedure can be used as a reusable sensing material for the detection of NH_3 gas operating at room temperature, although, the indicated parameters are sensitive for aging.

4. Conclusions

In this work, polypyrrole based gas sensor for ammonia detection has been studied. Polypyrrole was electrochemically synthesized from an aqueous solution of 0.1 M pyrrole and 0.1 M lithium perchlorate. That prepared sensor was cyclically studied (5 minutes gas-on step and 30 minutes air-on step) in response to different concentrations of ammonia gas. Such times were chosen to compare the results with the literature. It can be expected that measurement cycle may be reduced to about 5 minutes. However, for some applications requiring fast measurement time it might be too long. When the NH_3 gas was introduced into the chamber, it was noticed that the resistance of PPy sensor was increasing. Based on the sensor response curves, various parameters have been determined in order to find the best indicator of the gas concentration. The data was approximated to the exponential function (parameters t_{AN1}, t_{AN2}, NR_1 and NR_2), the linear function (parameters a_1, a_2, t_{AL1}, t_{AL2}) and the tangent function (parameters t_{S1}, t_{S2}). The calculations were carried out for both the gas-on step (t_{AN1}, NR_1, a_1, t_{AL1} and t_{AS1}) as well as air-on step (t_{AN2}, NR_2, a_2, t_{AL2} and t_{AS2},). The results show, that the t_{AN1}, NR_1 and NR_2 values were approximately linear with concentration for the gas-on step. However, the non-linear parameters had a high computational complexity. It was noticed that the linear method parameters (t_{AL1} and a_1) were linear with concentration for the gas-on step for both the flow rate of 50 sccm and 100 sccm. Based on calculations, it can be seen, that the air-on step brought less information about the gas concentration because of its non-linearity. The method based on t_{S1} and t_{S2} parameters calculated from the tangent function has proven to be too sensitive for the small disturbances in sensor response. The further investigations will include the influence of the condition of PPy synthesis on its gas sensing properties, sensor responses to the other volatile compounds, data processing and analysis to enable efficient recognition and identification of measured gas.

5. References

[1] Selampinar F, Toppare L, Akbulut U, Yalçin T and Şefik S 1995 *Synth. Met.* **68** 109-116.
[2] Dunst K, Jasinski G, Jasinski P 2014 *Metrol. Meas. Syst.* **21** 2 205–216.
[3] Lakshmi A, Anandha J R, Gopu G, Armugam P and Vedhi C 2013 *Electrochim. Acta* **92** 92 452-459.
[4] Patois T, Sanchez J B, Berger F, Rauch J Y, Fievet P and Lakard B 2012 *Sens. Actuators, B* **171-172** 431-439.
[5] Kanicki J 1986 *Handbook of Conducting Polymers* 1, (Dekker New York) 543.
[6] Sharma S, Hussain S, Singh S and Islam S S 2014 *Sens. Actuators, B* **194** 213–219.
[7] Agbor N E, Petty M C and Monkman P 1995 *Sens. Actuators, B* **28** 173–179.
[8] Chiang C J, Tsai K T, Lee Y H, Lin H W, Yang Y L, Shih C C, Lin C Y, Jeng H A, Weng Y H, Cheng Y Y, Ho K C and Dai C A 2013 *Microelectron. Eng.* **111** 409–415.
[9] Xie D, Jiang Y, Pan W, Li D, Wu Z and Li Y 2002 *Sens. Actuators, B* **81** 158–164.
[10] Dyreklev P, Granstrom M and Inganas O 1996 *Polym.* **37** 2609–2613.
[11] Liang W, Lei J and Martin C R 1992 *Synth. Met.* **52** 227–239.
[12] Otero T F and Rodriguez J 1994 *J. Electroanal. Chem.* **379** pp 513–516.
[13] Lee S, Sung H, Han S, Paik W, 1994 *J. Phys. Chem.* **98** 1250–1252.
[14] Carquigny S, Segut O, Lakard B, Lallemand F and Fievet P 2008 *Synth. Met.* **158** 453–461.
[15] Patois T, Lakard B, Monney S, Roizard X and Fievet P 2011 *Synth. Met.* **161** 2498–2505

IOP Conf. Series: Materials Science and Engineering **104** (2015) 012028 doi:10.1088/1757-899X/104/1/012028

[16] Kaynak A 1997 *Mater. Res. Bull.* **32** 271–285.
[17] Ogamasawara M, Funahashi K, Demyra T, Hagiwara T and Iwata K 1986 *Synth. Met.* **14** 61–69.
[18] Chmielewski M, Grzeszczuk M, Kalenik J and Kpas-Suwara A 2010 *J. Electroanal. Chem.* **647** 169–180.
[19] Sutar D S, Padma N, Aswal D K, Deshpande S K, Gupta S K, Yakhmi J V 2007 *Sens. Actuators, B* **128** 286–292.
[20] Li Q H, Wu J H, Tang Q W, Lan Z, Li P J, Lin J M and Fan L Q 2008 *Electrochem. Commun.* **10** 1299–1302.
[21] Carquignya S, Sanchez J -B, Berger F, Lakard B, Lallemand F 2009 *Talanta* **78** 1 pp 199-206
[22] Navale S T, Mane A T, Chougule M A, Sakhare R D, Nalage S R, Patil V B 2014 *Synth. Met.* **189** 94-99.

Acknowledgments
Financial support of the Polish Ministry of Science and Higher Education by the „Diamond Grant" programme for the project "Conducting polymers in environmental gas sensors" (0251/DIA/2013/42) and by the Statutory Funds for Research of Faculty of Electronics, Telecommunications and Informatics, Gdansk University of Technology is greatly appreciated.

IOP Conf. Series: Materials Science and Engineering **104** (2015) 012029 doi:10.1088/1757-899X/104/1/012029

Investigation of solid polymer electrolyte gas sensor with different electrochemical techniques

A Strzelczyk[1], G Jasinski[1,2], B Chachulski[1]

[1] Gdansk University of Technology, Dept. of Chemical and Process Engineering, Narutowicza 11/12, 80-233 Gdansk, Poland
[2] Gdansk University of Technology, Dept. Biomedical Engineering, Narutowicza 11/12, 80-233 Gdansk, Poland

E-mail: grzegorz.jasinski@eti.pg.gda.pl

Abstract. In this work solid polymer electrolyte (SPE) amperometric sulphur dioxide sensor is investigated. Nafion was used as a membrane electrode and *1M* sulphuric acid as an internal electrolyte. Sensor response to sulphur dioxide was measured. Besides traditional constant voltage amperometry also different electrochemical techniques were used. Results obtained by these methods are compared.

1. Introduction

Amperometric sensors due to high sensitivity and good dynamic properties are frequently used in toxic gas concentration monitoring [1]. Most of such sensors are based on liquid electrolytes. Up to now usage of solid polymer electrolyte (SPE) membrane with layer of porous, metal electrode in construction of amperometric sensors is the most successful approach [2]. Main advantages of this type of sensors are a very high active area of the working electrode (WE) and a short response time [3].

In this work performance of SO_2 amperometric sensor with SPE is investigated. The main goal of this work is to verify if usage of different electrochemical methods can improve sensor properties. Sensor responses were measured with constant potential amperometry (CPA), pulsed amperometry (PA), cyclic voltammetry (CV) and square wave voltammetry (SWV) techniques. Obtained results are compared.

These techniques are used with different sensors in the determination of different compounds. For example, PA technique was used with the glucose enzyme sensor [4]. CV method is commonly used in liquid solution electrochemistry and also with electrocatalytic gas sensors based on solid state electrolytes [5-9]. SWV technique can be used in the determination of volatile organic compounds [10]. In contrast with CPA, when only single value of the current is obtained, in case of the CV or SWV methods a response in a form of a curve with more information for a given gas mixture is obtained and analysed. These additional information obtained from CV and CWV response can be used to improve sensor properties. In order to analyse such responses a various methods of data analysis can be used [11]. To our knowledge, the PA and SWV techniques have not been used so far for the measurements of SPE sensors responses. For such sensors, a cyclic voltammetry technique is frequently used, however only to perform preliminary investigation in order to determine working potential. The CV is not employed for the determination of gas concentration and additional information contained in measured responses can not be used.

2. Experimental

2.1. Sensor structure

Housing of investigated sensor was made from polyether ether ketone (PEEK) which is highly resistant to acids and organic solvents. In the upper part of the housing a counter electrode (CE), a reference electrode (RE) and their contacts are placed. CE is in the form of a piece of gold sheet. Mercury sulphate reference electrode (MSE) is made of pure metallic mercury and mercury(I) sulphate obtained from the mercury(II) nitrate. Mercury/mercury sulphate system is in contact with internal electrolyte containing sulphuric acid. Between upper and middle part of the sensor housing a working electrode and its contact are placed. Contact of WE is in the form of a gold ring. In the middle part of the housing WE contact ending, gas inlet and outlet are placed. Bottom part of the sensor has reservoir for hygrostatic liquid. Solution of *1M* sulphuric acid in water was used as an electrolyte.

Nafion® 350 (Sigma-Aldrich), in the form of foil, thickness 0.01 inch (0.254 mm) strengthened by Teflon net was used. Layer of gold was prepared by using chemical plating method known as Takenaka-Torikai method [12]. Details of this process was described elsewhere [13]. Geometric surface of the WE is 0.636 cm^2.

Mixtures of high purity gases, namely synthetic air (mixture of nitrogen and oxygen), and 1000 ppm SO_2 in synthetic air, have been used. Aalborg or Brooks mass flow controllers were used in order to obtain the proper gas composition. All measurements were performed at room temperature.

Voltammetric and amperometric sensor responses were acquired by using electrochemical interfaces Gamry Interface 1000 and VersaSTAT 3 controlled by computer with suitable software. Constant gas flow rate of 50 millilitres per minute was maintained. For CPA measurements, working potentials 0.5 V and 0.7 V have been used. In case of the PA, a base potential (E_b) was 0.2 V and a measurement potential (E_m) was 0.5 V. Pulses of the same duration (τ = 1s) and a constant amplitude ($\Delta E = E_m - E_b = 0.3$ V) were applied. The intervals between pulses was also 1s. For the cyclic voltammetry measurements a linearly changing voltage (range -0.3 – 1.2 V) was used. The voltage sweep rate was 20 mV/s. In case of square wave voltammetry step size was 2 mV and the frequency was 10 Hz, what corresponds to the sweep rate of 20 mV/s. Pulse size has been fixed at 50 mV and the voltage range was the same as for CV.

3. Results

3.1. Constant potential amperometry

Results of CPA measurement for working potential of 0.5 V and of 0.7 V are presented in Figure 1 and Figure 2, respectively. In these figures an increase of current with increasing sulphur dioxide concentration is clearly visible.

Based on data showed in Figure 1 and Figure 2 calibration curves have been calculated (Figure 3). In case of amperometric sensor a measured response should linearly depend on the measured gas concentration. Experimental results obtained in particular concentration of SO_2 are presented in the Figure 3 (points). Effect of fitting experimental data to straight lines is presented in the same figure (lines). Sensor sensitivity at fixed potentials is the slope of these curves. In case of the Figure 3, the following linear equations were obtained (for working potential of 0.5 V and of 0.7 V the equations (1) and (2), respectively):

$$I[mA] = 1.15 \cdot 10^{-4}[mA/ppm] \cdot c[ppm] + 0.01065[mA] \tag{1}$$

$$I[mA] = 5.27 \cdot 10^{-4}[mA/ppm] \cdot c[ppm] + 0.07593[mA] \tag{2}$$

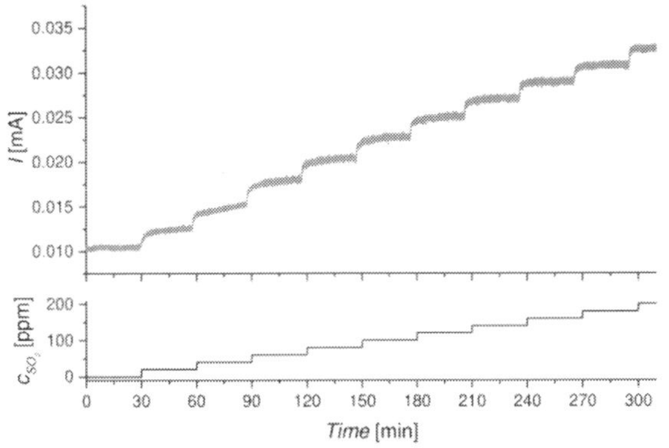

Figure 1. Current vs. time response of the sensor measured by the CPA technique at potential of 0.5 V (upper part of figure) and change of SO_2 concentration during the CPA measurement (lower part of figure).

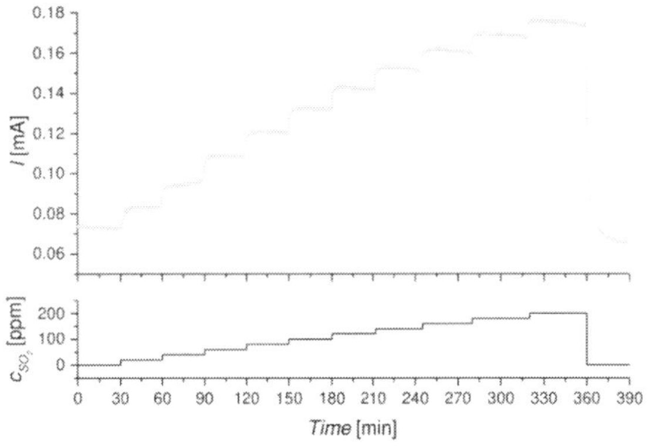

Figure 2. Current vs. time response of the sensor measured by the CPA technique at potential of 0.7 V (upper part of figure) and change of SO_2 concentration during the CPA measurement (lower part of figure).

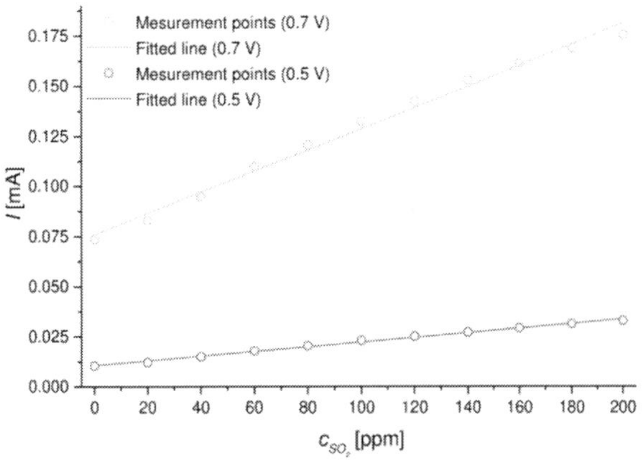

Figure 3. Calibration curves for measurement at 0.5 V and at 0.7 V made by the CPA technique.

The equations (1) and (2) shows that sensitivity at working potential of 0.7 V is almost 5 times higher than for the working potential of 0.5 V. Unfortunately value of the current registered for pure synthetic air (0 ppm SO_2) at potential 0.7 V is also much higher. This can have a negative influence on the sensor detection limit. In case of interfering gases present, a lower potential will be more desired, because for higher potentials more interfering gas components can be oxidized.

3.2. Pulsed amperometry

Pulsed amperometry is an enhancement of the constant potential amperometry technique. Shape of used excitation signal is presented in Figure 4. In case of the presented results, the measurement potential was 0.5 V. When potential was increasing to the 0.5 V, the oxidation reactions occurred on the WE surface, on the other hand, while potential was decreasing the reduction processes occurred. As a result, a surface of the sensor electrode was cleaned and refreshed before each measurement pulse. The current response of the sensor was measured 4 times per second. Namely, during each measurement, current of the pulse was measured 4 times at points A, B, C and D (as it is visible in the Figure 4). Next stairs-shaped characteristics were plotted. Current vs. time sensor response for two last measurement points C and D are presented in the Figure 5. Based on this data, the calibration curves were plotted similarly like in the case of the CPA measurements (results showed in the Figure 6).

Equations showing current-concentration relationship are presented below (equation (3) for point A, equation (4) for point B, equation (5) for point C and equation (6) for point D).

$$I[mA] = 3.2 \cdot 10^{-4}[mA/ppm] \cdot c[ppm] + 0.535[mA] \quad (3)$$

$$I[mA] = 3.11 \cdot 10^{-4}[mA/ppm] \cdot c[ppm] + 0.197[mA] \quad (4)$$

$$I[mA] = 3.45 \cdot 10^{-4}[mA/ppm] \cdot c[ppm] + 0.112[mA] \quad (5)$$

$$I[mA] = 3.46 \cdot 10^{-4}[mA/ppm] \cdot c[ppm] + 0.079[mA] \quad (6)$$

Coefficients of determinations (R^2) of these equation were 0.966 (equation 3), 0.994 (equation 4), 0.996 (equation (5)) and 0.997 (equation (6)). The current has the greatest value for points registered immediately after voltage increase (point A). It is because current increase immediately after the voltage disruption and decrease during duration of the voltage pulse. The capacitive current (unrelated to gas concentration) decrease slower than the faradic current (related to gas concentration). For this reason values of current registered for further points decrease. Value of sensitivity is the smallest for measurement made immediately after the voltage disruption and increase for further points. The best results are obtained for measurement made just before decreasing of the voltage (point D). The obtained sensitivity is in case of the PA sensitivity 3 times higher then for the CPA at the same working potential of 0.5 V.

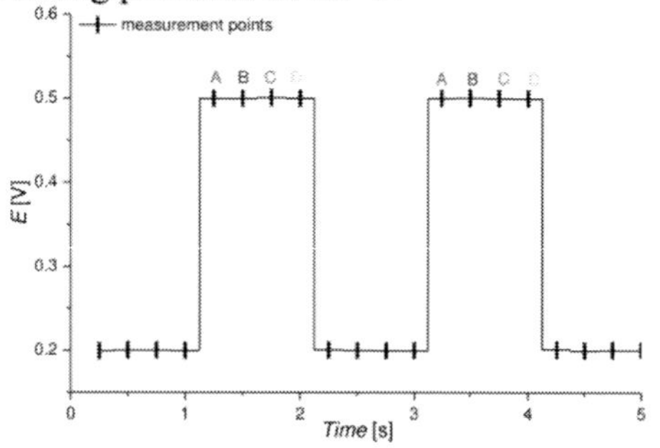

Figure 4. Shape of the voltage excitation signal used for the pulsed amperometry measurements.

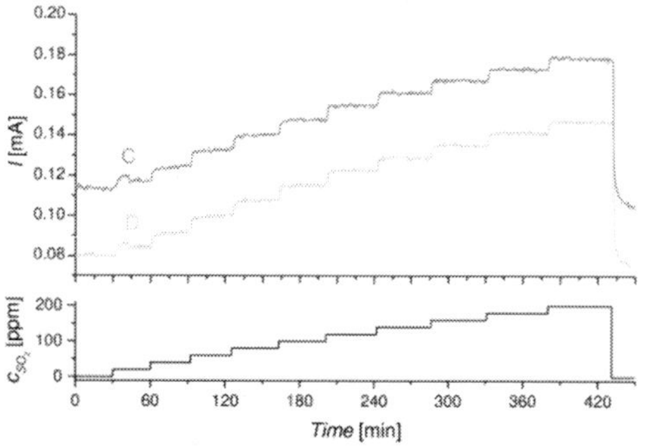

Figure 5. Current vs. time response of the sensor measured by the PA technique (upper part of figure) and change of SO_2 concentration during the PA measurement (lower part of figure).

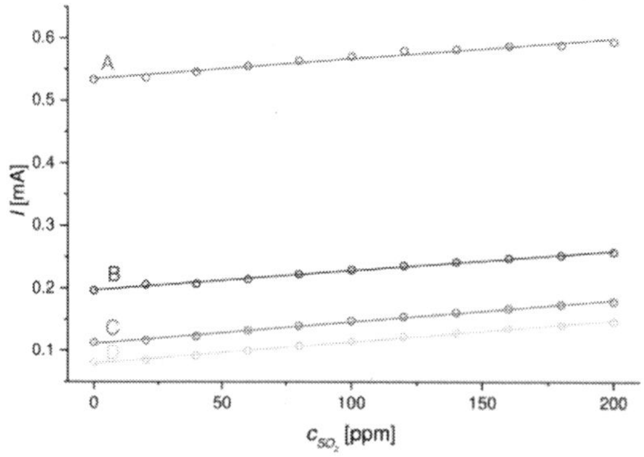

Figure 6. Calibration curves for measurement made by the PA technique.

3.3. Cyclic voltammetry

Another technique which can be used for determination of gas concentration of the SPE sensor is cyclic voltammetry. In Figure 7 two curves measured in the atmosphere of synthetic air and mixture of 1000 ppm SO_2 in synthetic air are presented. Usually, in case of the SPE sensor this type of

IOP Conf. Series: Materials Science and Engineering **104** (2015) 012029 doi:10.1088/1757-899X/104/1/012029

measurement is used for determination of the working potential used later in the CPA measurements. Cyclic voltammetry plots can be analysed with different multivariate signal processing methods [14, 15]. However, while the main goal of this article is only comparison of results obtained by different measurement techniques, so in this work only a sensitivity S was used as a representation of sensor response defined as:

$$S[mA/pm] = \frac{(I_{SO_2} - I_{sa})[mA]}{c_{SO_2}[ppm]} \tag{6}$$

where I_{SO_2} and I_{sa} are currents, obtained by the measurement in atmosphere of 1000 ppm SO_2 in synthetic air and synthetic air at the same voltage during anodic sweep, c_{SO_2} is concentration of SO_2 (1000 ppm). The sensitivity calculated for these measurements is about $6 \cdot 10^{-4}$ mA/ppm in the voltage range of 0.5-0.8 V.

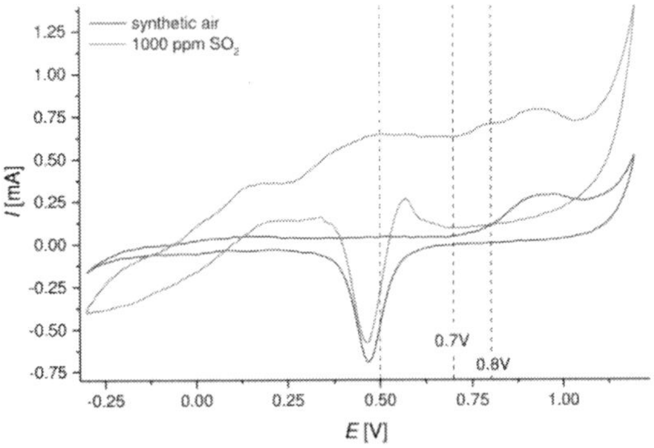

Figure 7. Current-voltage plots for synthetic air and 1000 ppm SO_2.

3.4. Square wave votlammetry

Parameters of the SWV measurements are presented in the experimental section. Plots obtained from measurements results are presented in Figure 8 and Figure 9. In the Figure 8 potential dependence of current forward (I_{fwd}) is presented. This is the current measured at the end of potential pulse. In the Figure 9 potential dependence of current difference is presented. This current is the value of difference between I_{fdw} and I_{rev}. The I_{rev} (current reverse) is the value of current measured just before the potential pulse. Usage of difference between I_{fdw} and I_{rev} allows to reduce capacitive current influence on the value of total current. Based on this data calibration curves were plotted (Figure 10).

In case of the I_{fdw} for presentation in the calibration curve, a voltage of 0.8 V was chosen as an optimal while in case of the I_{dif} value of 0.75 V was chosen. Following equations have been obtained (equation (7) for I_{fdw}, equation (8) for I_{dif}):

$$I[mA] = 1.59 \cdot 10^{-3}[mA/ppm] \cdot c[ppm] + 1.164[mA] \tag{7}$$

$$I[mA] = 6.57 \cdot 10^{-4}[mA/ppm] \cdot c[ppm] + 1.506[mA] \tag{8}$$

Coefficients of determinations (R^2) of these equation were 0.9996 and 0.9972 for equation (8) and equation (7), respectively. Sensitivity (slope of the curve) obtained for the I_{fdw} is almost 2.5 times higher than sensitivity obtained from I_{dif}. Both values of sensitivity calculated from results of SWV technique are higher than sensitivity obtained by CV technique.

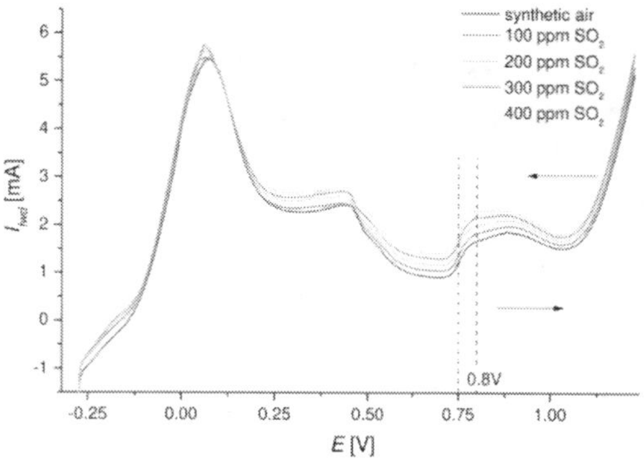

Figure 8. Current forward-voltage plots for different concentration of SO_2.

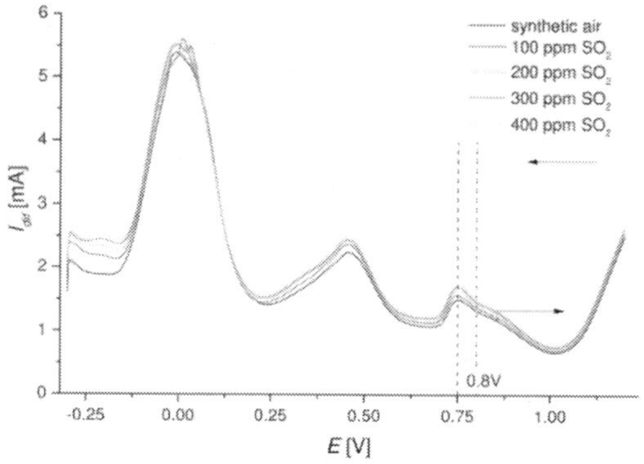

Figure 9. Current difference-voltage plots for different concentration of SO_2.

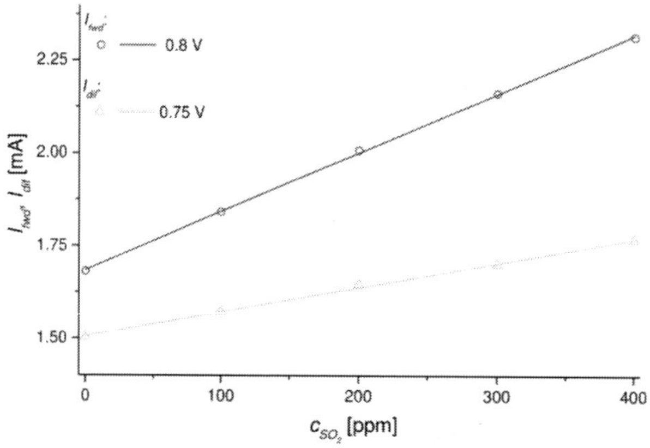

Figure 10. Calibration plots for SWV. Results obtained for the I_{fwd} and I_{dif}.

Table 1. Sensitivity of the SPE amperometric sensor obtained in measurements made by different electrochemical techniques.

Technique	Potential [V]	Sensitivity [mA/ppm]	R^2	Sensitivity [mA/(ppm·cm²)]
CPA	0.5	$1.15 \cdot 10^{-4}$	0.9945	$1.81 \cdot 10^{-4}$
	0.7	$5.27 \cdot 10^{-4}$	0.9903	$8.29 \cdot 10^{-4}$
PA	0.5	$3.46 \cdot 10^{-4}$	0.997	$5.44 \cdot 10^{-4}$
CV	0.5	$6.05 \cdot 10^{-4}$	N.A.	$9.51 \cdot 10^{-4}$
	0.7	$5.85 \cdot 10^{-4}$	N.A.	$9.20 \cdot 10^{-4}$
	0.8	$5.98 \cdot 10^{-4}$	N.A.	$9.40 \cdot 10^{-4}$
SWV (I_{fwd})	0.8	$1.59 \cdot 10^{-3}$	0.9996	$2.50 \cdot 10^{-3}$
SWV (I_{dif})	0.75	$6.57 \cdot 10^{-4}$	0.9972	$1.03 \cdot 10^{-3}$

For a better comparison, all sensitivities of the SPE amperometric sensor obtained with different electrochemical techniques were summarised in Table 1. These results clearly shows, that usage of the PA improve sensitivity in comparison with the CPA, when the same voltage is used. Further improvement is obtained by usage of the CV, but the best sensitivity can be obtained, when SWV is used.

4. Conclusions

In this article experimental results of the SPE sensor investigation are presented. Sensor response to sulphur dioxide was measured. It was proven that measurements of a gas concentration by this kind of sensor can be realised by other techniques, than constant potential amperometry which is usually used. Measurements with more sophisticated techniques, such as pulsed amperometry, cyclic amperometry and square wave amperometry were made and their results were presented. Sensitivity was calculated from all results. It can be concluded that proper choice of measurement method can improve sensor sensitivity without modification of its construction.

5. References

[1] Cao Z, Buttner W J and Stetter J R 1992 *Electroanalysis* **4** 253
[2] Jacquinot P, Hodgson A W E and Hauser P C 2001 *Anal. Chim. Acta* **443** 53
[3] Chachulski B 1998 *Analyst* **123** 1141
[4] Nagy L, Kálmán N and Nagy G 2006 *J. Biochem. Bioph. Methods* **69** 133
[5] Bard A J and Faulkner L R 2001 *Electrochemical Methods: Fundamentals and Applications* (New York: Wiley)
[6] Jasinski P, Nowakowski A and Weppner W 2000 *Sensor. Mater.* **12** 89
[7] Jasinski G, Jasinski P, Chachulski B and Nowakowski A 2005 J. *Eur. Ceram. Soc.* **25** 2969
[8] Jasinski G, Jasinski P, Nowakowski A and Chachulski B 2006 *Meas. Sci. Technol.* **17** 17
[9] Jasinski G, Jasinski P, Chachulski B and Nowakowski A 2006 *Mater. Sci. Poland* **24** 261
[10] Gebicki J and Kloskowski A 2010 *Metrol. Meas. Syst.* **XVII** 637-650
[11] Kalinowski P, Strzelczyk A, Wozniak L, Jasinski G and Jasinski P 2014 *Meas. Sci. Technol.* **25** 025101
[12] Takenaka H and Torikai E, inventors 1980 *Production of an ion-exchange membrane-catalytic electrode bonded material for electrolytic cells* (Tokyo, Japan patent no 55-38934)
[13] Strzelczyk A, Jasinski G and Chachulski B 2013 *PhD Interdisciplinary Journal* **3** 21
[14] Jasinski G, Jasinski P and Nowakowski A 2006 *Proc. SPIE* **6348** 63480E
[15] Kalinowski P, Wozniak L, Strzelczyk A, Jasinski P and Jasinski G 2013 *Metrol. Meas. Syst.* **20** 501

Acknowledgements

This work was partially supported by the National Centre for Research and Development, project LIDER No. 22/103/L-2/10/NCBiR/2011, and State Budget Funds assigned for statutory activity.

Improving the sensitivity of the ZnO gas sensor to dimethyl sulfide

P Suchorska-Woźniak, W Nawrot, O Rac, M Fiedot, H Teterycz

Wrocław University of Technology, Faculty of Microsystem Electronics and Photonics, Janiszewskiego 11/17, Wroclaw, Poland

E-mail: patrycja.wozniak@pwr.edu.pl

Abstract. This study was focused on how to improve the gas sensing properties of resistive gas sensors based on zinc oxide to dimethyl sulfide (DMS). The aim of this research was to investigate possible ways of improvement detection of dimethyl sulfide, such as volume doping with synthesized gold nanoparticles or applying sepiolite passive filter. The addition of noble metal into the gas sensing layer is a widely known method of increasing gas sensor response. Sepiolite is a clay mineral with highly porous structure consisting of nanotubes few micrometers long and water absorption abilities. In this work thick-film resistive gas sensors based on zinc oxide were made (pure ZnO, modified by gold nanoparticles, with the addition of filter) and tested for low concentration (2 ppm) of dimethyl sulfide. The sensitivities to DMS of developed sensors were compared. Attention was paid to the analysis of the impact of high humidity (90% RH) on the sensor time response.

1. Introduction

Gas detection is a common issue in environmental protection, work safety, automotive industry and alarm systems. Medical diagnostics is one of the more innovative application [1]. By detecting biomarkers, i.e. the compounds that are formed through metabolic processes in the body, we are able to notice potential disturbances in its functioning. More significant among such biomarkers, present in the exhaled air, are volatile sulfur compounds, such as dimethyl sulfide, whose increased concentration may indicate halitosis. It is a syndrome characterized by bad breath, which may be caused by liver or kidney disease, lung infection or the presence of undesirable bacteria in the mouth that cause damage to teeth and periodontal tissues [2, 3]. In order to detect small changes in concentration of dimethyl sulfide in exhaled air, a sensor sensitive to very low concentrations of this compound and resistant to the influence of high humidity is needed.

The most commonly used gas sensitive materials are metal oxides, such as zinc oxide, tin oxide and tungsten trioxide - semiconductors which, under the influence of particular gases, change their conductance. Among the most significant advantages are low production cost, high sensitivity, fast response time and the ability to detect different types of gases. Unfortunately, low selectivity is often mentioned as one of its disadvantages. In order to reduce these shortcomings filters [4] and dopants are used and the best operating temperature of sensor is matched up. The most popular gas sensitive layer dopants are metals such as gold [5, 6] and platinum [7].

In this paper the results of studies on chemical gas sensors implementation in halitosis detection are presented. The influence of dopants and filter on operating parameters of the sensors was examined. The study involved use of a thick film resistive gas sensor based on zinc oxide, material often used in detection of volatile sulfur compounds, carbon dioxide, acetone or ethanol, with high chemical and thermal stability. In order to improve sensitivity, synthesized gold nanoparticles were used as a dopant. Previous studies have shown that such additive, due to its high affinity for sulfur, significantly improves sensor sensitivity [8]. In result the lower detection threshold is decreased, which allows accurate measurement of e.g. dimethyl sulfide levels present in exhaled air. The nanometric dimensions of dopant guarantee its even distribution within the entire volume of gas sensitive layer and increases the active surface.

A great challenge during analysis of exhaled air is high humidity, above of 90% RH. It has a significant influence on sensors based on oxide materials, as adsorbed water molecules significantly impact conductance of the gas sensitive layer. To reduce distortions arising from presence of the water, a passive filter in form of sepiolite layer was used. Sepiolite is a clay mineral having a highly porous structure, consisting of nanotubes with a length of several micrometers. It is stable in the entire operating range of sensor [9] and has high adsorption capacity. Under the influence of temperature, various forms of water [10] are sequentially released from sepiolite: hygroscopic, zeolitic [11], molecules from nodes of octahedral structure and hydroxyl groups [12]. Preliminary studies have shown that presence of such filter significantly improves selectivity towards volatile organic compounds, rather than inorganic, and increases sensitivity. In this article methods of modifying sensor layer based on zinc oxide by gold nanoparticles dopant and sepiolite filter, in order to improve sensor sensitivity towards dimethyl sulfide, are presented.

2. Experimental part

2.1. Synthesis of materials

In preparation of the zinc oxide powders, commercially available reagents, characterized by the purity analytic grade 1M solution of zinc nitrate $Zn(NO_3)_2$ and a 1M solution of hexamethylenetetramine (HMT) were used. The solution was mixed in 1:1 proportion and then diluted to obtain concentration of 100 mM. Afterwards the solution was heated at 90°C for 8 hours. Resulting zinc oxide powder suspension was centrifuged, rinsed repeatedly with deionized water in the process of decantation and then dried using a lyophilizer.

Gold nanoparticles were prepared through reduction of chloroauric acid ($HAuCl_4$) by ascorbic acid. In reduction reaction of gold as stabilizers used low molecular weight polyethyleneimine (PEI) having an average molecular weight of 10 kDa, and Triton® X-100. In order to dope powder with gold nanoparticles, previously obtained ZnO powder was dispersed in isopropanol; then a solution of obtained gold nanoparticles with an average size of 17.5 nm was added to obtained mixture. The content of gold ions in solution based on weight of zinc oxide was 0.25% by weight. After stirring with a magnetic stirrer, solution was dried at 80°C and then heated in an oven at 450°C for 0.5 hours to remove organic residues.

Both pure and doped ZnO powders were processed into a paste was, which was screen printed onto alundum substrates. A series of sensors at different stages of modification was prepared:

- layer of zinc oxide;
- layer of zinc oxide doped with gold nanoparticles in an amount of 0.25 wt.%.;
- layer of zinc oxide doped with gold nanoparticles in an amount of 0.25 wt.%. with a sepiolite filter.

2.2. Sensors preparation

Sensors were formed on an alundum ceramics substrate 250 μm thick. On one side of the substrate, a meander-shaped platinum heater and gold conductive paths (electric supplies to platinum meander) were printed. On the other side of the substrate, gold electrodes with gas sensing layer and a filter, each 40 μm thick, were printed (Fig. 1).

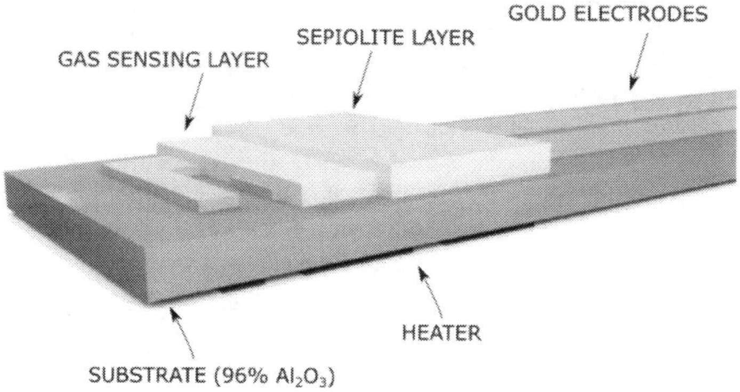

Figure 1. A schematic cross-section of the sensor structure with a sepiolite filter.

In paste preparation, an ESL-403 carrier manufactured by ESL Europe was used. The paste of given material was printed twice, with drying after each print, first at room temperature and second at 125°C for 10 mins. The complete structure of sensor was fired at 850°C for 2.5 hours. In the case of sensor with a sepiolite filter, the sensor layer was printed in such way that gas sensitive material had no direct contact with the atmosphere.

2.3. Electrical measurements

Sensor characterization was performed by a method known as Temperature Stimulated Conductance (TSC). The TSC method consists in measurements of sensor conductance as a temperature function. During measurements, tested sensor was placed in a chamber containing a gaseous atmosphere with a well-defined composition (synthetic air with humidity level at 30% RH or 90% RH and 2 ppm dimethyl sulfide). Working temperature of the heater was stimulated linearly with a constant increase of 2 deg/s in range from 150°C to 750°C (Fig. 2).

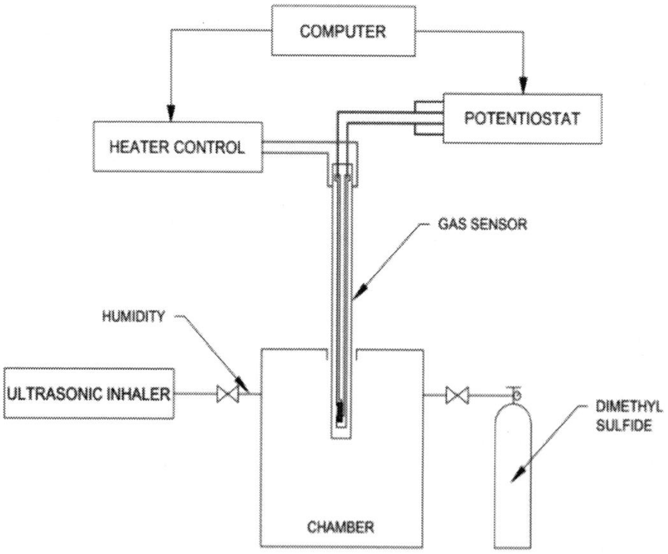

Figure 2. Schematic diagram of a measuring workstation.

Value of current flowing through the gas sensitive material was recorded during both growth and drop in temperature. Gas sensitive structure was polarized with a DC voltage, using Keithley 2400 current-voltage source and the electric current was measured. Then, using the Ohm's law, conductance of the gas-sensitive material was determined.

Furthermore, sensitivity changes in function of temperature were determined; the sensitivity was defined as a ratio of sensor conductance in an atmosphere containing particular gas to conductance in synthetic air with humidity level at 30% RH. A sensor response, i.e. time until steading of the output signal value after change from a steady value as a result of particular input signal (τ_{resp90}) and recovery time (τ_{rec90}), was also determined.

3. Results

The crystalline structure of obtained powders was determined by an X-ray Philips Materials Research Diffractometer equipped with a CuKα radiation source. Measurements were carried out by scanning the sample in an angle range Θ/2Θ. XRD studies have shown that the crystallographic structure of obtained oxide is the same as a model crystallographic structure of zinc oxide. The zinc oxide grains crystallized into a wurtzite-type structure with hexagonal unit cell and P63mc space group and lattice constants a = 3.2501 Å, b = 3.2501 Å and c = 5.2071 Å (Fig. 3).

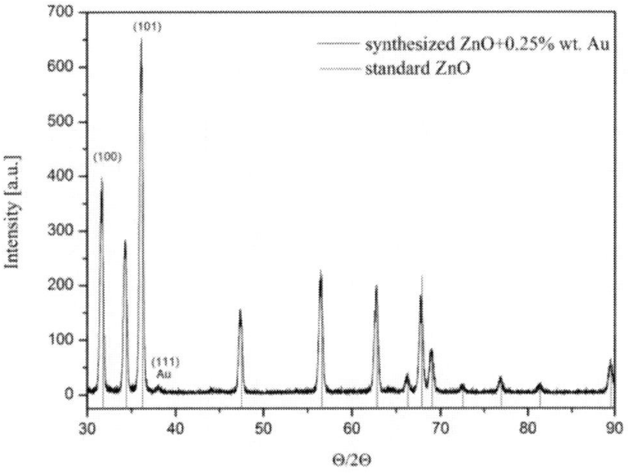

Figure 3. An X-ray diffraction pattern of synthesized zinc oxide doped with gold nanoparticles.

A peak characteristic for gold was observed on diffraction pattern. However, due to a small weight quantity of dopant (0.25 wt.%) and its large dispersion in ZnO volume, this peak is not highly visible. The average crystallite size of ZnO was determined by Scherrer equation (1):

$$L_{(101)} = \frac{K \cdot \lambda}{B \cdot \cos \Theta} \quad (1)$$

where: K – Scherrer constant (for ZnO was 0.9 based on [13]), θ – glancing angle a given interference band [rad]; λ – length of an X-ray radiation beam [Å], B – integral width (FWHM) of the peak (101) [rad].

As a result of an analysis of a diffraction peak corresponding to the plane (101) and based on the equation (1) the mean size of ZnO crystallites was 19 nm. Similarly, average size of gold crystallites was determined to approx. 5 nm.

Spectroscopic UV-vis measurements of colloidal gold solution were performed by Optizen α-Hybrid Mecasys spectrophotometer. Quartz cuvettes were used during measurements. Absorption spectra were recorded at room temperature and analyzed wavelength range was 200-700 nm. Analysis of colloidal gold sample showed a characteristic peak at a wavelength of 528 nm, which indicates presence of spherical gold nanoparticles (Fig. 4).

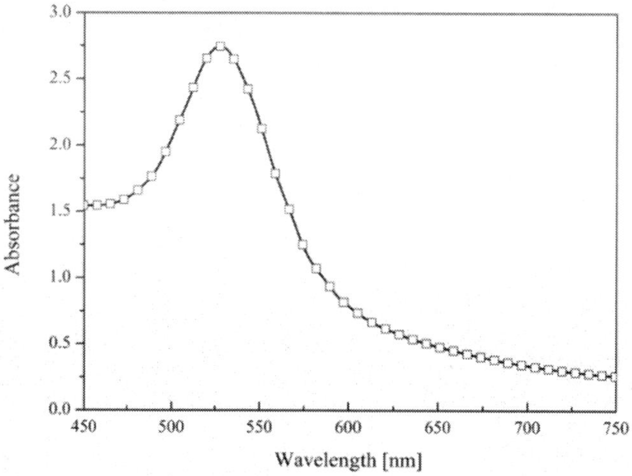

Figure 4. UV-vis spectrum of obtained gold nanoparticles.

The size distribution of gold nanoparticles was determined by Dynamic Light Scattering (DLS). Tests were performed using a Nicomp 380ZLS (Particle Sizing Systems, USA) apparatus, equipped with a 50 mW laser operating at 532 nm wavelength. The average gold particle size determined by this method was approximately 17.5 nm (Fig. 5).

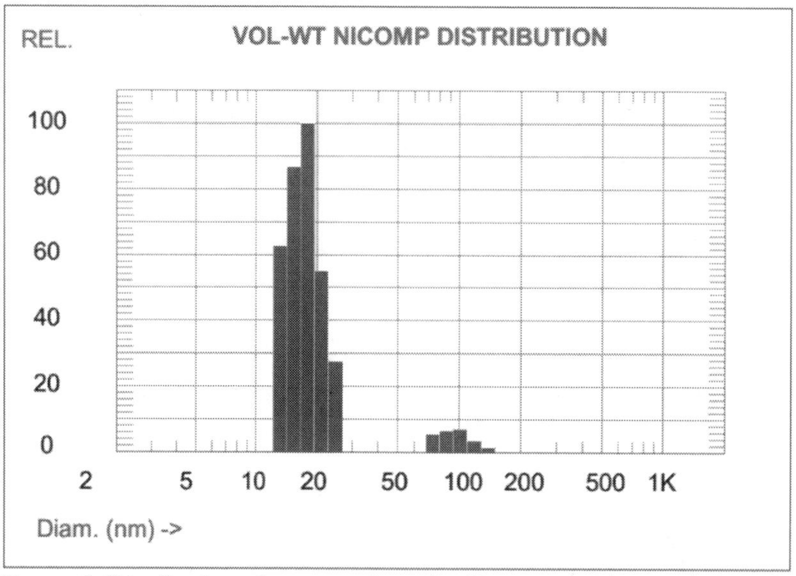

Figure 5. Distribution of average size of gold nanoparticles carried out by DLS.

The substructure of gold nanoparticles and sepiolite was observed using EM900 Transmission Electron Microscope (TEM) (Zeiss, Germany). TEM analysis of nanoparticles revealed that obtained nanoparticles have a spherical shape and sepiolite is composed of nanotubes having a length of several micrometers and a diameter of approx. 10-15 nm (Fig. 6).

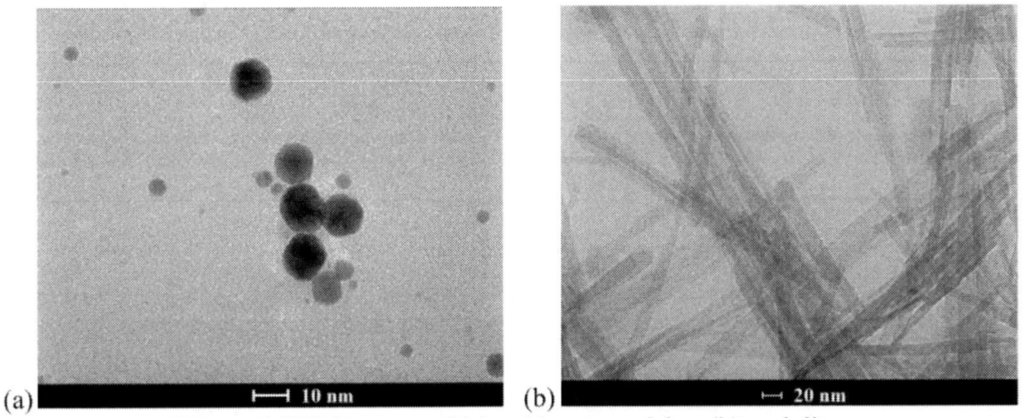

Figure 6. TEM images of (a) gold nanoparticles; (b) sepiolite.

In order to determine influence of gas sensing material, dopant and passive filter on sensor parameters during detection of dimethyl sulfide, electrical characterization of all three types of sensors was performed. Based on temperature stimulated conductance changes of sensor in humid dimethyl sulfide atmosphere, sensor sensitivity (Fig. 7) was determined. It has been found that use of dopant in form of gold nanoparticles significantly improved sensitivity towards examined sulfur compound. Furthermore, increase in sensitivity was observed after sepiolite filter was applied in addition to dopants. In this case the influence of water molecules desorption in whole temperature range is noticeable, which is a result of different water forms presence.

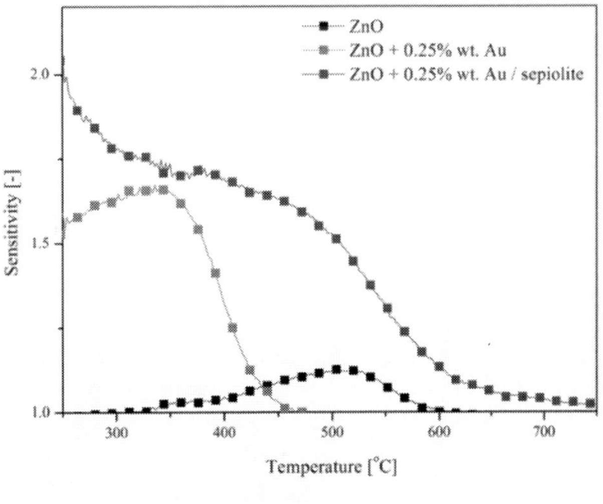

Figure 7. Sensor sensitivity in an atmosphere with a 2 ppm dimethyl sulfide and humidity of 90% RH.

The response rate and recovery time of sensors were examined. These are important parameters of sensors, as they determine required measuring period and interval before next measurement (Fig.8).

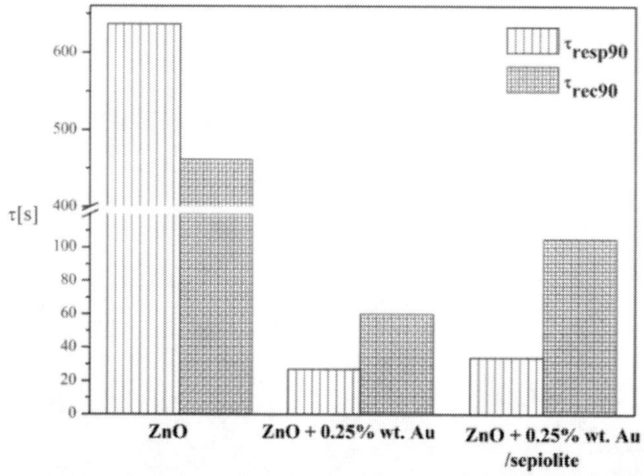

Figure 8. The response rate and recovery time of analyzed sensors after exposure to an atmosphere having a humidity of 90% RH containing 2 ppm dimethyl sulfide.

Noticeably, sensor based on pure zinc oxide has both longest response and recovery times. On the other hand, zinc oxide doped by gold nanoparticles responded fastest to presence of dimethyl sulfide. Presence of dopants not only improved sensitivity relative to determined gas, but also significantly affected pace increase. Additional presence of sepiolite is sort of barrier, extending adsorption and desorption of volatile compounds, which thus contributes to a slight increase in response and recovery times.

4. Summary

The article presents methods of modifying sensor layer based on zinc oxide by doping with gold nanoparticles and applying a sepiolite filter. Furthermore, effect of modifications on electrical characteristics of thick-layered resistive gas sensor during detection of dimethyl sulfide in atmosphere having a high humidity (90% RH) was determined.

All three sensors: based on pure ZnO, doped with gold nanoparticles and additionally coated with a sepiolite filter demonstrated good response relative to determined gas. Among analyzed sensors, ZnO+Au sensor was characterized by shortest response time (τresp90 = 26 sec.) and by shortest recovery time.

The results show that with every further modification of sensor material and structure (filter presence), sensitivity to dimethyl sulfide is increased. Improvement of sensitivity by introducing gold nanoparticles in volume of gas sensitive material is due to high affinity of gold for sulfur compounds.

On the other hand, temperature stimulated changes in the structure of sepiolite and elimination of high humidity influence in analyzed atmosphere yielded a sensor with the highest sensitivity.

Further research will focus on increasing quantity of gas-sensitive layer dopants in order to determine best concentration, thereby to obtain maximum sensitivity relative to determined gas. It is necessary to perform further research in order to unravel in detail the mechanism of dimethyl sulfide oxidation on the surface of developed sensors.

5. References

[1] Righettoni M, Amann A and Pratsinis S E 2015 Breath analysis by nanostructured metal oxides as chemo-resistive gas sensors *Mater. Today* **18** 163-171

[2] Krespi Y P, Shrime M G, Kacker A 2006 The relationship between oral malodor and volatile sulfur compound–producing bacteria *Otolaryngol. Head. Neck. Surg.* **135** 671-676

[3] Tangerman A and Winkel E G 2010 Extra-oral halitosis: an overview *J. Breath Res.* **4** 017003

[4] Pijolat C, Viricelle J P, Tournier G and Montmeat P 2005 Application of membranes and filtering films for gas sensors improvements *Thin Solid Films* **490** 7-16

[5] Guo J, Zhang J, Zhu M, Ju D, Xu H, Cao B 2014 High-performance gas sensor based on ZnO nanowires functionalized by Au nanoparticles *Sensor. Actuat. B Chem.* **199** 339-345

[6] Hosseini Z S, Mortezaali A, Iraji A and Fardindoost 2015 Sensitive and selective room temperature H_2S gas sensor based on Au sensitized vertical ZnO nanorods with flower-like structures *J. Alloy. Compd.* **628** 222-229

[7] Chang S J, Weng W Y, Hsu C L and Hsueh T J 2010 High sensitivity of a ZnO nanowire-based ammonia gas sensor with Pt nano-particles *Nano Commun. Netw.* **1** 283-288

[8] Suchorska-Woźniak P, Teterycz H 2013 Evaluation of the sensors for the detection of low concentrations of hydrogen sulfide *Proceedings of 37th International Microelectronics and Packaging IMAPS-CPMT Conference* (Kraków) pp 1-4

[9] Teterycz H, Günister E, Ece Ö I, Halek G and Klimkiewicz R 2008 The influence of temperature on the microstructure of sepiolite *Elektronika: konstrukcje, technologie, zastosowania* **49** 251-252

[10] Balci S 1996 Thermal decomposition of sepiolite and variations in pore structure without and with acid pre-treatment *J. Chem. Technol. & Biotechnol.* **66** 72-78

[11] Alkan M and Benlikaya R 2009 Poly(vinyl alcohol) nanocomposites with sepiolite and heat treated sepiolites *J. Appl. Polym. Sci.* **112** 3764-3774

[12] Yener N, Önal M, Üstünışık G and Sarıkaya Y 2007 Thermal behavior of a mineral mixture of sepiolite and dolomite *J. Therm. Anal. Calorim.* **88** 813-817

[13] Humayun Q, Kashif M, Hashim U and Qurashi A 2014 Selective growth of ZnO nanorods on microgap electrodes and their applications in UV sensors *Nanoscale Res. Lett.* **9** 1-5

Acknowledgements

The project was funded by the National Science Centre based on Decision Number DEC-2012/07/N/ST7/02304. The analysis of the influence of temperature and humidity on the time response of sensors was financed by Polish Ministry of Science and Higher Education from funds for young researchers and PhD students as project no. B40075 W-12.

IOP Conf. Series: Materials Science and Engineering **104** (2015) 012031 doi:10.1088/1757-899X/104/1/012031

Proximity sensors based on ball-lensed optical fibers

B Guzowski, M Lakomski, M Cywinski

Dept. of Semiconductor and Optoelectronics Devices, Technical University of Lodz, 90-924 Lodz ul. Wolczanska 211/215, Poland

Abstract. In this paper, proximity sensors based on ball-lensed single-mode fibers (SMF) are described. Two types of sensors are presented: (1) Type A with one transmitting and four receiving optical fibers and (2) Type B with one transmitting and eight receiving optical fibers. In both types ball-lensed optical fibers are used as a receiving line. Sensitivity of these sensors is compared to sensitivity of sensors with the same configurations, but involving cleaved optical fibers. All developed sensors were tested at two most popular in SMF wavelengths: 1310 nm and 1550 nm. As a refractive surface the silicon wafer was used.

1. Introduction

There is a great number of types of optical fiber sensors. They can be classified in many ways. Generally intrinsic and extrinsic sensors can be distinguished. In the intrinsic sensors, e.g. reflective sensor, the role of optical fibers is limited to distributing light to and from optical converter. In the extrinsic sensors, such as microbending or interferometric sensors, optical fiber distributes light and at the same time functions as an optical converter. The advantages of fiber optic sensors, such as the resistance to electromagnetic interference, high sensitivity or very small size result in wide range of application [1-5]. Reflective sensors are a group of intensity sensors. Due to high accuracy of processing, low cost and contactless acting they are often used [6-7].

In this type of sensors usually the multi-mode fibers (MMF) and are used. They have bigger core in comparison to single-mode fibers (SMF), therefore MMF can capture more reflected light. However typical wavelengths used in SMF have much lower attenuation than wavelengths transmitted in MMF and in SMF there is no mode dispersion. Therefore in some precision application SMF can be more suitable.

In this paper the results of the tests of four different proximity sensors are shown. To increase amount of reflected light captured by receiving optical fibers, the ball lenses were proposed. Obtained results were compared to the results received during the tests, where cleaved optical fibers were used. As a head of a sensor novel MTP C12405 ferrule was used. This small element allowed us to precisely distribution and adjustment of optical fibers.

2. Theoretical models of sensors

2.1. Cleaved optical fibers

In literature different models of optical fiber reflective sensors are described. In some models the receiving optical fibers are parallel to transmitting optical fibers [8]. There are also models with inclined optical fibers [9,10]. In this paper the theoretical model based on [11] is presented. Kaczmarek in his model assumed that optical fibers are in contact (figure 1a). In our model there is a gap *d* between optical fibers (figures 1b-1d). To simplify the analysis, it is assumed, that reflective surface reflects all incoming

IOP Conf. Series: Materials Science and Engineering **104** (2015) 012031 doi:10.1088/1757-899X/104/1/012031

light and there is no absorption of the light between optical fibers and the surface. To analyze reflective sensor operation, conical shape of the light beam emitted by the transmitting optical fiber and circular cross-section of all optical fibers are assumed. The light intensity of the image of receiving optical fiber is determined. As the analysis method, geometrical optics method was used. It assumes that the intensity of illumination in cross-section of the cone is constant and given by the formula:

$$I(z) = \frac{P_O}{\pi w^2(z)} \tag{1}$$

where: P_o is output power at the tip of transmitting optical fiber, w – cone radius as a function of z.

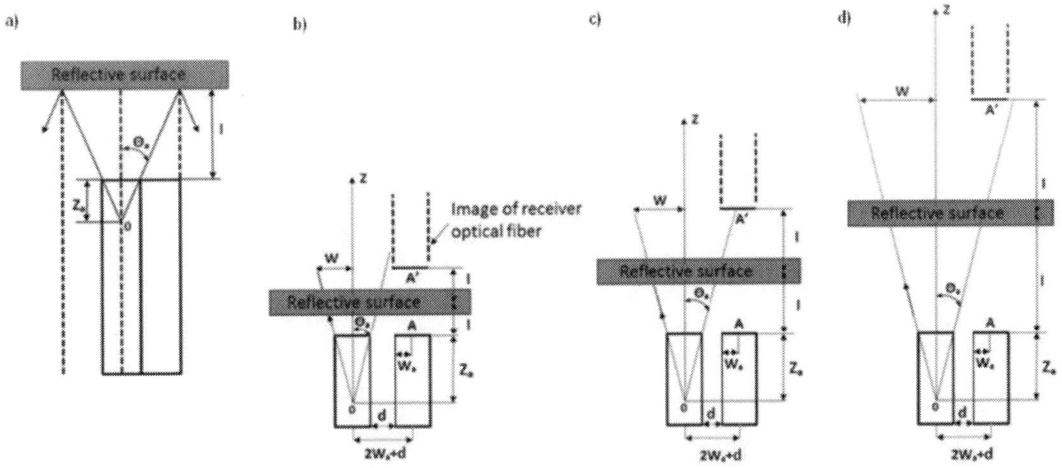

Figure 1. The idea of reflective sensor: a) optical fiber are in contact [11], b) out of operating range, c) partially illuminated, d) fully illuminated

Formula of $w(z)$ function is given below:

$$w(z) = z\,\mathrm{tg}\theta_a = z\frac{w_a}{z_a} = z\frac{w}{l+z_a} \tag{2}$$

For very small displacements ($l < \frac{dz_a}{2w_a}$) the reflected light is not captured by receiving fibers.

When $\frac{dz_a}{2w_a} < l < \frac{z_a\left(w_a+\frac{d}{2}\right)}{w_a}$ surface of receiving optical fiber $S(z)$ is illuminated only partly as shown in figure 2a). For bigger displacements $\left(l > \frac{z_a\left(w_a+\frac{d}{2}\right)}{w_a}\right)$ surface $S(z)$ is illuminated fully – figure 2b).

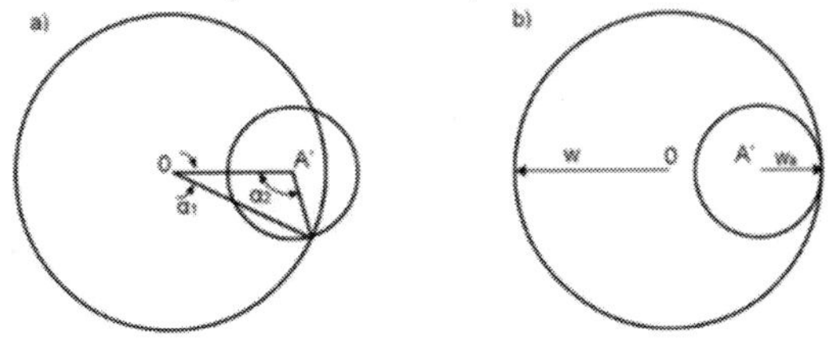

Figure 2. Reflected surface of the receiving optical fiber: a) partially for $\frac{dz_a}{2w_a} < l < \frac{z_a\left(w_a+\frac{d}{2}\right)}{w_a}$, b) fully for $l > \frac{z_a\left(w_a+\frac{d}{2}\right)}{w_a}$

When $\frac{dz_a}{2w_a} < l < \frac{z_a\left(w_a+\frac{d}{2}\right)}{w_a}$ surface $S(z)$ is described by formula:

$$S(z) = w^2(z)\alpha_1(z) + {w_a}^2[\alpha_2(z) - \sin\alpha_2(z)] - w(z)w_a\sin\alpha_1(z) \quad (3)$$

Thus, the power of light incident on the end of the receiving optical fiber is described by formula:

$$P(z) = \frac{S(z)}{\pi w^2(z)} P_o \quad (4)$$

When $l > \frac{z_a\left(w_a+\frac{d}{2}\right)}{w_a}$ surface $S(z) = \pi {w_a}^2$ therefore the power $P(z)$ is described as:

$$P(z) = \frac{{w_a}^2}{w^2(z)} P_o \quad (5)$$

2.2. Theoretical model of ball-lensed optical fibers

To increase the amount of received light and thereby the sensor sensitivity we proposed using ball-lensed optical fiber (BLOF) [12], shown in figure 3, as a receiving optical fibers.

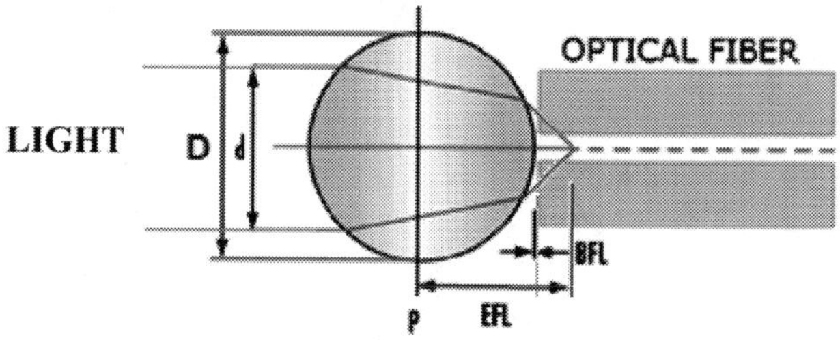

Figure 3. Ball-lensed optical fiber (BLOF), where D – diameter of ball lens and d – diameter of input source, EFL – effective focal length, BFL – back focal length of ball lens

The EFL of a ball lens is determined using the equation (6), while the BFL is represented using the equation (7):

$$EFL = \frac{nD}{4(n-1)} \quad (6)$$

$$BFL = EFL - \frac{D}{2} \quad (7)$$

where n is the refractive index of the material.
The numerical aperture NA of the lens is represented by equation (8):

$$NA = \frac{2d(n-1)}{nD} \quad (8)$$

The maximum surface (when $d = D$) of the ball lens that can be illuminated is given by the formula $S(z) = 2\pi r^2$. In this special case, when we assumed that the ball lens is made of glass ($n = 1.5$) the NA of lens is 0.67. It is five times more than for cleaved SMF. However it must be taken into account that not all of the incoming light will be focused to the core of the optical fiber in this case. Some of the light will be focused to the cladding and will be lost.

3. Construction of the sensor

3.1. Ball-lensed optical fibers

In the Optical Fibers Techniques Laboratory at Lodz University of Technology four types of BLOF were realized. The lenses diameters were in the range of 140 µm to 230 µm.

The process of realization of BLOF was made with the use of Fitel S153 by Furukawa fusion splicer. To realize micro-lens, cleaved optical fiber SMF class G.652 was placed in V-groove, as it is presented on figure 4. The end of the optical fibers was located between the V-groove and two electrodes.

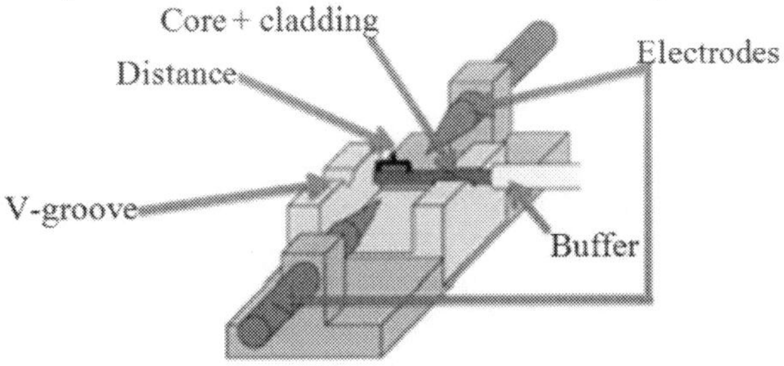

Figure 4. Schematic of fiber alignment inside the splicer.

The manual mode in optical splicer allow to set some parameters which are responsible for making the lens. The most important of them are shown in the Tab 1. First of all, desired operating parameters of splicer was set and SMF fiber was positioned relative to the two electrodes, according to "Initial length of fiber". Then the two electrical discharges of electrodes were induced and caused melting of SMF fiber end. In this process ball micro-lens was formed. During these discharges the optical fiber was moved in accordance with the parameter indicated as the " Z push distance".

Table 1. Programming parameters.

Parameter	Names of the micro-lenses			
	Ball-lensed 140 µm	Ball-lensed 170 µm	Ball-lensed 200 µm	Ball-lensed 230 µm
Diameter of lens (µm)	140 ± 4	170 ± 3	200 ± 4	230 ± 4
First arc power (mW)	50	100	125	200
Electrodes discharge time (s)	3	2	6	3
Second arc power (mW)	80	167	167	200
Initial length of fiber (µm)	150	200	200	200
Z push distance (mm)	0.02	3	16.4	32.767

3.2. Head of the sensor

The most crucial element in proximity sensor is the head of the sensor. It must provide the proper and precise alignment of the optical fibers. Due to this fact this element can be complicated and expensive. In our construction of the sensor as a head of the sensor, the MTP C12405 optical fiber ferrule shown in figure 5 was used. This type of ferrule, developed by US Conec in 2014, allows to place up to 48 optical fibers.

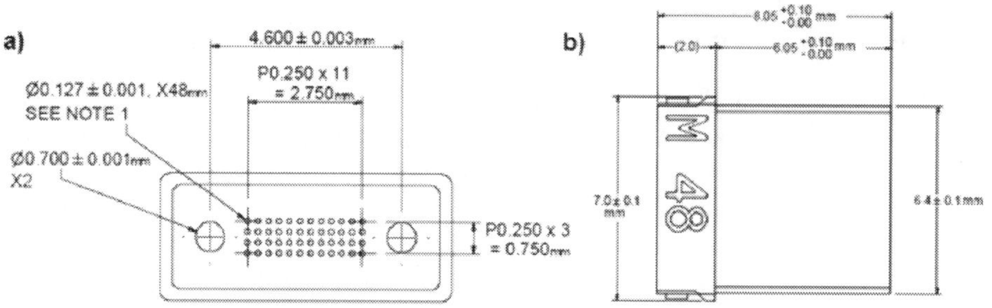

Figure 5. Head of sensor - MTP C12405 optical connector: (a) front view, (b) view from the top [13]

Two different configurations of sensors were designed, called type A and type B. The difference is between receiving optical fibers, respectively 4 and 8 optical fibers. Moreover, for these two types sensors involving also cleaved optical fibers as receiving optical fibers were prepared. All four configurations of the developed proximity sensors are shown in figure 6. In each configuration, the transmitting fiber was always the central fiber (red color in figure 6) and it was always cleaved. All of the used optical fibers were SMF class G.652.

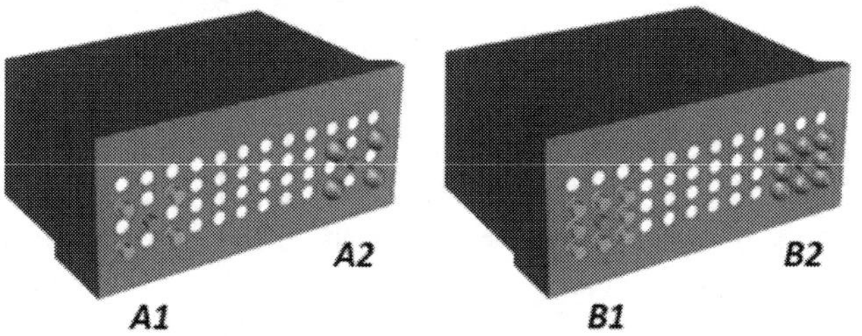

Figure 6. The distribution of receiving optical fibers in sensors: a) Type A sensors with 4 cleaved optical fibers and 4 ball-lensed optical fibers, b) Type B sensors with 8 cleaved optical fibers and 8 ball-lensed optical fibers.

The photos of developed sensor heads, taken by a scanning electron microscope Carl Zeiss model EVO MA10, are shown in figure 7 and figure 8.

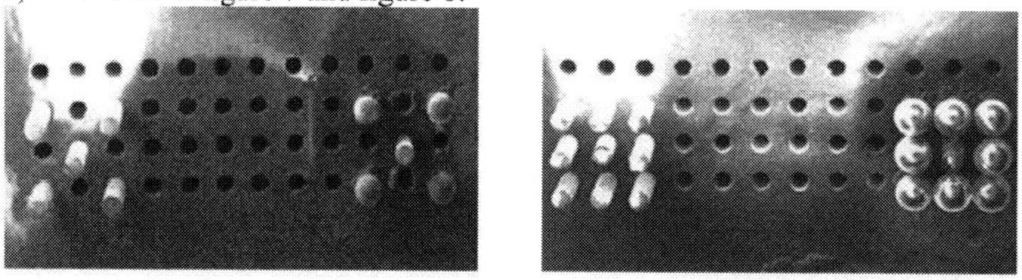

Figure 7. Real photo of head of sensors enlarged 88x: a) Type A with 4 receiving optical fibers, b) Type B with 8 receiving optical fibers.

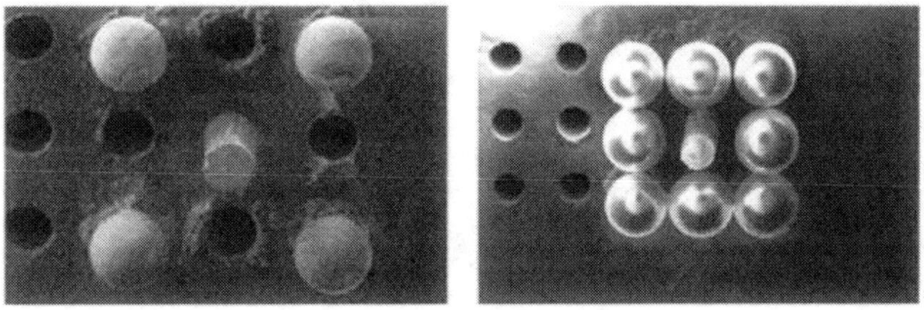

Figure 8. a) Type A sensor with 4 ball-lensed optical fibers enlarged 284x, b) Type B sensor with 8 ball-lensed optical fibers enlarged 170x.

4. Measuring setup

A measuring setup used to carry out all the measurements is presented in figure 9a). Additionally, the Thorlabs MAX373D fiber launch system (5) was used as shown in figure 9b). In this setup the light source FLS-600 by EXFO (1) was connected through SCPC/APC patchcord (2) to transmitting fiber (3). This fiber is the part of the sensor head (4) which was placed on a fixed plate with a holder (6). The reflecting surface was silicon wafer (7) fixed to a moving plate of MAX373D system. The receiving fibers (8) was connected to the optical power meter AF-ORL-3.1 (9).

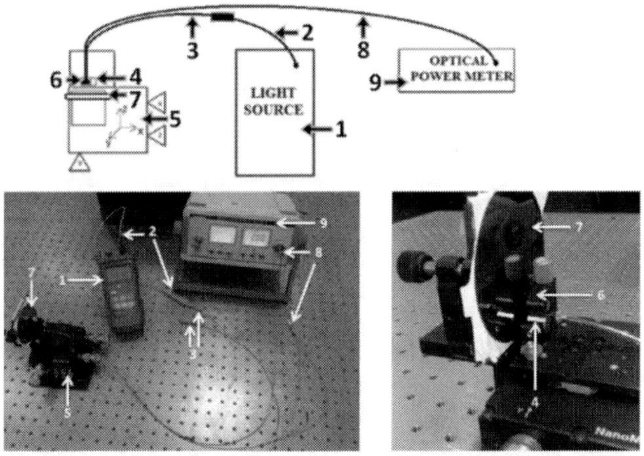

Figure 9. a) Scheme of the measuring setup b) Measuring setup and head of sensor mounted in front of the reflecting surface.

5. Results

All 4 configurations were tested for two wavelengths: 1310 nm and 1550 nm ± 20 nm. Output characteristics $P_R/P_T = f(h)$ (for constant optical power source) have been designated. Operating range of the sensor was set to 4000 µm with 25 ± 1 µm resolution. In order to eliminate the influence of external light, each configuration was examined five times, in a darkened room at the temperature of 22°C. To remove all possible light from the cladding, the receiving optical fibers were wound five times around a coil with a diameter of 10 mm.

Figure 10 shows the comparison of Type A configuration both for cleaved and ball-lensed optical fibers for both used wavelengths. Similarly results obtained for Type B sensor are shown in figure 11. On the other hand, figure 12 and figure 13 show the characteristics of both wavelengths, 1310 nm and 1550 nm. These figures show the differences between Type A and Type B sensors.

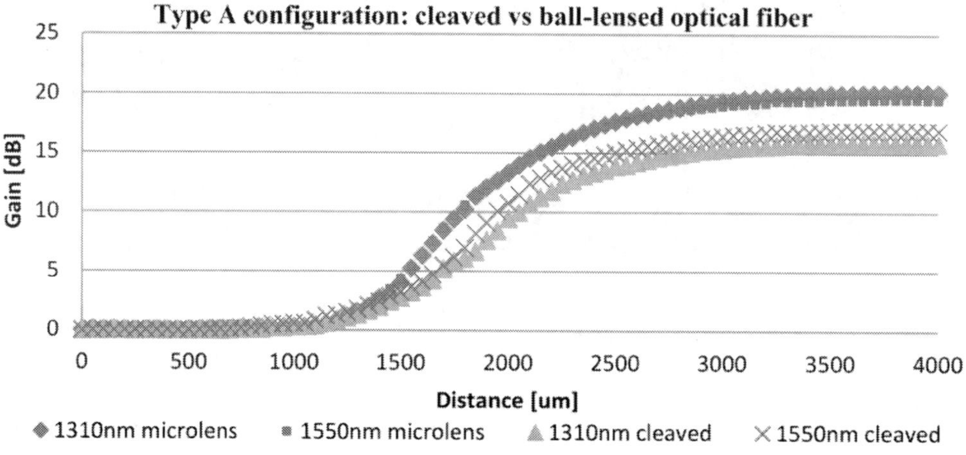

Figure 10. Comparison of cleaved and ball-lensed optical fiber in Type A sensor

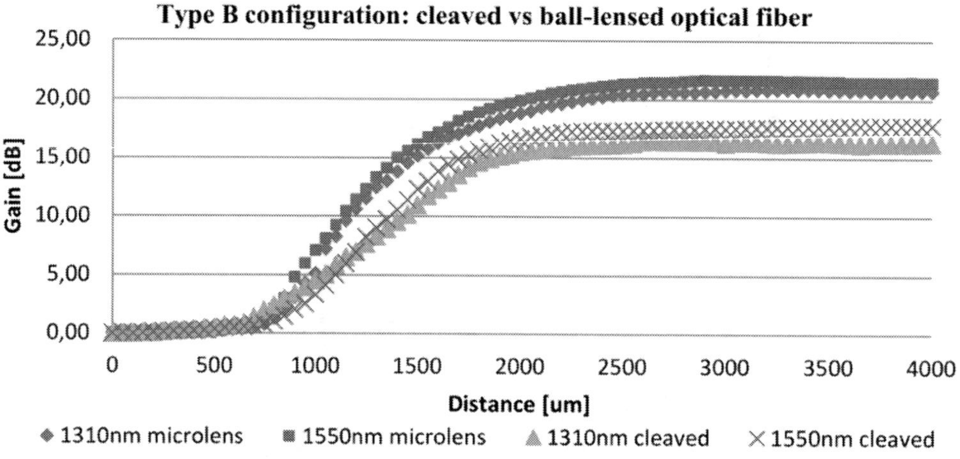

Figure 11. Comparison of cleaved and ball-lensed optical fiber in Type B sensor

On the basis of the results presented in figure 10 and figure 11, it can be stated that for the distance around 1.45 - 4 mm for both wavelengths (1310 nm and 1550 nm), developed sensors (type A and type B) with BLOF are more sensitive than sensors with cleaved optical fibers. For 1310 nm wavelength (figure 10) sensors with BLOF, received power of reflected light is almost 50% bigger in comparison to sensors with cleaved optical fiber (in distance 1.5–2 mm). For bigger distance using BLOF in sensors results in increased sensors sensitivity up to 33%. For the wavelength 1550nm using ball-lenses increases received power proportionally by c.a. 40% in distance 1.5–2 mm, and for more than 2mm by 20%.

More receiving optical fibers in type B sensors (figure 11) results in increased operating range of sensors. The operating range starts 5mm earlier (in distance 750 μm) compared to type A sensor. Also in this configuration, sensors with ball-lensed optical fibers collect and distribute more (about 30-40 %) reflected light than sensors with cleaved optical fibers.

Another correlation was noticed on the basis of the research presented in figure 12 and figure 13. Type B sensor with 8 receiving optical fibers, called B2, is the most sensitive sensor in the whole tested range and both wavelengths. For bigger distance (2.2 - 2.5 mm and more) sensor with 4 BLOF as receiving fibers (type A2) collect more light than sensor B1 with 8 cleaved optical fiber. These result clearly show the better effectiveness of the sensors with ball-lensed optical fibers.

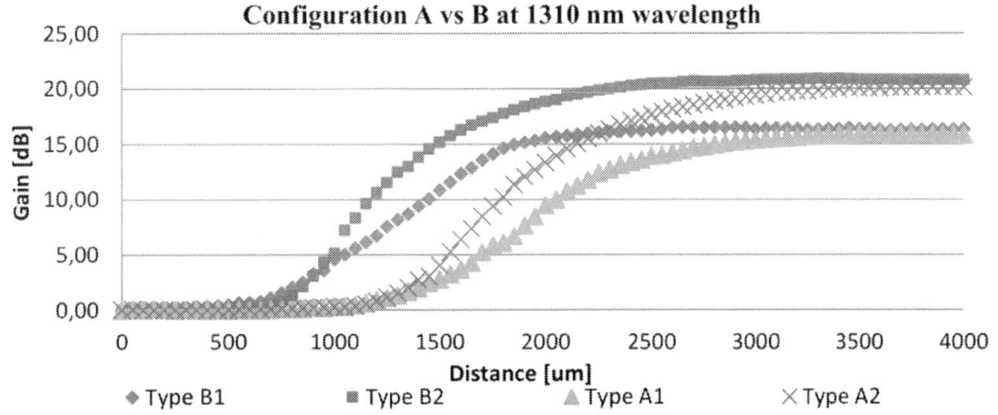

Figure 12. Comparison of two configurations in both type of sensors – wavelength 1310 nm.

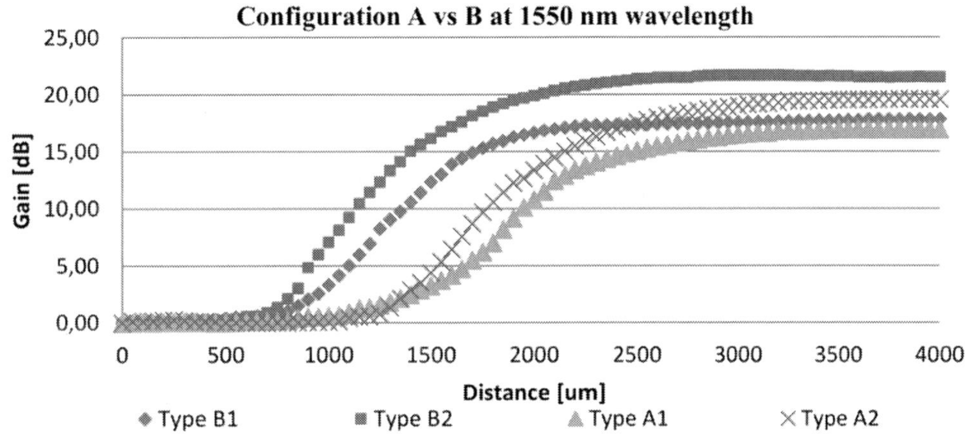

Figure 13. Comparison of two configurations in both type of sensors – wavelength 1550 nm.

6. Conclusions

In this paper, two configurations of proximity sensors using BLOF were presented and compared with similar sensors using cleaved optical fibers. As a head of a sensor novel MTP C12405 ferrule was used, which allowed us to precisely distribution of optical fibers. Based on our study, we show that using BLOF as a receiving optical fibers, significantly increase the amount of collected light. Type A sensor with 4 BLOF (A2 sensor) for the distance bigger than 2.4 - 2.5 mm collects more light than 8 cleaved optical fibers (B1 sensor). Moreover the operating range of the sensors with BLOF is wider due to increased collecting surface. Number of receiving optical fibers placed in the sensors also change the sensitivity of proximity sensor. Both types of sensors: A and B with 8 receiving optical fibers are more sensitive than sensors with 4 receiving optical fibers.

The obtained results may contain small errors due to unequal positioning ± 5 μm of the cleaved fibers into the MTP connector, or due to the lenses being of different sizes ± 4 μm. Those errors have been found to be within the range of 2 - 3 μm, which certainly does not affect how the proximity sensor works or the results obtained from the measurements. To make the theoretical model more precisely the Beer–Lambert law must be taken in account, as well as the absorption of the light by the reflective surface.

7. References

[1] Walker S Loewke K et al 2007 An optical fiber proximity sensor for haptic exploration with a robotic finger (*IEEE Robotics and Automation*) pp 473-478

[2] Sasaki M Ando T et al 2001 Lens fabrication on optical fiber end using photolithography pp 258-259

[3] Nabeel R A et al 2006 Harsh environments minimally invasive optical sensor using free-space targeted single-crystal silicon carbide (*IEEE Sensors Journal* **6(3)**) pp 672-685

[4] Quero G et al 2013 Two-dimensional hybrid metallo-dielectric nanostructures directly realized on the tip of optical fibers for sensing applications (*Proceedings of the SPIE* **8774** id. 877402) pp 1-10

[5] Consales M Ricciardi A et al 2012 Lab-on-Fiber Technology: Toward Multifunctional Optical Nanoprobes (*ACS Nano* **6(4)**) pp 3163–3170

[6] Yuan J et al 2014 A Fresnel reflection-based optical fiber sensor system for remote refractive index measurement using an OTDR (*Photonic Sensors, Vol* ***4****, Issue 1*) pp 48-52

[7] Chen C et al 2012 Reflective Optical Fiber Sensors Based on Tilted Fiber Bragg Gratings Fabricated With Femtosecond Laser, (*Journal of Lightwave Technology Vol* ***31****, Issue 3*) pp 455-460

[8] Wanninayake B Indika 2015 Modeling and Optimizing Output Characteristics of Intensity Modulated Optical Fiber-Based Displacement Sensors, (*IEEE Transactions on Instrumentation and Measurement Vol* ***64****, Issue 3*) pp 758-767

[9] Woo-Seong C et al 2011 An intensity function for inclined multimode optical fiber sensors, (*10th International Workshop on Electronics, Control, Measurement and Signals ECMS*) pp 1-6

[10] Puangmali P et al 2010 Mathematical Modeling of Intensity-Modulated Bent-Tip Optical Fiber Displacement Sensors, (*IEEE Transactions on Instrumentation and Measurement Vol* ***59****, Issue 2*) pp 283-291

[11] Kaczmarek Z 2006 Swiatlowodowe czujniki i przetworniki pomiarowe, ISBN 83-87982-07-5 pp 105-111

[12] Kato D 1973 Light coupling from a stripe-geometry GaAs diode laser into an optical fiber with spherical end (*Journal of Applied Physics* **44**) pp 2756-2758

[13] Yount M 2009 Customer Drawing http://www.usconec.com/images/drawings/C12405.pdf

Acknowledgment

The authors would like to thank the US Conect company for sharing with us the MTP ferrules.

Computational complexity and length of recorded data for fluctuation enhanced sensing method in resistive gas sensors

Ł Lentka, J Smulko

Department of Metrology and Optoelectronics, Faculty of Electronics, Telecommunications and Informatics, Gdańsk University of Technology, Gdańsk, Poland

E-mail: lukasz.lentka@gmail.com

Abstract. This paper considers complexity and accuracy of data processing for gas detection using resistance fluctuation data observed in resistance gas sensors. A few selected methods were considered (Principal Component Analysis – PCA, Support Vector Machine – SVM). Functions like power spectral density or histogram were used to create input data vector for these algorithms from the observed resistance fluctuations. The presented considerations are important for proposing relatively cheap and mobile gas detection devices of limited computations abilities and utilizing fluctuation enhanced gas sensing method.

1. Introduction

Resistive gas sensors are commonly used in various practical applications [1-3]. Their most important task is to detect gas of interest or determine its concentration at sufficiently high accuracy because this feature can be often crucial for living standard and health of human beings. A single resistance gas sensor is often insufficient to reach high concentration accuracy or high detection rate. Therefore a set of gas sensors is applied with selected data processing algorithm. There are many well-known algorithms and methods [4-5] which can be used to detect various gases utilizing resistive sensors or other, even more expensive, gas sensors types (e.g. electrochemical or optical [6, 7]). Nowadays there is strong pressure on proposing inexpensive and miniaturized gas sensors to detect various gases in portable systems, accessible for wide range of customers (e.g. in mobile phones). Thus, the system which applies a set of resistive gas sensors cannot be used in portable applications because of high energy consumption (each resistive gas sensor has to be heated up). Another proposed method can utilize fluctuation phenomena observed in resistive gas sensors (fluctuation enhanced sensing – FES) at low frequency range to gather more data from much limited number of gas sensors. Unfortunately, saving energy for heating requires more energy for data recording and processing. That issue has been addressed in the paper. Selected algorithms are considered due to their computational complexity and accuracy. These algorithms are applied in gas sensing area and can use the FES method as well. The conclusions should help us to decide what kind of statistical parameters characterizing resistance fluctuations would be most efficient in practical applications of the FES method.

Power spectral density and histogram characterizing noise phenomena together with the selected algorithms of data processing were taken into account. Both functions characterizing fluctuations were considered by analysing their random error due to limited number of recorded noise samples used for their estimation.

2. Power spectral density

Fluctuations can be characterized by power spectral density (PSD), determining their power distribution versus frequency. The power spectral density $S(f)$ of random signal can be defined by equation:

$$S(f)=\frac{1}{2\pi N}\sum_{i=1}^{N}\left|X_i(f)\right|^2 \tag{1}$$

where $X_i(f)$ is a Fourier transform (spectrum) of time series being a part of the analysed signal $x(t)$. Power spectrum density is estimated by recording noise samples, computing Fourier transform for the consecutive time series and averaging over the set N of the estimated spectra to reduce its random error. The FES method utilizes low frequency noise which depends on ambient atmosphere of the applied gas sensor [2, 3]. Resistance fluctuations are observed as voltage fluctuations across the sensor biased by DC voltage [8-11]. It has been confirmed that low frequency noise power spectrum is often much more sensitive to any changes of ambient atmosphere of gas sensor than a change of its DC resistance only [12]. Moreover, the PSD is a function of frequency and gathers information about presence of the investigated gases by changing its intensity and slope versus frequency. Usually, the product $f{\cdot}S_u(f)/U^2$ of voltage power spectral density multiplied by frequency f and normalized by squared DC voltage bias of the sensor is used. That product is independent from actual measurement circuit and depends on the sensor and measured gas only. Thanks to multiplying the measured spectra by frequency, visibility of any spectra shape changes is improved [12].

Averaging over the spectra set reduces variance of the estimated PSD. When N independent spectra are averaged a random error ε_r of the PSD is given by equation:

$$\varepsilon_r=\left(\frac{1}{N}\right)^{1/2} \tag{2}$$

For example, if we average $N = 40$ spectra, the random error is reduced to $\varepsilon_r \approx 16\%$ which can be accepted for noise measurements. Strong increase of the averaged spectra number is often not recommended because it requires much longer noise records. Assuming a sampling frequency $f_s = 4$ kHz necessary to observe $1/f$ noise in the gas sensors and 1024-point spectrum we need to record noise in time of 10.24 s to estimate $N = 40$ independent spectra. That time can be accepted for practical applications. Twice lower random error means four time longer time of noise recording. It means that further increase of noise recording time cannot be often accepted.

The PSD function requires application of fast Fourier transform algorithm which is very popular and can be implemented even in a cheap portable gas sensing devices optimized for signal processing operations, like FPGA arrays or signal processors. The estimated PSD can be used as input data vector for regression or detection algorithms to determine gas concentration or its presence only. Such algorithms require additional computation. Another possibility is to reduce dimension of the PSD function by estimating its slope only to use it for gas detection. Unfortunately, the gas detection based on the PSD slope only gives much worse detection results than PCA or SVM algorithms using the PSD function [13].

3. Histogram

Histogram of noise record is one of the statistical methods characterizing the observed fluctuations and is independent from the presented PSD function. Unfortunately, in numerous cases the observed noise has Gaussian distribution only. Then, we cannot get any additional information about the present gases. Histogram is a good tool when we observe non-Gaussian noise component, characteristic for the investigated gas. Such noise was observed in some resistive gas sensors and characterized by various statistical measures [2, 14]. We can claim that such situation should be more common for new gas sensor materials (e.g. nanoparticles functionalized by organic ligaments) because of low dimension of the grains (about a few nm only) composing the gas sensing layer.

In order to achieve an acceptable random error ε_r of estimated histogram we have to record sufficient number of noise samples. When the investigated signal time series $x(t)$ has probability density $p(x)$ it is necessary to collect sufficient number K of noise samples. The random error depends on probability of occurring the observed signal value $x(t)$. That probability is determined by the product of probability density $p(x)$ and range Δx where the histogram was estimated. Thus, the random error ε_r is defined by:

$$\varepsilon_r = \left(\frac{1}{K\Delta x p(x)}\right)^{1/2} \quad (3)$$

We can suppose that any differences from normal distribution should be visible only at relatively high signal values $x(t)$, typically 2-3 times greater than its standard deviation σ_x as reported in literature [14]. When we consider that $p(x)$ is Gaussian distribution and that we divide the range $\pm 3\sigma_x$ into 20 equal intervals we have to collect about $2\cdot 10^6$ of noise samples to limit random error ε_r of the extreme ranges to a few percent only. It means that we have to prolong time of noise recording about 50 times ($2\cdot 10^6/(40\cdot 1024)$) when compared with the data collected for estimating PSD function at the same sampling frequency. Such increase can be difficult in practical applications. Even much longer recording time was used when instead of histogram a bispectrum function was estimated to determine intensity of non-Gaussian component [14].

Intensity of non-Gaussian component can be determined by other measures, like skewness or kurtosis [14]. Both measures can be used to determine asymmetry (skewness) or peakedness (kurtosis) of the observed distribution from the normal one. These measures can be estimated using much shorter noise records than histogram. For example, a skewness γ_1 of the signal $x(t)$ having standard deviation σ_x and mean value μ_x is defined by formulae:

$$\gamma_1 = E\left[\left(\frac{x-\mu_x}{\sigma_x}\right)^3\right] \quad (4)$$

where the operator E means averaging. The estimated value γ_1 is a parameter only instead of function as it was in a case of histogram. It means that its usefulness for gas detection algorithms will be certainly more limited but does not require so long noise recording.

4. Principal Component Analysis

Very popular method applied for data processing in electronic nose applications is Principal Component Analysis (PCA). The method considers data collected from the gas sensors array as an input vector and performs their linear transformation to expose the most informative components (principal components). That method reduces size of the input data and makes further data processing much easier. The model created by the PCA method can be used to predict gas concentration (regression algorithm) or detect its presence (detection algorithm). Both algorithms require linear algebra operations only and can be performed by relatively cheap and popular embedded systems like smartphones or dedicated microcontrollers. The same method can be applied when the input data is a data vector composed of PSD function or histogram.

In general the regression algorithm is performed during three steps. Firstly, the PCA method transforms the sensors data matrix to determine the most informative principal components for further analysis. Secondly, a model of linear regression is prepared using the input data vector obtained in the first step and gas concentration measurements using the reference method. The model comprises of a set of regression coefficients having the same dimension as the number of applied principal components. The same method is used in various chemometric applications [4,15]. The created model can be used in the third step to predict gas concentration by the input data representing answer of the gas sensors (e.g. the estimated PSD of resistance fluctuations in a gas sensor) to their ambient atmosphere.

If a low number of principal components, typically limited to 5-10 only, is selected we can assure relatively low computational complexity and accuracy of gas concentration estimation equal to a few ppm. Such conditions were observed for the FES method for some gases [13]. Thus, it is possible to use the PCA algorithm in a portable device.

5. Support Vector Machine

Gas sensors response to presence of various gases is often nonlinear (e.g. dependence between DC resistance of gas sensor and gas concentration is logarithmic). Therefore we have to consider other algorithms than the presented PCA as a more advanced alternatives. Such algorithms should be more efficient in gas concentration prediction by the cost of more complicated computations. Support vector machine (SVM) algorithm is a developed method of the mentioned PCA algorithm because it includes a nonlinear case which should be more efficient for gas detection using the FES method [16]. The SVM algorithm for regression or detection aims was proposed by Vapnik [16] and further developed into its least square version (LS-SVM) by Suykens to reduce necessary computations when compared with the original SVM algorithm [5]. Again the LS-SVM regression algorithm requires three steps as was mentioned for the PCA algorithm.

The LS-SVM algorithm transforms an input data into a new space using nonlinear function (figure 1). When this function is linear the algorithm simplifies to the PCA method. The LS-SVM algorithm finds the optimal solution by solving the system of linear equations and using least square criterion. The applied nonlinear function provides the conversion of the input feature space into the transformed space where data are linearly separable to assure the highest margin between the data.

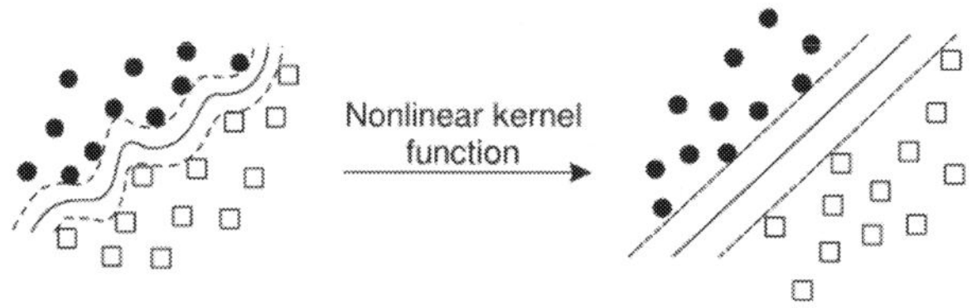

Figure 1. Illustration of support vector machine algorithm using nonlinear transformation.

The most intense computations in the LS-SVM algorithm are performed during model creation using a set of input vectors x_k ($k = 1, \ldots N$) recorded during N measurements at different ambient gas atmospheres. When the necessary computations are performed by PC computer the LS-SVM algorithm requires at least a few times longer time computing when compared with the PCA algorithm.

The established model for the LS-SVM method estimates a vector of coefficients α_k ($k = 1, \ldots N$) defining the set of hyperplanes which maximizes the margin between the input data after nonlinear transformation. Then we can predict gas concentration $y(x)$ by utilizing the input data vector x (e.g. data vector of PSD or histogram when the FES method is applied) and the created model, defined by coefficients α_k and parameter b:

$$y(x) = \sum_{k=1}^{N} \alpha_k K(x_k, x) + b \tag{5}$$

where $K(x_k, x)$ is a nonlinear kernel function applied in the LS-SVM algorithm. The most popular kernel function is polynomial, sigmoidal or Gaussian radial basis function (RBF). Equation (5) does not require intense computations. Thus, data processing of similar complexity has to be performed as in a case of previously mentioned PCA algorithm. The more intense computations are performed only during model creation. That step has to be done only once at the first stage before applying regression method. It means that we can use regression method based on the LS-SVM algorithm than on the PCA even in portable gas detection systems.

The main privilege of the LS-SVM method is its better accuracy when compared with the PCA. It has been observed that the same input data can give a few percentage better accuracy of gas concentration prediction when the LS-SVM algorithm and Gaussian radial basis function were used instead of the PCA method.

6. Conclusions

Problems of data processing for gas detection using fluctuation enhanced sensing method were considered. It has been underlined that power spectral density of the observed resistance fluctuations requires lower number of recorded noise samples to be estimated at given accuracy (random error) than histogram. Both functions can be uses as input data vectors for gas detection or gas concentration regression algorithms. Next, two popular algorithms for gas sensing were presented and compared (PCA and LS-SVM). It was concluded that the slightly better results are observed when the LS-SVM algorithm is applied due to better prediction of gas concentration. Unfortunately, that algorithm means more computations during creation of the regression model.

7. References

[1] Heszler P, Ionescu R, Llobet E, Reyes L F, Smulko J, Kish L B, and. Granqvist C G 2007 On the selectivity of nanostructured semiconductor gas sensors *Physica Status Solidi* B **244(11)** 4331–4335

[2] Kwan C, Schmera G, Smulko J, Kish L B, Heszler P, and Granqvist C G 2008 Advanced agent identification with fluctuation-enhanced sensing *IEEE Sens. J.* **8(6)** 706–713

[3] Kotarski M and Smulko J 2010 Hazardous gases detection by fluctuation-enhanced gas sensing *Fluct. Noise Lett.* **9** 359–371

[4] Barman I, Dingari N C, Singh G P, Soares J S, Dasari R R, and Smulko J M, 2012 Investigation of noise-induced instabilities in quantitative biological spectroscopy and its implications for noninvasive glucose monitoring *Analytical Chemistry* **84.19** 8149–8156

[5] Suykens J, Van Gestel T and De Brabanter J 2002 *Least squares support vector machines* (Singapore: World Scientific)

[6] Wojtas J, Rutecka B, Popiel S, Nawała J, Wesołowski M, Mikołajczyk J, Bielecki Z 2014 Explosives vapors–concentrating and optoelectronic detection *Metrol. Meas. Syst.* **21** 177–190

[7] Dunst K, Jasinski G, Jasinski P 2014 Potentiometric oxygen sensor with solid state reference electrode *Metrol. Meas. Syst.* **21** 205–216

[8] Kotarski M and Smulko J 2009 Noise measurement set-ups for fluctuations-enhanced gas sensing *Metrol. Meas. Syst.* **16** 457–464

[9] Ederth J, Smulko J, Kish L B, Heszler P and Granqvist C G 2006 Comparison of classical and fluctuation-enhanced gas sensing with Pd x WO 3 nanoparticle films *Sens. Actuators* B **113** 310–315

[10] Kish L B, Li Z, Solis J L, Marlow W H, Vajtai R, Granqvist C G, Lantto V, Smulko J and Schmera G 2005 Detecting harmful gases using fluctuation-enhanced sensing with Taguchi sensors *IEEE Sens. J.* **5** 671–676

[11] Kubersky P, Sedlák P, Hamácek A, Nespurek S, Kuparowitz T, Sikula J, Syrovy J T 2014 Quantitative fluctuation-enhanced sensing in amperometric NO 2 sensors *Chem. Phys.*, http://dx.doi.org/10.1016/j.chemphys.2014.10.021

[12] Kish L, Vajtai R and Granqvist C 2000 Extracting information from noise spectra of chemical sensors: single sensor electronic noses and tongues *Sens. Actuators B* **71.1** 55–59

[13] Lentka Ł, Smulko J, Jonescu R 2015 *Efficiency of Gas Detection Algorithms Using Fluctuation Enhanced Sensing* The 23rd Int. Conf. on Noise and Fluctuations, 2-6 June 2015, Xi'an China

[14] Smulko J, Granqvist C G, Kish L B 2001 On the statistical analysis of noise in chemical sensors and its application for sensing *Fluctuation and Noise Letters*, **1.03**, L147-L153

[15] Lorber A, Kowalski B R Estimation of prediction error for multivariate calibration 1988 *Chemom.* **2**, 93-109

[16] Vapnik V 2000 *The nature of statistical learning theory* (Springer Science & Business Media)

Acknowledgments
This work was supported by the National Science Center, Poland, the grant decision: DEC-2012/06/M/ST7/00444 "Detection of gases by means of nano-technological resistance sensors".

IOP Conf. Series: Materials Science and Engineering **104** (2015) 012033 doi:10.1088/1757-899X/104/1/012033

Gas sampling system for matrix of semiconductor gas sensors

Grzegorz Jasinski [1], Anna Strzelczyk [2], Piotr Koscinski [1]

[1] Faculty of Electronics, Telecommunication and Informatics,
Gdansk University of Technology, Narutowicza 11/12, 80-233 Gdansk, Poland
[2] Faculty of Chemistry, Gdansk University of Technology,
Narutowicza 11/12, 80-233 Gdansk, Poland

E-mail: grzegorz.jasinski@eti.pg.gda.pl

Abstract. Semiconductor gas sensors are popular commercial sensors applied in numerous gas detection systems. They are reliable, small, rugged and inexpensive. However, there are a few problem limiting the wider use of such sensors. Semiconductor gas sensor usually exhibits a low selectivity, low repeatability, drift of response, strong temperature and moisture influence on sensor properties. Sample flow rate is one of the parameters that influence sensors response what should be considered in the measurement system. This paper describes low cost module for controlling measured gas flow rate. The proposed equipment will be used as a component of electronic nose system employed for classifying and distinguishing different levels of contamination in air.

1. Introduction

In recent years, gas sensors are used in more and more applications. Gas sensors, which are commercially available or under development, have a lot of merits, such as rapid detection speed, high sensitivity and durability [1]. What is the most important, they are very inexpensive. However, there are a few problems limiting the wider use of such sensors. Semiconductor gas sensors usually exhibit a low selectivity, low repeatability, drift of a response and strong temperature and moisture influence on sensor properties [2, 3]. Some shortcomings can be minimized to some extent by using an array of sensors and by adopting some pattern recognition techniques [4]. Still overcoming all the problems is difficult.

One of the problems that have to be taken into account is sample flow rate that influence sensors response. The easiest solution, usually chosen in the measurement systems, is to stabilize gas flow rate over gas sensor at desired level. In laboratory condition a mass flow controllers along with some air pumps are commonly used for this purpose. Due to large size and high price such solution is not convenient.

In this paper we propose electronic module for sampling gas into measuring cell under controlled gas flow rate. The module is simple and inexpensive. Emphasis has been placed on universality and modularity. This module can be easily incorporated into system that measures response of a gas sensors matrix. The developed equipment will be used as a component of electronic nose system employed for classifying and distinguishing different levels of contamination in air. The system was prepared for semiconducting gas sensors, but it can be used with any kind of a gas concentration

IOP Conf. Series: Materials Science and Engineering **104** (2015) 012033 doi:10.1088/1757-899X/104/1/012033

measuring equipment. To our best knowledge, a similar system has not yet been proposed and examined.

2. Effect of air flow rate on sensor response

There is a lot of research on different sensor constructions working based on various concepts [5]. Influence of gas flow-rate is, however, unusually neglected. There are only few studies investigating this effect [6], thus it is worth to determine relation between gas flow speed and response of some most common gas sensors types.

Amperometric sensors due to high sensitivity and good dynamic properties are frequently used in toxic gas concentration monitoring [7]. Sulphur dioxide amperometric sensor with solid polymer electrolyte of our own design was investigated in this paper. Details of the experiment and sensor construction can be found elsewhere [8, 9]. The response (value of current) of this sensor at various gas flow-rates in two mixtures of high purity gases, namely synthetic air and 100 ppm SO_2, is presented in Figure 1. The gas flow was controlled with Brooks mass flow controllers and changed in steps from 10 mL per minute (mLPM) to 90 mLPM. In both gas compositions, the increase of gas flow resulted in the increase of current response. In synthetic air the instability effect is visible, but relatively small. In 100 ppm of SO_2, however, this effect is more significant. In case of the amperometric sensor it is a diffusion of gas through the porous electrode membrane that limits the measured current. Therefore, increasing flow of the gaseous analyte over the sensor electrode surface results in the current increase.

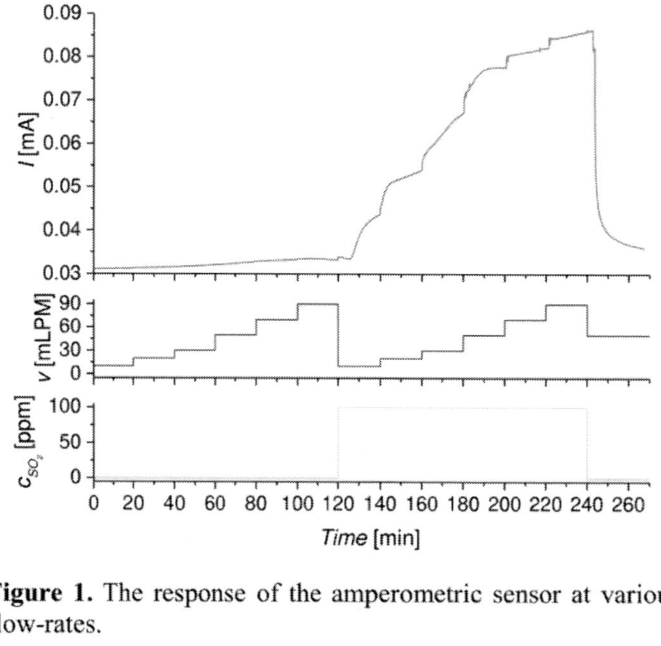

Figure 1. The response of the amperometric sensor at various gas flow-rates.

Similar experiment have been performed with semiconductor gas sensor. This time commercially available sensors GGS1331, GGS2331, GGS3331 (all from Umweltsensortechnik GmbH), TGS2600 and TGS880 (both from Figaro Engineering Inc.) have been used. Each sensor work at the operation temperature suggested by manufacturer in the application note. Such sensors, also referred as metal oxide sensors, in the presence of reducing or oxidizing gases change their resistance. The sensors response to the step change of gas flow from 0 mLPM to 80 mLPM is presented in Figure 2. It can be seen that the gas flow rate influence all sensors response. For each sensor, an introduction of gas flow causes increase of sensors resistance. It can be explained by the lowering of the sensor operating

IOP Conf. Series: Materials Science and Engineering **104** (2015) 012033 doi:10.1088/1757-899X/104/1/012033

temperature by gas flowing over sensor surface. The resistance change is different for each sensor, in case of the GGS 2331, the resistance for 100 mLPM is almost 5 times higher than for gas not flowing.

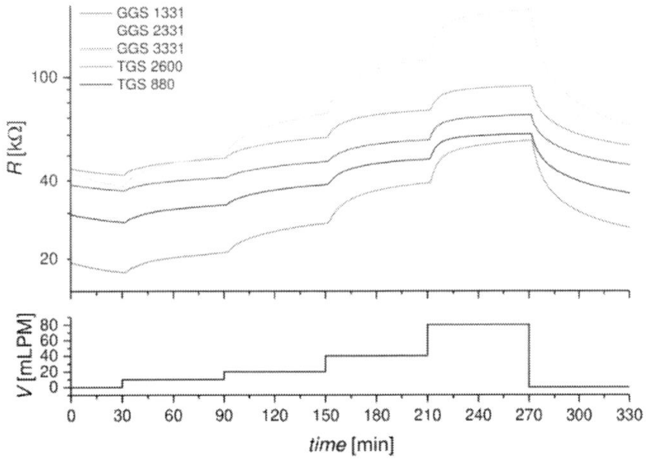

Figure 2. The response of the commercially available common semiconductor gas sensor at various gas flow-rates.

Based on above results, it can be concluded, that the gas flow rate influence different gas sensor response and it should be taken into consideration during measurements. Maintaining known and constant gas flow rate seems to be a good strategy. Alternatively, more complex procedure can be introduced, e.g. stop-flow mode of operation [10]. This mode can be accomplished by the alteration of gas flow in measuring chambers in controlled way.

3. Experimental set up

Preliminary studies have shown that a system for controlling gas flow rate requires feedback in the form of gas flow measuring circuit. Honeywell's Zephyr™ Digital Airflow Sensors seems to be good candidates for this purpose. HAF Series of such sensors are designed to measure mass flow of air and other non-corrosive gases. The Microelectronic and Microelectro-Mechanical System (MEMS) sense die is located in a precise and calculated airflow channel to provide repeatable flow response. These sensors provide a digital interface for reading airflow. With thermally isolated heater and temperature sensing elements, they provide a fast response to the air or gas flow.

They are available in standard flow ranges, from ±50 standard cubic centimeter per minute (sccm) to ±750 sccm, and are fully calibrated and temperature compensated with an on-board Application Specific Integrated Circuit (ASIC). All presented result are for two HAF sensors, namely HAFBLF0750CAAX5 and HAFBLF0100CAAX5, having the ranges of ±750 sccm and ±100 sccm, respectively. The main advantages of HAF sensors are their low price, high accuracy and easy integration with electronics using digital interface.

First, we checked HAF sensors accuracy and usefulness in developed system. As a reference source of gas flow a cylinder of compressed synthetic air and Tylan mass flow controllers (MFC) were used. Desired gas flux was set with MFC, and the readings from both HAF sensors were measured. Each measurement was repeated 20 times. Statistical analysis have been done. Measurements are gathered in Table 1. The obtained result prove that both sensor are very accurate and stable. In all cases, error during measurements with HAF sensor is smaller than 1%.

IOP Conf. Series: Materials Science and Engineering **104** (2015) 012033 doi:10.1088/1757-899X/104/1/012033

Table 1. HAF sensor calibration with Tylan MFC.

Flow set with MFC [sccm]	HAF sensor with 100 sccm range [sscm]	Standard deviation [sscm]	HAF sensor with 750 sccm range [sscm]	Standard deviation [sscm]
100.0	99.7	0.02	99.9	0.09
90.0	90.0	0.04	88.9	0.26
80.0	79.2	0.11	79.1	0.15
70.0	69.2	0.05	68.9	0.06
60.0	59.7	0.10	58.8	0.00
50.0	49.4	0.06	49.0	0.25
40.0	39.1	0.08	39.3	0.00
30.0	29.5	0.06	29.5	0.06
20.0	19.5	0.06	19.5	0.14
10.0	9.5	0.06	9.3	0.06
0.0	0.1	0.04	0.2	0.00

Considering the above, an electronic module was proposed to control and stabilize the gas flow rate at a desired value based on AVR microcontroller. Microcontroller is responsible for communication with the rest of the electronic nose system (I2Cmain). Microcontroller build-in Two Wire Interface (TWI) bus, which is I²C compatible, is used. Microcontroller is acting on this bus as a SLAVE device with individually set slave address. For configuration of the HAF sensor a two unused pins of the microcontroller are controlled using software procedures. This way communication of ATmega8 and HAF sensor is manually realized in MASTER mode. As a result, it is possible to read digital value of gas flow.

Another task of the microcontroller is to control the air intake. Two solutions have been proposed, usage of pump from blood pressure meter, powered with nominal 3 V voltage, and DC fan (Comair Rotron Flight II FS12B3), powered with nominal 12 V. Both approaches have been investigated in more details. In each case, microcontroller PWM signal was used for obtaining different average voltages and a result a gas flow management. In case of the fan, a TACH signal was measured as a feedback of rotation speed. Designed module block structure is shown in Figure 3.

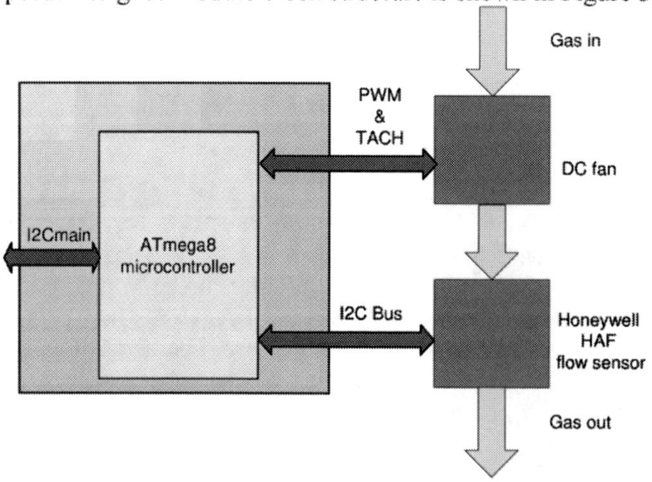

Figure 3. Block diagram of the proposed module

4. Results of the tests

Obtained result for the blood pressure meter pump are collected in Table 2. All measurements were repeated 20 times, averaged and standard deviation was calculated. Measurements have been made using two lengths of the silicone tube with an inner diameter 3 mm, which connects the pump to the measuring cell. Relatively high maximum flow rates have been obtained. Unfortunately, the pump is very unstable source of air flow, and as a result, the gas flow is highly irregular. Value of standard deviation is very high in relation to obtained gas flow. Large fluctuations in flow make the pump useless as part of the proposed system.

Table 2. Pump calibration with HAF sensor for different tube length.

	140 cm tube length		490 cm tube length	
Pump supply voltage [V]	Flow [sccm]	Standard deviation [sccm]	Flow [sccm]	Standard deviation [sccm]
1.0	171.7	70.73	148.4	95.60
1.5	280.7	113.01	297.9	91.74
2.0	347.7	107.98	366.0	106.77
2.5	429.9	93.90	443.5	121.72
3.0	276.5	119.88	440.4	115.69
3.5	627.6	91.26	399.8	103.89

Table 3. DC fan calibration with HAF sensor for different tube length.

	140 cm tube length		490 cm tube length	
Fan supply voltage [V]	Flow [sccm]	Standard deviation [sccm]	Flow [sccm]	Standard deviation [sccm]
3	10.7	0.24	5.2	0.27
6	44.1	0.29	18.5	0.23
9	100.5	1.21	39.1	0.96
12	169.8	2.26	67.7	1.11
15	250.5	3.10	101.1	1.82
18	308.2	3.93	126.7	2.28

Obtained result for the DC fan are collected in Table 3. The data are presented in a similar manner to the one described earlier in case of the pump. Flow fluctuation measured in case of the fan are in acceptable level. The obtained data show that there is no basic relation between supply voltage and gas flow rate. Additionally, values of air flow depend on the length of the tubes. It is showing the need to use of the flow sensor in the proposed system.

Finally, DC fan was chosen as a most suitable solution. Proposed module is controlled with simple dedicated software. Microcontroller repeatedly polls the HAF sensor and correct PWM factors according to error value of feedback loop (difference between value of desired flow set by supervising module and current flow reading).

IOP Conf. Series: Materials Science and Engineering **104** (2015) 012033 doi:10.1088/1757-899X/104/1/012033

5. Conclusions

The paper presents inexpensive module for measuring and controlling gas flow rate. It is used as a component of electronic nose system employed for classifying and distinguishing different levels of contamination in air. The hardware part of the system consists of a number of independent, single-board modules working in parallel. They are connected each other using I^2C bus, which can be controlled by software of supervising module [11].

6. References

[1] Nenov T G and Yordanov S P 1996. *Ceramic sensors: technology and applications* (Lancaster, Technomic Pub)

[2] Kalinowski P and Jasinski G 2013 *PhD Interdisciplinary Journal* **1** 99

[3] Smulko J M, Trawka M, Granqvist C G, Ionescu R, Annanouch F, Llobet E and Kish L B, 2015 *Sensor Review* **35** 340

[4] Strike D J, Meijerink M G H and Koudelka-Hep M 1999 *Fresenius J. Anal. Chem.* **364** 499

[5] Sberveglieri, G 1992 *Gas Sensors: Principles, Operation and Developments* (Springer)

[6] Liu S-J, Shen H-X and Feng J-X 1995 *Anal. Chim. Acta* **313** 89

[7] Jasinski P 2006 *Mater. Sci. Poland* **24** 269

[8] Chachulski B and Gebicki J 2006 *Sens. Actuator B-Chem.* **119** 435

[9] Strzelczyk A, Jasinski G and Chachulski B 2013 *PhD Interdisciplinary Journal* **3** 21

[10] Szczurek A, Maciejewska M, Flisowska-Wiercik B and Bodzoj L 2010 *Sens. Actuator B-Chem.* **148** 522

[11] Jasinski G, Koscinski P, Strzelczyk A, Wozniak L and Lentka G 2015 Low cost measurement module for matrix of amperometric gas sensor *Proc. of XXI IMEKO World Congress: Measurement in Research and Industry* (Prague, Czech Republik, 30 August-4 September)

Acknowledgments

This work was partially supported by the National Centre for Research and Development, project LIDER No. 22/103/L-2/10/NCBiR/2011, and State Budget Funds assigned for statutory activity.

Low cost electrochemical sensor module for measurement of gas concentration

Grzegorz Jasinski [1], **Anna Strzelczyk** [2], **Piotr Koscinski** [1]

[1] Faculty of Electronics, Telecommunication and Informatics,
Gdansk University of Technology, Narutowicza 11/12, 80-233 Gdansk, Poland
[2] Faculty of Chemistry, Gdansk University of Technology,
Narutowicza 11/12, 80-233 Gdansk, Poland

E-mail: grzegorz.jasinski@eti.pg.gda.pl

Abstract. This paper describes a low cost electrochemical sensor module for gas concentration measurement. A module is universal and can be used for many types of electrochemical gas sensors. Device is based on AVR ATmega8 microcontroller. As signal processing circuit a specialized integrated circuit LMP91000 is used. The proposed equipment will be used as a component of electronic nose system employed for classifying and distinguishing different levels of air contamination.

1. Introduction

In recent years, gas sensors are used in more and more applications. Gas sensors, which are commercially available or under development, have a lot of merits, such as rapid detection speed, high sensitivity and durability. What is the most important, they are very inexpensive. There are many sensor constructions working based on various concepts [1]. Electrochemical gas sensors can be divided into three main classes according to the operating principle: amperometric, potentiometric, and conductometric sensors. Gas sensors working based on amperometric principle provide relatively high analytical performance at modest cost [2, 3].

Amperometric gas sensors develop a current response signal by reaction of a measured gas at an electrode. This response can be measured at a fixed or variable electrode potential, although a fixed potential is usually used. The reaction rate, reflected by the current at the sensing electrode, occurs at a thermodynamically determined potential for any given reaction and, when operated under appropriate diffusion-limited conditions, is simply proportional to the concentration of the measured gas concentration. The relationship between current and concentration is linear, typically over 3 orders of magnitude, and measurements with high sensitivity are possible.

Amperometric sensors, in order to work properly at a fixed potential, require to be controlled by a potentiostatic circuit. In the worldwide market such devices in the form of professional instruments are available (eg. Solartron Analytical 1287 electrochemical interface) and usually used in stationary conditions for purposes of science. These devices usually are very sophisticated and offer excellent capabilities. However, due to a high complexity they are very expensive. There is a strong need to develop a portable and low-cost devices for measuring gas concentration with amperometric sensors.

In this paper a relatively simple and inexpensive module for measurement of gas concentration with amperometric gas sensors is proposed. Lack of selectivity is usually a common shortcoming of most of sensor available. Such sensors reacts not only with the selected gas but also with other gases. Various techniques for selectivity improvement can be used such as special sensor construction [4], sophisticated mode of operation [5-7], sensor operating temperature modulation or sensor response fluctuations measurement [8]. It seems that improvement of selectivity can be obtained using sensor arrays combining several sensors along with some pattern recognition algorithms as well. Extensive research based on concept of improving selectivity by combining a few not selective sensors response analysis have been done since the 80-ties of 20th century. A new type of device called an electronic nose have been proposed [9]. Thus a proposed system is designed in such way that several modules can be easily connected into one system for measuring response of amperometric gas sensors matrix. Other modules already designed include gas sampling controlling module or electrochemical impedance spectroscopy analyzer.

2. Amperometric sensor module

A potentiostat circuit required for amperometric sensor proper operation can be designed from analogue components. However, the continuing progress in integrated chip technologies results in combing many systems on a single chip. Today, a system on chips (SoC), an integrated circuit that integrates all components of a computer or other electronic system into a single chip are becoming more common in the market. Such components may contain digital, analog or mixed-signal functions—all on a single chip substrate. This technology provides new possibilities, especially due to their relatively low cost and low power consumption.

SoC microsystems can realize many functions. One interesting application is impedance measurement with AD5933 chip [10]. Such chips not only provides possibility for miniaturized and low-power applications, but greatly simplifies system. Thus, in this paper as signal processing circuit a specialized integrated circuit LMP91000 produced by Texas Instrument is used. The LMP91000 is a programmable Analog Front End for use in micro-power electrochemical sensing applications. It provides a complete signal path solution between a sensor and a microcontroller that generates an output voltage proportional to the sensor 3-elecrode cell current. The LMP91000's adjustable cell bias and transimpedance amplifier gain are programmable through the I^2C interface. Based on this chip a module for measurement of gas concentration based on AVR ATmega8 microcontroller is proposed (Figure 1).

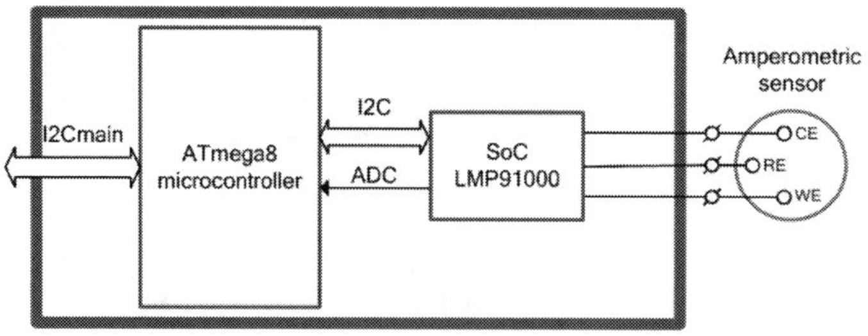

Figure 1. Block diagram of the amperometric sensor module.

One module measures response of single amperometric gas sensors. ATmega8 is a heart of proposed module. Microcontroller is responsible for communication with the rest of the system (I2Cmain). Microcontroller build in Two Wire Interface (TWI) bus, which is I^2C compatible, is used. Microcontroller is acting on this bus as a SLAVE device with individually set slave address. The ATmega8 also responds to I^2C General Call Address. This solution allows connecting multiple analysers to the same bus and control them separately (using address) or all at the same time (using General Call Address).

For configuration of the SoC, two unused pins of microcontroller are controlled using software procedures. This way communication of ATmega8 and SoC is manually realized in MASTER mode. It is possible to configure various parameters such as: the range of the output voltage, the voltage of potentiostat circuit, gain of transimpedance amplifier, load of gas sensors and many others which are more described in datasheet for LMP91000 [11]. Generally, one set of adjustments correspond to certain type of electrochemical sensor. If it is necessary to use other type of sensor, it is possible to do it by changing the programmable parameters within the LMP91000 integrated circuits.

Last task of the module is to measure the voltage at the SoC output pin, which is proportional to the concentration of the target gas. The voltage is converted to digital format via a 10-bit ATmega8 internal analog digital converter (ADC). Software averaging was used, thus low ADC resolution did not affect the measurements accuracy.

3. Tests and results

A prototype board of proposed amperometric sensor module was realized. Dedicated software for PC and AVR microcontroller have been developed. An ammonia - sulphur dioxide amperometric sensor with solid polymer electrolyte of our own design was used for a test. Details of the experiment and sensor construction can be found elsewhere [12]. The response of this sensor in mixture of high purity gases, namely synthetic air and 100 ppm SO_2, is presented in Figure 2. The sensor response in both forms is presented, voltage measured by module and calculated current. Continuous measurements were made with averaging of 100 ADC conversions. Change in the composition of the gas mixture was carried out at a period of 20 minutes.

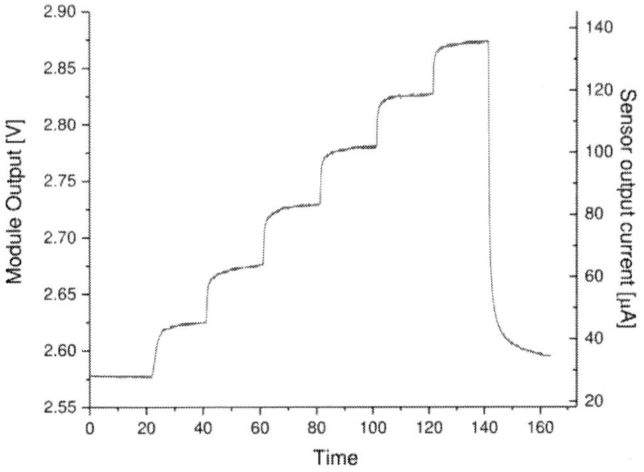

Figure 2. The response of the amperometric sensor to different concentrations of SO_2 measured by module

It can be seen that sensor reaction to the toxic gas concentration change is clearly visible. The obtained curve is smooth. The measured current by module related to almost stable sensor response for given SO_2 concentration is presented in Figure 3. Obtained relationship is almost linear. Slope of the curve, that correspond to the sensor sensitivity, equals to 0.73µA/ppm and is similar to results obtained with laboratory stationary electrochemical interfaces. The obtained results allow concluding that the work proposed in the framework of the measuring module is suitable for use with amperometric gas sensors.

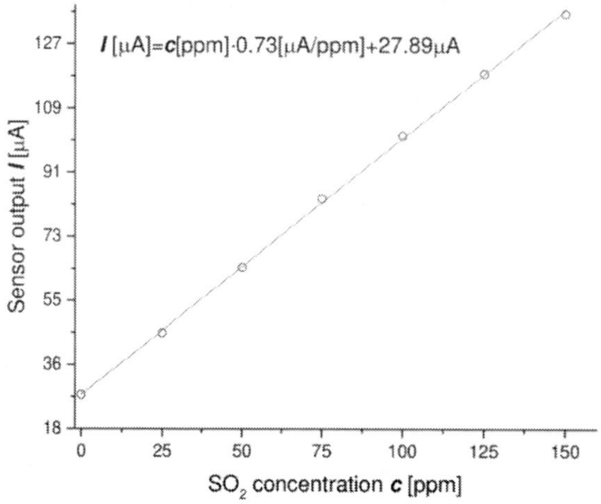

Figure 3. Calibration curve for measurement in Figure 2.

4. Conclusions

The paper presents scalable, flexible measurement system designed for measurement of amperometric gas sensors response. The hardware part of the system consists of a number of independent, single-board modules working in parallel. They are connected each other using I^2C bus, which can be controlled by software of supervising module. Flexibility and scalability as well as a low price of proposed system have been achieved.

5. References

[1] Nenov T G and Yordanov S P 1996. *Ceramic sensors: technology and applications* (Lancaster, Technomic Pub)
[2] Jasinski P 2006 *Mater. Sci. Poland* **24** 269
[3] Stetter J R and Li J 2008 *Chem. Rev.* **108** 352
[4] Lorenc P, Strzelczyk A, Chachulski B and Jasinski G 2015 *Solid State Ionics* **271** 48
[5] Jasinski G, Jasinski P, Nowakowski A and Chachulski B 2006 *Meas. Sci. Technol.* **17** 17
[6] Kalinowski P, Wozniak L, Strzelczyk A, Jasinski P and Jasinski G 2013 *Metrol. Meas. Syst.* **20** 501
[7] Kalinowski P, Strzelczyk A, Wozniak L, Jasinski G and Jasinski P 2014 *Meas. Sci. Technol.* **25** 025101
[8] Kotarski M M and Smulko J M 2010 *Fluct. Noise Lett* **9** 359
[9] Gardner J W and Bartlett P N 1993 *Sens. Actuator B-Chem.* **18-19** 211
[10] Hoja J and Lentka G 2013 *Metrol. Meas. Syst.* **20** 43
[11] LMP91000 Sensor AFE System: Configurable AFE Potentiostat for Low-Power Chemical Sensing Applications, Texas Instruments, 2011
[12] Strzelczyk A, Jasinski G and Chachulski B 2013 *PhD Interdisciplinary Journal* **3** 21

Acknowledgments

This work was partially supported by the National Centre for Research and Development, project LIDER No. 22/103/L-2/10/NCBiR/2011, and State Budget Funds assigned for statutory activity.

IOP Conf. Series: Materials Science and Engineering **104** (2015) 012035 doi:10.1088/1757-899X/104/1/012035

Energy requirements for methods improving gas detection by modulating physical properties of resistive gas sensors

M Trawka, M Kotarski

Department of Metrology and Optoelectronics, Faculty of Electronics, Telecommunications and Informatics, Gdańsk University of Technology, Gdańsk, Poland

E-mail: maciej.trawka@pg.gda.pl

Abstract. One of the most important disadvantage of resistive gas sensors is their limited gas selectivity. Therefore, various methods modulating their physical properties are used to improve gas detection. These methods are usually limited to temperature modulation or UV light irradiation for the layers exhibiting photocatalytic effect. These methods cause increased energy consumption. In our study we consider how much energy has to be supplied to utilize such methods and what kind of additional information can be gathered. We present experimental results of selected resistive gas sensors, including commercial and prototype constructions, and practical solutions of modulating their physical properties.

1. Introduction

The majority of resistive gas sensors start to be gas sensitive at elevated temperatures only, reaching even a few hundred Celsius. It means that such sensors require continuous energy supply for their heating. The electrical power necessary for heating commercial resistive gas sensors (e.g. Taguchi gas sensor TGS 823) is close to 0.7 W for more stable constructions, giving reliable gas detection results. Less precise sensor constructions (e.g. TGS 2600) require about 0.2 W only [1, 2]. Their physical properties, like intensity of adsorption-desorption processes of ambient gas molecules, will change respectively to their operating temperatures. This means that a change of heating power can modulate sensor DC resistance and this phenomenon can be utilized to improve gas detection. We can expect that resistance fluctuations ($1/f$-like noise) change at different temperatures as well as DC resistance. Thus, a method of fluctuation enhanced sensing (FES) used for gas detection since a decade can be also applied at different temperatures to gather more information [3, 4]. This study will compare experimental results of both mentioned methods observed for a few investigated commercial sensors and selected gases.

Some resistive gas-sensing layers (e.g. WO_3, TiO_2 [5, 6]) exhibit a photocatalytic effect when irradiated by UV light. Application of that effect is often limited to sensor cleansing only but it can be utilized to modulate gas sensitivity and selectivity as well. We can suppose that such method should be quite efficient because of limited energy consumption and measurement time, especially when compared with the results obtained by temperature modulation. Both issues will be considered in the paper to give an overview of which method could be more efficient for portable gas detection systems to improve their detection efficiency when energy consumption is one of the most important parameter.

2. Effectiveness of temperature modulation

An elevated temperature of the resistive gas sensor is secured by supplying DC voltage to its heater (typically 5 V for commercial gas sensors). Thus, any change of that voltage will result in a change of sensor temperature determined by the thermal inertia of the sensor. Experimental results for exemplary commercial gas sensor and selected gas mixtures suggest that changes of the sensor's DC resistance at stepwise change of DC voltage depends on gas sensor ambient atmosphere (Figure 1). Similar results were published by other authors [7, 8]. Time constants related to dynamic changes of the sensor DC resistance (initial increase and further drop) or position and level of its maximum would secure additional parameters, determined by ambient gas atmosphere (the chemical time constant is slower). Such parameters can be estimated by observing DC resistance changes during time interval at least about 60 s according to exemplary experimental results (Figure 1) and applied in popular detection algorithms [9]. It means that we have to heat up the sensor and have to spend energy equal to at least about 0.2 W·60 s = 12 J. In a case of a more stable commercial gas sensor (e.g. TGS 823) the required energy will be even a few times greater.

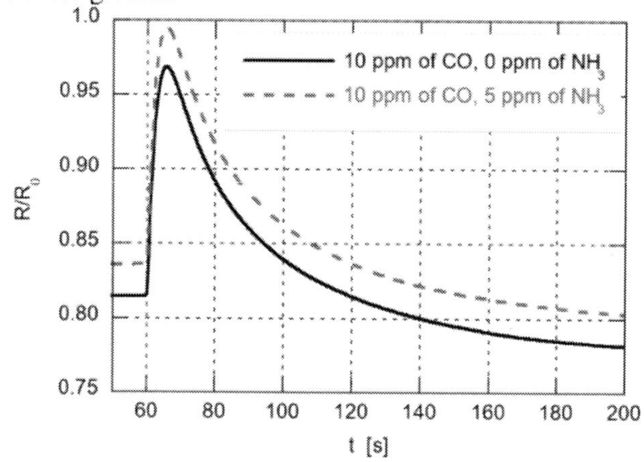

Figure 1. Change of DC resistance R of commercial gas sensor TGS 2600 at ambient atmosphere of the selected gas mixtures (CO and NH_3 diluted in synthetic air) normalized to its resistance R_0 observed at synthetic air versus time t; a stepwise drop of heating voltage from 5 V to 4.7 V was applied at $t = 60$ s.

The presented method of applying temperature modulation for better gas detection utilizes rather a small change of heating voltage and therefore additional time of measurement is relatively short and energy necessary for the prolonged heating is limited as well. The method is quite efficient to observe changes of DC resistances characteristic for different gases. When we observed $1/f$-like resistance noise at the same heating voltage drop the results cannot be used for better gas detection. There was almost no change in intensity of $1/f$-like noise at the drop of heating voltage from 5 V to 4.7 V, at least in a case of investigated gases (CO, NH_3, CH_4). Thus we have to apply other measurement conditions which could utilize the FES method. That issue has been considered recently by changing sensor temperatures at wider range and resulted in much prolonged measurement time [10, 11]. One of these methods estimates power spectral density of $1/f$-like sensor resistance noise at different temperatures and observes a change of maximal noise intensity (power spectrum density at the selected frequency) at different temperatures due to various rates of adsorption-desorption processes [10]. That change is

IOP Conf. Series: Materials Science and Engineering **104** (2015) 012035 doi:10.1088/1757-899X/104/1/012035

characteristic for the ambient gas and can be utilized to determine its presence. The method requires noise measurements at a few temperatures of the gas sensor. We have to continue such measurements at least an hour and to heat the applied sensor all time [10]. It means that the supplied energy will be many times greater than the energy necessary to record DC dynamic changes as presented in Figure 1. Moreover, recording of resistance noise requires more advanced measurement setup than for measurements of DC resistance only and additional energy consumed by that system, even if the system is limited to selected frequency bands (e.g. as proposed for toxic gases detection [12]).

Another method when resistance noise can be utilized to improve gas sensing by its measurements at different temperatures is a method called sample-and-hold [13]. The method keeps the gas sensor at its operating temperature at ambient atmosphere of the investigated gas and cools to room temperature to keep the "frozen" gas species inside the porous gas-sensing layer structure. Next, the sensor is slowly heated when placed in atmosphere of synthetic air only. Changes of resistance noise intensity during heating can be used to detect the type of the "frozen" gas species. Especially, temperature value when the "frozen" gas species evaporate from the porous gas sensing layer due to desorption processes depends on energy activation of the investigated gas [11]. That method ensures high selectivity for the detected gas but requires at least 1-2 hours of slow gas sensor heating. Thus, both presented methods [10, 11] utilizing resistance noise measurements at various temperatures require many times more energy than the method recording DC resistance changes induced by a change of heating voltage. It means that only the last method could be applied in portable applications due to energy consumption even if the FES methods at modulated temperatures ensure a higher selectivity for the detected gas.

3. Effectiveness of UV irradiation

Some gas sensing materials (e.g. WO_3, TiO_2) exhibit a photocatalytic effect when irradiated by UV light [5, 14, 15]. That effect is often used to cleanse the gas sensor instead of using heating pulse because it is often faster and more energy efficient than the heating pulse [5]. The main disadvantage of that method is the need for an additional UV LED-diode. These diodes are characterized by different wavelengths of maximal optical power emission but require relatively small power consumption to be used (e.g. maximally 4 V × 10 mA = 0.4 J/s for exemplary diode T5F36).

In our experimental studies we have applied the diode T5F36, having maximal emission at about 362 nm (Figure 2). The UV diode could be biased up to 10 mA but a photocatalytic effect present in the investigated material was observed at lower currents.

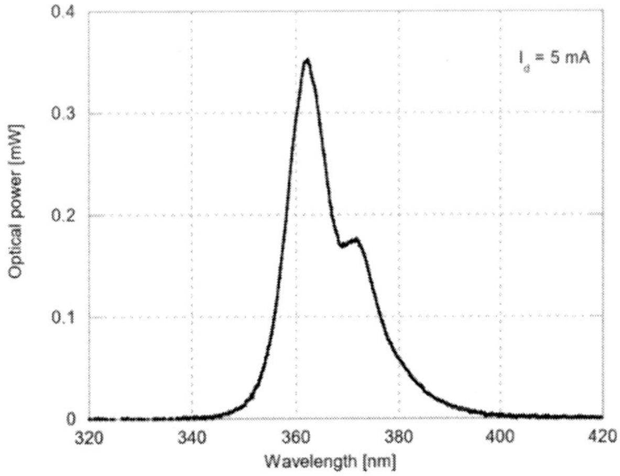

Figure 2. Optical power of the applied UV LED-diode T5F36 when biased by DC current I_d = 5 mA; the diode was placed at the same distance from the power meter as in a case of measurements of gas sensing WO_3 layer

The prototype gas sensing layer of WO_3 was investigated. All details of their preparation and parameters are presented elsewhere [15]. The gas sensing layer was irradiated by the diode placed about

5 mm from the gas sensing surface. We have observed that the diode current I_d limited to a few mA was sufficient to saturate effects of the DC resistance changes induced by the emitted UV light (Figure 3). A relative change of DC resistance induced by UV irradiation depended on the ambient gas and reached even 35%. That change was more intense in comparison with relative changes of DC resistance induced by drop of heating voltage (e.g. change of its maximum value – Figure 2). The DC resistance has stabilized about 10-15 min after switching on the UV LED-diode. It means that the UV-light can replace temperature modulation to improve gas detection, at least for some gases and for the gas sensing layer exhibiting photocatalytic effect. Moreover, that method requires similar amount of energy (600 s × 4 V × 5 mA = 12 J) as in a case of temperature modulation method (Figure 1) but means a use of additional element – UV LED-diode.

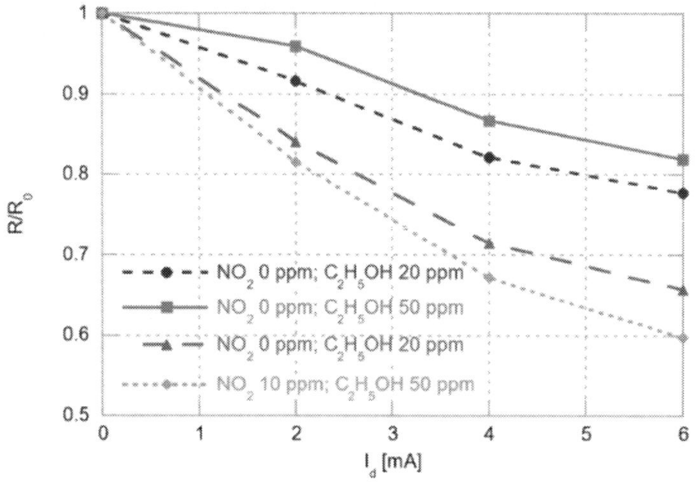

Figure 3. Changes of the normalized DC resistance R of WO_3 gas sensing film [15] versus UV LEDdiode DC current I_d at two different gas mixture concentrations of NO_2 and C_2H_5OH, at working temperature of 200°C; R_0 – DC resistances observed for the selected gas mixtures at $I_d = 0$ mA.

Our preliminary studies of 1/*f*-like resistance noise changes induced by UV light in the investigated WO_3 gas sensing layer suggest that noise measurements depend strongly on UV light and can be used to improve gas detection [15]. Noise intensity has changed even ten times when UV light was used to modulate the gas sensing layer. Unfortunately, the FES method is not so energy efficient than DC resistance measurements only. It requires energy for noise amplification, sampling, recording and necessary computing but can potentially assure much more information for detection algorithms when compared with the DC resistances at modulated conditions [16-18]. The 1/*f*-like noise is often used as a source of information about various materials and their qualities [19, 20]. Thus, effectiveness of noise measurements for gas detection has to be considered separately for each case. We can only conclude that the method of gas detection based on measurements of DC resistances modulated by temperature or UV light consume less energy when compared with the FES method.

4. Conclusions

In our experimental study we have presented and analysed exemplary data of DC resistance changes at different temperatures or modulated by UV light. The results suggest that UV light can be utilized for better gas detection in portable systems because of energy consumption. We can suppose that even better gas detection can be obtained when the *1/f*-like noise is used for gas detection when the sensor is modulated by UV light. Unfortunately such improvement requires more energy and more sophisticated measurement system. Further experimental studies are necessary to answer this question for gases of interest. Moreover, smaller gas sensors means their faster response and shorter measurement time. We can conclude that even without applying smaller sensors the measurement time can be reduced by cost of lower accuracy. The compromise between accuracy and measurement time requires more detailed studies.

5. References

[1] Macku R, Smulko J, Koktavy P, Trawka M, Sedlak P 2015 Sens. Actuators B **213**, 390–396.

[2] Figaro Taguchi gas sensors data sheets. Available: e.g., www.figarosensor.com

[3] Kish L, Vajtai R and Granqvist C 2000 Sens. Actuators B **71** 55–59

[4] Ederth J, Smulko J, Kish L B, Heszler P and Granqvist C G 2006 Sens. Actuators B **113** 310– 315

[5] Korotcenkov G, Cho G B K 2013 Sens. Actuators B **188** 709–728

[6] Zakrzewska K, Radecka M, Rekas M 1997 Thin Solid Films **310.1** 161–166

[7] Ngo K A, Lauque P, Aguir K 2007 Sens. Actuators B **124** 209–216

[8] Ionescu R., Hoel A., Granqvist C G , Llobet E, Heszler P 2005 Sens. Actuators B **104** 132–139

[9] Dunst K, Jasinski G, Jasinski P 2014 *Metrol. Meas. Syst.* **21** 205–216

[10] Macku R, Smulko J, Koktavy P, Trawka M, Sedlak P 2015 Sens. Actuators B **213** 390–396.

[11] Smulko J, Trawka M. 2015 Sens. Actuators B **219** 17–21

[12] Kotarski M, Smulko J 2010 Fluctuation and Noise Letters, **9** 359–371

[13] Solis J L, Kish L B, Vajtai R, Granqvist C G, Olsson J, Schnurer J, Lantto V 2001 Sens. Actuators B **77** 312–315

[14] Topalian Z, Smulko J, Niklasson G A, Granqvist C G 2007 IOP Publishing In Journal of Physics: Conference Series **76.1** 012056–012060

[15] Trawka M, Smulko J, Hasse L, Ionescu R, Annanouch F E, Llobet E 2014 Fluctuation enhanced gas sensing using UV irradiated Au-nanoparticle-decorated WO_3-nanowire films. The 8th International Conference on Sensing Technology September 2-4, 2014

[16] Heszler P, Ionescu R, Llobet E, Reyes L F, Smulko J, Kish L B, Granqvist C G 2007 *Physica Status Solidi* B **244(11)** 4331–4335

[17] Kotarski M, Smulko J. 2009 Metrol. Meas. Syst. **16** 457–464

[18] Kwan C, Schmera G, Smulko J, Kish L B, Heszler P, Granqvist C G 2008 Sensors Journal IEEE, **8.6** 706–713

[19] Stadler A W, Zawiślak Z, Dziedzic A, Nowak D 2014 Metrol. Meas. Syst., **21** 15–26

[20] Granqvist C G, Green S, Jonson E K, Marsal R, Niklasson G A, et al. 2008 Thin Solid Films, **516** 5921–5926

Acknowledgments

This work was supported by the National Science Center, Poland, the grant decision: DEC2012/06/M/ST7/00444 "Detection of gases by means of nano-technological resistance sensors".

IOP Conf. Series: Materials Science and Engineering **104** (2015) 012036 doi:10.1088/1757-899X/104/1/012036

Non-invasive blood glucose monitoring with Raman spectroscopy: prospects for device miniaturization

M.S. Wróbel

Gdańsk University of Technology, Faculty of Electronics, Telecommunications and Informatics, Department of Metrology and Optoelectronics, Narutowicza 11/12, Gdańsk 80-233, Poland

E-mail: maciejswrobel@gmail.com

Abstract. The number of patients with diabetes has reached over 350 million, and still continues to increase. The need for regular blood glucose monitoring sparks the interest in the development of modern detection technologies. One of those methods, which allows for non-invasive measurements, is Raman spectroscopy. The ability of infrared light to penetrate deep into tissues allows for obtaining measurements through the skin without its perforation. This paper presents the limitations and possibilities of non-invasive blood glucose monitoring with Raman spectroscopy. Especially focusing on the possibilities for device miniaturization. Such device incorporates a Raman spectrometer, a fiber-optical probe, and a computing device (microcontroller, smartphone, etc.) which calculates the glucose concentration using specialized algorithms. Simplification of device design, as well as turbidity correction technique and a new proposed method of synchronized detection are described.

1. Introduction

The number of diabetics in the world equals about 347 millions (in 2013), and is estimated to increase up to about 438 millions by the year 2030. In the EU countries alone, more than 30 m. people suffer from diabetes and over 55 m. in all European countries, according to the International Diabetes Federation [1]. Treatments costs of diabetes and complications resulting from it are estimated to be between 10% and 18% of national health care budgets, especially in the developed countries.

Continuous blood glucose levels monitoring is required for patients with diabetes mellitus to establish the quantity of glucose in blood [2]. Then, the required amount of insulin to be administered is assessed. Widely accepted methods for blood glucose monitoring require sampling of patients' blood and performing measurements on a blood drop. With at least one finger pick per day (and usually three or more) such measurements cause serious health concerns, as well as great inconvenience in patients' daily lives. Complications related to blood draws include weakening of skin which leads to problems with fast healing, risk of infections, pain, etc. Therefore, there is a need to develop new methods, which would allow for continuous, non-invasive detection of blood glucose levels [3].

This papers provides a short review of possible non-invasive blood glucose measurement techniques, with focus on Raman spectroscopy. Furthermore, prospects for miniaturization of such device, based on Raman spectroscopy and subsequent signal processing, is described. A new method, synchronized Raman detection, is proposed to solve the most limiting factor in non-invasive detection, the problem of transferring calibration model between patients due to changes in skin compositions.

2. Non-Invasive blood glucose monitoring methods

Currently, the gold standard method for blood glucose detection is the enzymatic reaction of the sample with glucose oxidase to form gluconic acid and hydrogen peroxide, and subsequent readout of the change in current between electrodes on a test strip, or after reaction with oxygen, with a colorimetric readout [4]. A wide range of possible techniques have been applied to the problem of non-invasive blood glucose monitoring, from which we will focus only on the optical techniques.

2.1. Fluorescence

Fluorescence is a highly sensitive technique for detection of analytes in general, however, several drawbacks with its application to non-invasive measurements are limited specificity and scattering of tissues [5–7]. Several approaches have been undertaken for blood glucose monitoring, including excitation with UV light and subsequent detection of fluorescence from skin [8], detection of glucose in tears [9], or change in autofluorescence of enzymes, such as NAD(P)H [10].

2.2. Optical coherence tomography

Optical coherence tomography detects backscattered light from the samples. Although this method operates mainly in near infrared biological window, allowing for light penetration of about 1 mm into the tissues, it lacks chemical specificity. The scattering changes induced by change of refractive index of the medium due to glucose interfere with other tissue components, thus reducing this method's usefulness [11–14].

2.3. Near infrared spectroscopy

Near infrared spectroscopy (NIRS) relies on the measurements of transmitted or reflected light at wavelength in the range between about 750 nm – 2500 nm. This allows for a very deep penetration into the skin, which is use in the measurements of absorption of light by specific molecules [15], as well as in laser therapy applications [16]. This is a widely accepted method for the measurements of oxygenation-deoxygenation of hemoglobin and blood volume pulse known as photoplethysmography (PPG) [17]. The complexity of background spectra arising from the presence of other tissue components (water, hemoglobin, proteins, fat, etc.) is a major obstacle in the development of a NIRS-based blood glucose sensor. Moreover, changes in skin absorption and scattering properties (pigment, disease, etc.) as well as environmental factors play significant roles [5, 18].

2.4. Mid-infrared spectroscopy

In mid-infrared spectroscopy (2500 – 10 000 nm range) the effects of scattering are largely decreased, whereas the absorption is greatly increased. Thus, up to a few mm of tissue can be penetrated. Spectral bands of chemical substances are much stronger and sharper, than those in case of NIRS, which permits greater chemical sensitivity. The interference of hemoglobin and blood greatly obstruct the measurements [19].

3. Raman spectroscopy

Raman spectroscopy has been previously used for non-invasive blood glucose monitoring, and is most promising due to its superb chemical stability, good penetration depth with near-infrared sources, as well as the amount of developed methods of quantitative data analysis [20, 21]. Raman spectra contain information about the change in molecule energy levels under the influence of laser light irradiation, and subsequent detection of light inelastically scattered from the sample at different wavelengths than the excitation laser wavelength. The difference between wavelengths depends on the specific chemical bond energy. This yields a set of characteristic peaks in the spectrum, which is a "fingerprint" of that specific chemical substance.

Raman spectra exhibit numerous lines specific for glucose molecular vibrations (Figure 2), such as the skeletal vibrational modes (anomeric ν -C-O and ν -C-C stretch at carbon #1 at 820–950 cm^{-1}, the ν C-hydroxyl and δ C-O-H stretch at 1000–1180 cm^{-1}) and the CH_2 vibrations and C-O-H bending at 1200–1500 cm^{-1}. It is noteworthy that aqueous glucose spectra differ from its powder form.

Application of this method is particularly difficult in the case of biological materials, because of high absorption μ_a, and scattering μ_s coefficients of tissues (plasma: $\mu_a \approx 0.03\ mm^{-1}$, $\mu_s \approx 0.06\ mm^{-1}$; RBCs: $\mu_a \approx 0.45\ mm^{-1}$, $\mu_s \approx 30\ mm^{-1}$; skin/tissue: $\mu_a \approx 0.5\ mm^{-1}$, $\mu_s \approx 1.2\ mm^{-1}$) [22–24], which causes low levels of Raman scattered signal. Moreover, the random noise, as well as the interference of signals from other substances in the tissues, and strong fluorescence background are another limiting factors [25].

3.1. Measurement setup

The basic Raman spectrometer setup consists of components: a laser light source, an optical probe, a spectrometer , and a PC or other data-processing device (Figure 1). The light from a laser source is transmitted through a fiber-optic cable to the optical probe, which contains an optical filter, to select a narrow laser line. Then the laser light irradiates the sample, and the Raman effect occurs, causing the emission of the elastically scattered Raman light back to the probe. The probe collects the Raman light, filters if with a high-pass optical filter, and focuses it on a fiber-optic cable. The light then is transmitted onto a spectrometer where it is split into separate wavelengths and detected on a CCD camera.

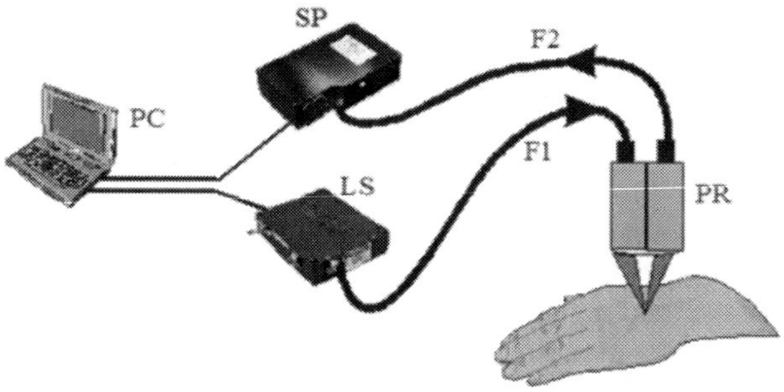

Figure 1. Laboratory RS setup: LS – laser, SP – spectrometer, F1 – lead-in fiber, F2 – lead-out fiber, PR – optical probe, PC – personal computer.

3.2. Detection algorithms

Due to the nature of the Raman spectroscopy, the detected light is often of very low intensity, thus the introduction of random noise is evident, as well as often strong optical instrument background, as well as background coming from fluorescence of the sample. This requires the use of spectral pre-processing techniques and application of de-noising algorithms, background correction algorithms, as well as the removal of cosmic rays and other spectral conditioning methods [26, 27]. In the case of biological samples measurements the amount of noise, as well as inter-patient variability [28] requires use of elaborated multivariate calibration techniques. Complexity of the spectra, and aforementioned patient-to-patient changes in skin optical properties as well as chemical signature, causes largely non-linear response on the examined analytes concentrations, thus the linear methods, such as principal component analysis (PCA) [29] are being replaced by their nonlinear counterparts such as Support Vector Machines (SVM) [30, 31].

With the advent of these technologies, the weight of the problem of blood glucose monitoring is no longer caused by the measurements itself, but is shifted more towards the data analysis. Thus, the actually achievable measurement precision, and its further improvement, (although still important) seems to introduce less progress and be won't bring the ultimate solution of the problem. However, the improvement in data processing will be more valuable and more impactful. Thus, the issue of device miniaturization can be brought to greater attention, when the most of the work remains to be done in the software rather than hardware to make non-invasive glucose measurements a possibility.

4. Prospects for device miniaturization

A vast majority of the Raman studies are performed using benchtop laboratory systems. Such devices are often bulky, stationary, and costly. However, a mobile, miniature RS systems are comparable in their metrological performance to their counterparts [32–34]. With the use of several techniques, it should be possible to allow for further miniaturization of such devices, without the loss in their performance. At the same time, some of the most significant challenges in non-invasive measurements may be additionally overcome.

4.1. Set-up simplification

Upon the identification of spectral bands where the most significant changes are due to the content of investigated analyte, it is unnecessary to measure and process the whole spectrum [35]. It is suggested to focus on those selected areas during measurements, thus to reduced the complexity of a measurement setup by a great deal. First of all, a simplification in the detection setup would require the use of a set of photo detectors and complimentary optical filters in the place of a spectrograph, diffraction grating, and a CCD camera. A miniature IR diode laser modules can be used to reduce space requirements. A schematic of this approach is presented (Figure 2).

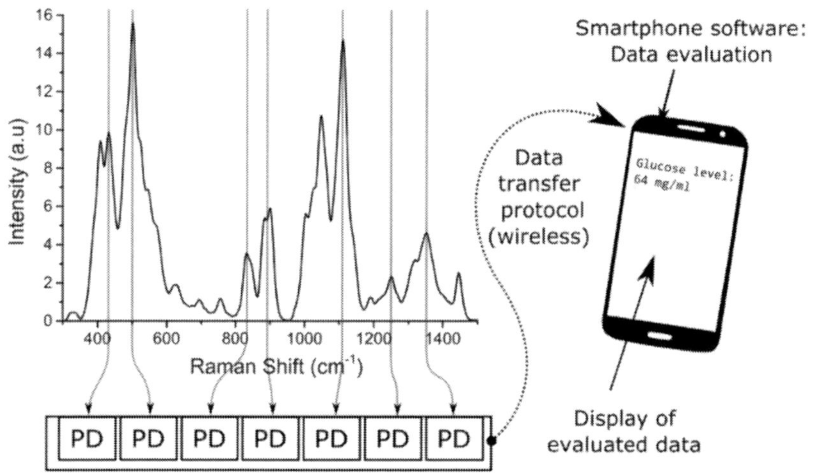

Figure 2. A schematics of a simplified device set-up, where a set of photodiodes (PD) with band-pass filters replaces CCD camera, and the data analysis and projection can be performed on a smartphone.

During first experimental analysis, specific Raman band intensity correlated with the change in glucose quantity (here, a sample aqueous glucose Raman spectrum) will be detected using chemometric data processing. For this purpose, Principal Component Analysis (PCA) can be used as an exploratory analytical method. It is possible to detect the bands which carry the informational content related to the investigated analyte, which can be seen on the specific principal components from transformed complex spectra. Furthermore, a PLS-Regression or other more elaborated techniques (PLS-DA, SVM-regression) can be used for quantification. Then, only Raman signal in those selected bands should be detected by a set of photodetectors, such as photodiodes (PD). Then, the received data is transferred to an auxiliary device where the data processing can be performed and the data projected to the end-user; possible devices include: a PC/laptop, microcontroller, or a smartphone. The ever-increasing capabilities and subsequent drop in prices, versatility, as well as a simplicity in use and familiarity for end-users makes smartphone the optimal choice for both a data processing unit and user interface device.

4.2. Turbidity-correction

The biggest issue with the non-invasive blood glucose detection, is the spot-to-spot and patient-to-patient differences in skin optical properties and slight differences in chemical composition. This in turn causes huge errors in the detection algorithms and calibration models, and enable their transfer between. Several methods for the correction of that problem have been investigated, and their performance greatly reduces the prediction errors of multivariate calibration methods. These methods apply a light-propagation model of elastic light scattering, to predict photon migration paths in tissues with various optical properties [36, 37]. Then, upon the measurements of diffuse reflectance at the Raman laser excitation wavelength, as well as the broadband Raman emission wavelength range, a correction factor is derived to minimize turbidity-induced spectral baseline variability [38]. Such approach does not require the use of tissue simulating phantoms (which mimic the μ_s and μ_a of tissues at specific wavelength range [39, 40]) as well as the knowledge of the absorption and scattering coefficients of tissues for multivariate calibration purposes. However, it is relevant only in a limited range of skin optical properties, but the introduction of diffuse reflectance into the setup is rather simple. Use of this method greatly reduces the prediction error of multivariate calibration techniques.

4.3. Synchronized Raman Detection

Other methods used for the reduction of tissue variability on Raman spectra have been proposed, such as tissue modulation. In this method, the fingertip is gently pressed to later released, to accumulate as much blood as possible in the measurement spot. This results in a *"modulated"* signal with high and low volumes of tissue blood perfusion [41]. The subsequent analysis proved that both the elastically (laser light and Rayleigh scattering) and inelastically (i.e. Raman) scattered light intensity follows the change of tissue blood perfusion [42].

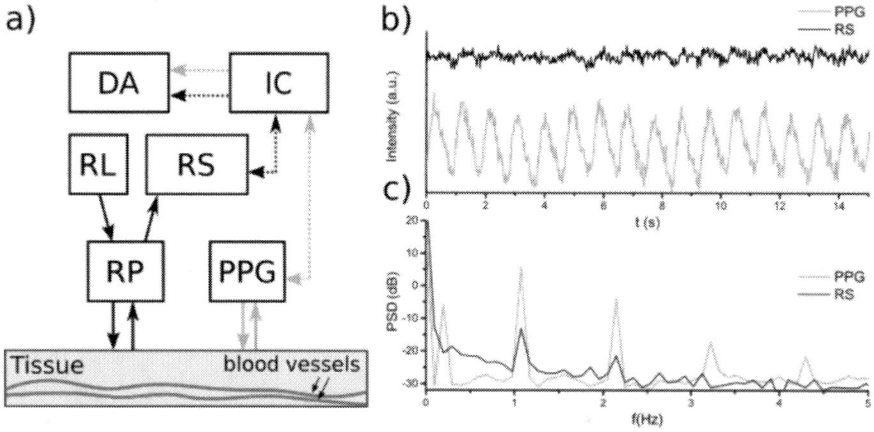

Figure 3. a) Synchronized detection set-up with data flow (optical signals – sold lines), (electrical signals – dotted lines). RL – Raman laser, RS – Raman spectrometer, RP – optical Raman probe, PPG – Photopletysmographic detector, IC – instrument control unit, DA – Data analysis unit; b) a simulation of Raman signal correlated with blood volume pulse from PPG; c) Power spectrum density of simulated PPG and Raman signals.

Our proposed solution to the problem of individual variability due to tissues could potentially be achieved by application of synchronous detection of optical Raman signal. In this method the Raman signal obtained from a fingertip can be treated as a composition of two components: a DC (or static) and AC (or dynamic) components. The static component contains the Raman signal which originates from the solid tissues (skin, muscle, vessel walls, lipids, etc.), as well as the small amount of blood which always remains in the blood vessels. In other words, it contains the information from the whole volume of irradiated tissue which is stationary in time. The dynamic component originates only from the objects that change with time, that is the volume of blood which is associated with the pulse. Due to the pumping action of heat, additional amount of blood fills the vessels, changing blood volume at the rate of the heartbeats. In the investigated case, only the dynamic component is of interest, as it carries the Raman spectrum of the blood itself. The static component varies with tissue composition between measurement spots and on a patient-to-patient basis. This causes random and unpredictable changes in baseline, thus the static component can be treated as noise, as it does not carry the informational content. Therefore, the detection of only the AC Raman signal and elimination of a DC Raman signal (associated with the variance in tissue composition) would greatly reduce the prediction error and allow to translate the calibration model to all patients. The synchronization of inelastically scattered Raman signal with the elastically scattered signal from a photoplethysmography (PPG) near-infrared measurement of blood volume, will allow for selection of only the useful (AC) Raman signal. This method can be efficient due to identical scattering properties of red blood cells between people of different age, race and sex. Moreover, scattering on red blood cells is almost two orders more intense when compared with other tissue constituents. The schematic of a system for synchronized Raman detection is presented on Figure 3.a). The Raman detection part comprises of a (RL) Raman laser, (RS) Raman spectrometer, (RP) optical Raman probe. The Raman light will be collected with high temporal resolution, to detect the rapid change in blood perfusion of a tissue, caused by a natural blood volume pulses from cardiac rhythm. Then, simultaneously with the Raman signal, a blood volume signal is detected by a (PPG) photopletysmographic detector, which is controlled by a (IC) instrument control unit. Both signals from those two modes of detection are transferred to (DA) data analysis unit.

The resulting spectra of our simulations are presented on Figure 3.b). The PPG signal was modeled as a sawtooth/triangle waveform with additional white noise, baseline variability, and additional *dicrotic notch* [17, 43–45]. The resulting Raman signal consists of a DC baseline and a PPG signal with greatly diminished intensity and additional white and pink noise, as well as another baseline variability. The frequency characteristic of these signals were checked with the use of Wiener power spectral density (PSD) presented on Figure 3.c). The PPG signal follows typical spectral distribution of a triangle wave, with basic frequency and power reduced at subsequent harmonics. The basic frequency and first harmonic is retained in the RS spectrum, while the rest of the harmonics is obstructed by noise, and a significant DC component. This preliminary analysis shows, that by application of such method, the inter-patient variability may be overcome, with a simple set-up modification and subsequent software analysis of signals.

5. Conclusion

This paper presented current trends in noninvasive blood glucose measurements, with special focus on the Raman spectroscopy. The greatest challenges in the application of this method were described, with the inter-patient variability being the most influential. With the advent of advance multivariate calibration methods, the weight of the problem of commercialization of such devices lies in the software part, thus permitting their miniaturization. Several signal-processing techniques as well as a set-up simplification were described. Additionally, a new method of synchronized Raman detection, correlated with a natural blood volume pulse was initially analyzed. The frequency response of simulated signals retains blood pulse information. The elimination of individual variability influence on the precision of estimating the blood glucose concentration is a key advantage of the proposed method. Therefore it is concluded that the combination of advanced multivariate techniques, and synchronized detection scheme will solve the inter-patient variability and allow for device miniaturization for commercial purposes. So far such use of a synchronous detection for non-invasive Raman spectroscopy has not been investigated in-depth.

6. References

[1] International Diabetes Federation 2013 IDF Diabetes Atlas, 6th ed. http://www.idf.org/diabetesatlas

[2] Vashist S K 2013 Continuous Glucose Monitoring Systems: A Review *Diagnostics* **3** 385–412

[3] So C-F, Choi K-S, Wong T K and Chung J W 2012 Recent advances in noninvasive glucose monitoring *Med. Devices Auckl. NZ* **5** 45–52

[4] McMillin J M 1990 Blood Glucose *Clinical Methods: The History, Physical, and Laboratory Examinations* ed H K Walker, W D Hall and J W Hurst (Boston: Butterworths)

[5] Tura A, Maran A and Pacini G 2007 Non-invasive glucose monitoring: Assessment of technologies and devices according to quantitative criteria *Diabetes Res. Clin. Pract.* **77** 16–40

[6] Klonoff D C 2012 Overview of Fluorescence Glucose Sensing: A Technology with a Bright Future *J. Diabetes Sci. Technol.* **6** 1242–50

[7] Pickup J C, Hussain F, Evans N D, Rolinski O J and Birch D J S 2005 Fluorescence-based glucose sensors *Biosens. Bioelectron.* **20** 2555–65

[8] Hull E L, Ediger M N, Unione A H T, Deemer E K, Stroman M L and Baynes J W 2004 Noninvasive, optical detection of diabetes: model studies with porcine skin *Opt. Express* **12** 4496

[9] Badugu R, Lakowicz J R and Geddes C D 2003 A Glucose Sensing Contact Lens: A Non-Invasive Technique for Continuous Physiological Glucose Monitoring *J. Fluoresc.* **13** 371–4

[10] Evans N D, Gnudi L, Rolinski O J, Birch D J S and Pickup J C 2003 Non-invasive glucose monitoring by NAD(P)H autofluorescence spectroscopy in fibroblasts and adipocytes: a model for skin glucose sensing *Diabetes Technol. Ther.* **5** 807–16

[11] Gabbay R A and Sivarajah S 2008 Optical coherence tomography-based continuous noninvasive glucose monitoring in patients with diabetes *Diabetes Technol. Ther.* **10** 188–93

[12] Alarousu E, Hast J T, Kinnunen M T, Kirillin M Y, Myllyla R A, Plucinski J, Popov A P, Priezzhev A V, Prykari T, Saarela J and Zhao Z 2004 Noninvasive glucose sensing in scattering media using OCT, PAS, and TOF techniques *Proc. SPIE* vol 5474, ed V V Tuchin pp 33–41

[13] Kinnunen M, Myllylä R, Jokela T and Vainio S 2006 In vitro studies toward noninvasive glucose monitoring with optical coherence tomography *Appl. Opt.* **45** 2251

[14] Popov A P, Bykov A V, Toppari S, Kinnunen M, Priezzhev A V and Myllyla R 2012 Glucose Sensing in Flowing Blood and Intralipid by Laser Pulse Time-of-Flight and Optical Coherence Tomography Techniques *IEEE J. Sel. Top. Quantum Electron.* **18** 1335–42

[15] Korhonen V O, Myllyla T S, Kirillin M Y, Popov A P, Bykov A V, Gorshkov A V, Sergeeva E A, Kinnunen M and Kiviniemi V 2014 Light Propagation in NIR Spectroscopy of the Human Brain *IEEE J. Sel. Top. Quantum Electron.* **20** 1–10

[16] Wróbel M S, Jedrzejewska-Szczerska M, Galla S, Piechowski L, Sawczak M, Popov A P, Bykov A V, Tuchin V V and Cenian A 2015 Use of optical skin phantoms for preclinical evaluation of laser efficiency for skin lesion therapy *J. Biomed. Opt.* **20** 085003

[17] Myllylä T, Korhonen V, Vihriälä E, Sorvoja H, Hiltunen T, Tervonen O and Kiviniemi V 2012 Human Heart Pulse Wave Responses Measured Simultaneously at Several Sensor Placements by Two MR-Compatible Fibre Optic Methods *J. Sens.* **2012** 1–8

[18] Yadav J, Rani A, Singh V and Murari B M 2015 Prospects and limitations of non-invasive blood glucose monitoring using near-infrared spectroscopy *Biomed. Signal Process. Control* **18** 214–27

[19] Liakat S, Bors K A, Xu L, Woods C M, Doyle J and Gmachl C F 2014 Noninvasive in vivo glucose sensing on human subjects using mid-infrared light *Biomed. Opt. Express* **5** 2397–404

[20] Dingari N C, Barman I, Singh G P, Kang J W, Dasari R R and Feld M S 2011 Investigation of the specificity of Raman spectroscopy in non-invasive blood glucose measurements *Anal. Bioanal. Chem.* **400** 2871–80

[21] Pandey R, Dingari N C, Spegazzini N, Dasari R R, Horowitz G L and Barman I 2015 Emerging trends in optical sensing of glycemic markers for diabetes monitoring *TrAC Trends Anal. Chem.* **64** 100–8

[22] Friebel M, Roggan A, Müller G and Meinke M 2006 Determination of optical properties of human blood in the spectral range 250 to 1100 nm using Monte Carlo simulations with hematocrit-dependent effective scattering phase functions *J. Biomed. Opt.* **11** 034021

[23] Bashkatov A N, Genina E A and Tuchin V V 2011 Optical properties of skin, subcutaneous, and muscle tissues: a review *J. Innov. Opt. Health Sci.* **04** 9–38

[24] Wróbel M S, Popov A P, Bykov A V, Kinnunen M, Jędrzejewska-Szczerska M and Tuchin V V 2015 Measurements of fundamental properties of homogeneous tissue phantoms *J. Biomed. Opt.* **20** 045004–045004

[25] Smulko J M, Dingari N C, Soares J S and Barman I 2014 Anatomy of noise in quantitative biological Raman spectroscopy *Bioanalysis* **6** 411–21
[26] Kwiatkowski A, Gnyba M, Smulko J and Wierzba P 2010 Algorithms of Chemicals Detection Using Raman Spectra *Metrol. Meas. Syst.* **17**
[27] Morawski R Z 2012 Measurement Data Processing in Spectrophotometric Analysers of Food *Metrol. Meas. Syst.* **19** 623–52
[28] Barman I, Dingari N C, Singh G P, Soares J S, Dasari R R and Smulko J M 2012 Investigation of Noise-Induced Instabilities in Quantitative Biological Spectroscopy and Its Implications for Noninvasive Glucose Monitoring *Anal. Chem.* **84** 8149–56
[29] Bispo J A M, de Sousa Vieira E E, Silveira J Landulfo and Fernandes A B 2013 Correlating the amount of urea, creatinine, and glucose in urine from patients with diabetes mellitus and hypertension with the risk of developing renal lesions by means of Raman spectroscopy and principal component analysis *J. Biomed. Opt.* **18** 087004–087004
[30] Barman I, Kong C-R, Dingari N C, Dasari R R and Feld M S 2010 Development of Robust Calibration Models Using Support Vector Machines for Spectroscopic Monitoring of Blood Glucose *Anal. Chem.* **82** 9719–26
[31] Barman I, Dingari N C, Rajaram N, Tunnell J W, Dasari R R and Feld M S 2011 Rapid and accurate determination of tissue optical properties using least-squares support vector machines *Biomed. Opt. Express* **2** 592–9
[32] Gnyba M, Smulko J, Kwiatkowski A and Wierzba P 2011 Portable Raman spectrometer - design rules and applications *Bull. Pol. Acad. Sci. Tech. Sci.* **59** 325–9
[33] Wood B R, Heraud P, Stojkovic S, Morrison D, Beardall J and McNaughton D 2005 A Portable Raman Acoustic Levitation Spectroscopic System for the Identification and Environmental Monitoring of Algal Cells *Anal. Chem.* **77** 4955–61
[34] Brennan J F, Wang Y, Dasari R R and Feld M S 1997 Near-Infrared Raman Spectrometer Systems for Human Tissue Studies *Appl. Spectrosc.* **51** 201–8
[35] Dingari N C, Barman I, Kang J W, Kong C-R, Dasari R R and Feld M S 2011 Wavelength selection-based nonlinear calibration for transcutaneous blood glucose sensing using Raman spectroscopy *J. Biomed. Opt.* **16** 087009–087009
[36] Shih W-C, Bechtel K L and Feld M S 2008 Intrinsic Raman spectroscopy for quantitative biological spectroscopy part I: theory and simulations *Opt. Express* **16** 12726–36
[37] Bechtel K L, Shih W-C and Feld M S 2008 Intrinsic Raman spectroscopy for quantitative biological spectroscopy part II: experimental applications *Opt. Express* **16** 12737–45
[38] Barman I, Singh G P, Dasari R R and Feld M S 2009 Turbidity-Corrected Raman Spectroscopy for Blood Analyte Detection *Anal. Chem.* **81** 4233–40
[39] Lamouche G, Kennedy B F, Kennedy K M, Bisaillon C-E, Curatolo A, Campbell G, Pazos V and Sampson D D 2012 Review of tissue simulating phantoms with controllable optical, mechanical and structural properties for use in optical coherence tomography *Biomed. Opt. Express* **3** 1381–98
[40] Wróbel M S, Popov A P, Bykov A V, Kinnunen M, Jędrzejewska-Szczerska M and Tuchin V V 2014 Multi-layered tissue head phantoms for noninvasive optical diagnostics *J. Innov. Opt. Health Sci.* **08** 1541005
[41] Chaiken J, Knudson P E, Bussjager R J, Peterson C M, Peterson K, Hagrman D, Weinstock R S, Finney W, Khan M, Hagrman P and Zhao Y 2005 Effect of hemoglobin concentration variation on the accuracy and precision of glucose analysis using tissue modulated, noninvasive, in vivo Raman spectroscopy of human blood: a small clinical study *J. Biomed. Opt.* **10** 031111–03111112
[42] Chaiken J, Goodisman J, Deng B, Bussjager R J and Shaheen G 2009 Simultaneous, noninvasive observation of elastic scattering, fluorescence and inelastic scattering as a monitor of blood flow and hematocrit in human fingertip capillary beds *J. Biomed. Opt.* **14** 050505
[43] Yamakoshi K and Yamakoshi Y 2006 Pulse glucometry: a new approach for noninvasive blood glucose measurement using instantaneous differential near-infrared spectrophotometry *J. Biomed. Opt.* **11** 054028
[44] Shamir M, Eidelman L A, Floman Y, Kaplan L and Pizov R 1999 Pulse oximetry plethysmographic waveform during changes in blood volume. *Br. J. Anaesth.* **82** 178–81
[45] Korpas D, Halek J and Dolezal L 2009 Parameters describing the pulse wave. *Physiol. Res.* **58** 473

Acknowledgements

This study was partially supported by DS Programs of the Faculty of Electronics, Telecommunications and Informatics, Gdańsk University of Technology; as well as the Polish National Science Center under the grant 2011/03/D/ST7/03540.

IOP Conf. Series: Materials Science and Engineering **104** (2015) 012037 doi:10.1088/1757-899X/104/1/012037

A novel concept for tissue-metal detection and differentiation using an inductive proximity sensor

A Gorczewska[1], A Poliński[1], J Wtorek[1], B Truyen[2]

[1]Gdansk University of Technology, Faculty of Electronics, Telecommunications and Informatics, ul. Narutowicza 11/12, 80-233 Gdansk, Poland
[2]Vrije Universiteit Brussel, Faculty of Engineering, Boulevard de la Plaine 2, 1050 Ixelles, Belgium

E-mail: jerzy.wtorek@eti.pg.gda.pl

Abstract. In this paper a novel application of inductive proximity sensors for detection of living tissue by means of measurements of the coil impedance changes at different frequencies is described. The mathematical analyses utilizing Bessel function estimation include detected object size and its distance from a sensor. The main aim of this study is to prove the possibility of distinguishing between metal objects and living tissues. The movement of an object can be also detected, assuming it is in a distance comparable with a size of the induction sensor.

1. Introduction

Proximity sensors are utilized in many applications in the industry, new technologies or home appliances. One of the fields of utilization of proximity sensors is the branch known as the human-system interaction. However, methods that are applied to living tissue detection are not realized only by proximity sensors but involve also other measurement techniques. A brief overview of the existing systems for the living tissue detection, as well as a short introduction to the principles of operation of such methods is presented below.

Typically, the living tissue can be detected by utilization of optical, thermal, ultrasounds, electric or electromagnetic methods. Generally, sensors can be divided due to the range of object detection. When a disturbance occurs close to the sensor the inductance or capacitance sensors are applied. When the distance between them is larger (e.g. in industry applications) it is better to use ultrasounds, photoelectric or radar sensors.

One of the types of sensors, which are able to detect a living tissue, is the ultrasound sensor. It is widely utilized for a diagnostic and treatment purposes of a human tissue. Their different interactions with tissue depend on their types (generally a different velocity of ultrasounds in a medium). Some of the emitted waves reflect, others propagate and partially scatter in a medium. Impact on it has an essential parameter called the acoustic impedance, which is a measure of a medium reaction to the acoustic wave. The unit of the specific acoustic impedance, Z_a, is Rayl [kg m^{-2} s^{-1}]. This parameter for living tissue and a metal object differ about an order of magnitude. For an air the acoustic impedance is yet four times smaller than for the living tissue [1]. The ultrasound sensors which are utilized on a large distance (even to 300 m) are applied only to metal object. Thus a transmitted wave is entirely reflected from the metallic object, which causes a larger difference of the specific acoustic impedance between air and the metal (e.g. aluminium Z_a=17*10^6 Rayl, air Z_a=4*10^2 Rayl).

Living tissues are more similar to an air than to metallic materials. They have lower value of reflection coefficient. It is a result of the fact that a boundary between air and soft tissue is not so sharp (difference between acoustic impedance of the air and the tissue is not so big).

Optical measurements of the living tissue are generally led in the regions of the greatest optical transparency, i.e., the tissue optical window (600 to 1300 nm) [2]. This is a range where blood components have the most impact on the absorption coefficient. Utilizing optical methods it is possible to detect and visualize biological tissue with a high-resolution. Due to the scattering of a light the light source-object distance must be short and does not exceed a few centimeters. The same situation concerns such medium as a fabric clothes which makes detection of a human body from some larger distance impossible. Other materials, e.g. metal, have a lower absorption coefficient with maximum value for shorter wave (200-400 nm) than for the biological tissue. It is difficult to detect the properties of metals and tissues by optic techniques because the wave has to penetrate into the human body, to reveal the differences between the optical characteristics (e.g. absorption coefficient). Measurement only the reflection coefficient does not allow for distinguishing different types of materials such as living tissues or metals.

In electromagnetic methods the proximity sensor always emits the electromagnetic field, which is disturbed (absorbed, etc.) by the object placed near the sensor. The disturbances of the electromagnetic field can be then recorded and utilized for detecting and recognizing of the object. Depending on the electromagnetic properties of the different types of proximity sensors, specific sensors are always oriented on detecting slightly different objects. For example impedance measurements can be used to detect of the person in the bathtub [3].

In general, a coil-object interaction is based on the fact, that the alternating magnetic field of the coil is inducing eddy currents in (conducting) the object. Induced eddy currents create a magnetic field that opposes the magnetic field that created it. The magnetic fields of eddy currents, deriving from the object, induced the electromagnetic field in the coil which resulting a change in its impedance [4]. Additionally, the inductance of the coil depends on the frequency, and this dependence will also have an impact on the impedance of the coil. Those principles are utilized in inductive sensing method, whereas capacitive sensing mechanism is based on the measurement of displacement currents [5].

There are two known types of coil interaction with the nearby object, namely, capacitive (electric) mechanism and inductive (magnetic) mechanism. The electrical properties of the object determine which of these mechanisms will be dominating. Capacitive proximity sensors are usually used for detection of the non-magnetic objects, which conductivity is low, thus the human – system interaction requires contact with the senor's surface. Wimmer in [6] presents a game in which user controls a ball by walking on the top of a metal sensor antenna. The measuring parameter is a change of a capacitance. Touchless 3D gesture recognition described in [7] allows in near-field of a sensor ambient detection a single human finger or a whole hand. However, this approach has not been checked in the case of detecting a fingers living tissue. Likewise in [8], when a RF – capacitive proximity sensor was used to detect the presence of human hand equipped in special gloves nearby a chainsaw, and the system turn off the power.

Inductive proximity sensors are usually applied for the recognition of magnetic objects of high conductivity. An example of such device is inductive proximity sensor designed to locate metal shrapnel during surgery, reported in [9]. A combination of inductive and capacitive properties was utilized in conductive-capacitive proximity sensor for seat occupancy sensing [10].

Living tissue has a low conductivity. The technique in which the inductive method is used for measurements of electromagnetic properties of biological tissue is called magnetic induction spectroscopy. In [11] the requirements and limitations of developing a system for biological tissue characterization were described. The estimation of the dielectric parameters (conductivity, dielectric constant) of biological tissue is realised by measurements of the magnetic field at the wide range of frequencies. However, there are some limitations of such system, e.g. the requirement of the

placement of an object between transmitting and receiving coils, which demands the adequate place and specific measurements procedure. Thus, it cannot be utilized for measurements of living objects anywhere.
Almost all presented methods require the separation of the living tissue and metal object from their natural environment. In real ambient exist not only those two kind of materials as metal or living tissue but also plastics, cements, plants, organic etc. in various configurations (e.g. a person with a dog, a motorcyclist on a motorcycle, a tree, a street light). They all cause that description of an object utilizing only electrical properties in one frequency is not enough. A sensor for detection of the living tissue and metal materials from real environment should also be able to recognize the geometry such as shape, distance from sensor and obviously kind of material.
In this paper we present the preliminary concept of utilization of the simple inductance proximity sensor for living tissue detection. The differentiation between living tissue and the metallic (conductive) objects is analyzed. As a differentiating parameter, the ratio of real to imaginary part of the reflected impedance is considered. Impedance model is analyzed in a wide range of frequency in order to enable recognition from complex surrounding living objects, their location, size and shape.

2. Material and methods

2.1. Impedance model

In terms of electrical properties the living tissue is a dielectric, which in contrast to metals, has a low conductivity. Properties of a material can be described by two passive electromagnetic properties, namely conductivity and permittivity.
The parameter which determines the dielectric properties of materials is the complex permittivity (inclusive the relative permittivity and conductivity), ε^* expressed as [12]:

$$\varepsilon^* = \varepsilon' - j\varepsilon'' - j\frac{\sigma_s}{\omega\varepsilon_0}, \tag{1}$$

where ε' and ε'' are respectively real and imaginary part of relative electric permittivity, σ_s is the static conductivity conductors, ε_0 - the permittivity of free space and ω – the angular frequency.
Another description of dielectric properties is presentation them as the complex conductivity:

$$\sigma^* = \sigma(\omega) + j\omega\varepsilon'(\omega), \tag{2}$$

In biological tissue when frequency is increasing the conductivity has the same trend, and the permittivity has opposite behavior [12]. These changes are not the same in a wide range of frequency.
The different situation is observed in metal materials [4], where there are large values of conductivity (e.g. copper has σ = 58.6*10^6 Sm^{-1}) and the relative permittivity is equal to 1. Further, these parameters for metal materials are independent of frequency. Therefore using several different frequencies for electromagnetic measurements living tissue and metal materials may make possible to distinguish them.
For living tissue, one of three ranges called α–, β– or γ–dispersion is observed in the multifrequency measurements. A dielectric dispersion, which is termed β-dispersion, is shown at a certain frequency of electric field (i.e. characteristic frequency) [13]. About 90% of the composition of human body is water. At the frequency range 10 kHz – 10 MHz (β-dispersion) this hydration state is well reflected by the frequency-dependence of the conductivity [11]. With increasing frequency the additional contribution of the intracellular fluid volume causes a significant increase of the conductivity compare to the lower frequencies.
When an object appears in the vicinity of a sensing coil a circuit can be treated as a contactless transformer. The presence of the object changes values of parameters, which is reflected in the impedance of a sensing coil [14].

The issue of measuring a change of coil impedance for the detection of materials of different conductivities is schematically presented in Figure 1.

Mortarelli [15] based on Geselowitz considerations [16] proposed the equation describing mutual impedance changes of ΔZ. This relationship has applied Lorentz reciprocity relation. The formula is presented below:

$$\Delta Z = -\frac{1}{I^2}\int_V \Delta\sigma^* E_a E_b dv, \tag{3}$$

where the complex conductivity change $\Delta\sigma^*$ is determined by:

$$\Delta\sigma^* = \Delta\sigma + j\omega\Delta\varepsilon$$

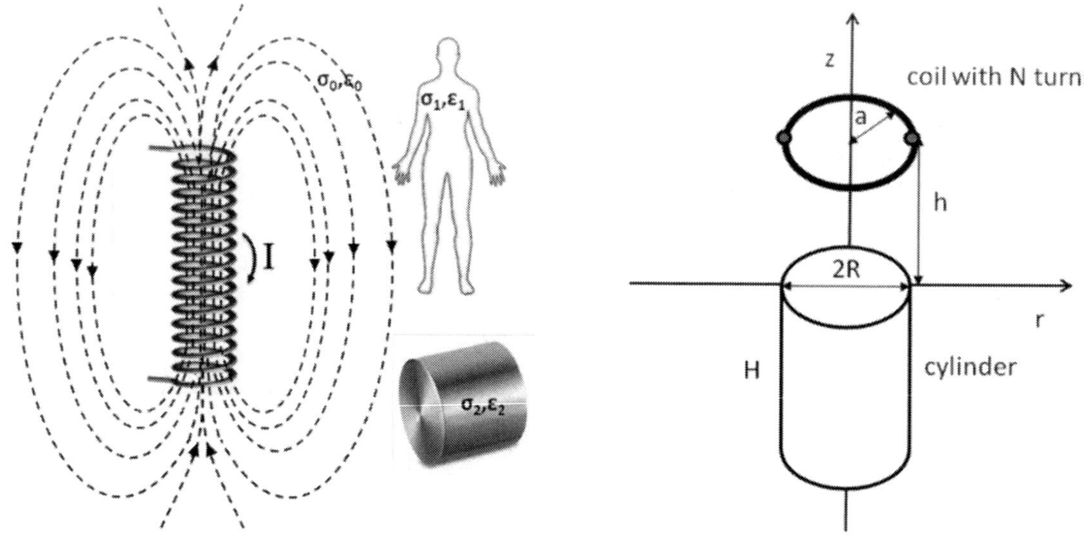

Figure 1. Scheme of a coil-object (living tissue or metal) interaction (a), and modeling geometry – ring coil above a cylinder (b).

$\Delta\sigma$ ($\Delta\varepsilon$) represents conductivity (permittivity) changes in the considered volume, which causes change in the field distribution from E_a to E_b. The volume of integration V includes the region where properties of the medium are changing. For biological tissues and metals there is no change in permeability so magnetic field component H has no influence on impedance change [17, 18].

In some situations it is useful to simplify calculations and assume that $E_a \approx E_b$ i.e. the field distribution is assumed not to be disturbed by conductivity change (or this disturbance is small). We have compared this approach with assumption that the fields are not the same and difference is caused by different properties of material for half space $z>0$ (region of coil) and $z<0$ (region of disturbance). This is also simplification, but allows using explicit formula for vector potential, and relation with the electromagnetic fields E_a and E_b:

$$E = -j\omega A_\phi, \tag{4}$$

thus:

$$\Delta Z = \frac{\omega^2}{I^2}\Delta\sigma^* \int_V A_{\phi,a} A_{\phi,b} dv, \tag{5}$$

$\Delta\sigma$ is independent of the region of integration so is treated as a constant. The definition of A_ϕ is taken from [19, 20]:

$$A_{\phi,a}(\rho,z) = \frac{\mu_0 I a}{2}\int_0^\infty J_1(\xi\rho)J_1(\xi a)e^{-\lambda_0|z-h|}\frac{\xi d\xi}{\lambda_0}, \tag{6}$$

where

$$\lambda_0 = \sqrt{\xi^2 - k_0^2}, \qquad k_0^2 = \omega^2\mu_0\varepsilon_0,$$

and μ_0 is permeability of free space and a is radius of the coil.
While for a heterogeneous medium:

$$A_{\phi,b}(\rho,z) = \frac{\mu_0 I a}{2}\int_0^\infty J_1(\xi\rho)J_1(\xi a)e^{\lambda_1 z-\lambda_0 h}\frac{2\xi d\xi}{\lambda_0+\lambda_1}, \tag{7}$$

where

$$\lambda_1 = \sqrt{\xi^2 - k_1^2}, \qquad k_1^2 = \omega^2\mu_0\varepsilon_1 - j\omega\mu_0\sigma_1,$$

Eqs.6-7 are true for $z<0$, i.e. in the region where the conductivity changes are expected. The region of $z>0$ is neglected in this approach. A graphic representation of the considered model is shown in Figure 2 where a is radius of the coil, h is a coil – cylinder distance, R and H are radius and height of the cylinder, respectively. The cylinder is coaxially located under the sensing coil. By modifying the complex conductivity σ^*, the cylinder radius R and height H, a frequency operation f, and the distance h (the dimensions of the coil have been constant), it is possible to calculate impedance changes of the coil for both types of media (dielectric or conductive) appearing in the proximity of the inductive sensor.
Estimation of the impedance change of the coil induced by the presence of the human tissue or a conductive material in the sensor proximity was realized by numerical simulations according to the formulas presented above. Subsequent values of the complex conductivity and the relative proximity for each frequency were taken from Gabriel tables for ovine muscle tissue [12].
Apart from the impedance modeling, measurements of the living tissue were carried out as well. The measuring set up was composed of a single coil, a metal cylinder, the impedance analyzer and its dedicated software to control measurement parameters. Measurements of living tissue were executed on the volunteer arm. The circular sensor coil was constructed by winding 10 turns (0.15 mm in diameter) of the total inductance 12 μH and the 4 cm radius of the coil. Measurement protocol included measurement the impedance of: empty coil, coil in the proximity of metal cylinder or human arm localized above the sensor. Measurements were performed using Solartron SI 1260 Impedance/Gain-Phase Analyzer in the frequency range from 10 kHz up to 1 MHz using 11 points per decade spaced uniformly in logarithmic scale (1 second for each frequency for metal cylinder, while for volunteer arm it was increase to 5 seconds).
The measurement results were verified using the simple transformer model [21]. However the secondary winding of the transformer is heavy to calculate since its values depend on the object location and parameters. Thus we use simplification that load impedance is equal to metal and arm properties, while secondary inductance is equal to primary one. The coupling factor k was estimated to obtain results similar to measurement one and was set to 0.55. To obtain more realistic results, transformer model include capacitance coupling also. The value of this capacitor was set to 1nF.

3. Results

The results were analyzed for single turn of the radius coil $a = 0.5$ m and four changing parameters: a frequency f (from 100 kHz to 10 MHz), the coil – cylinder distance h, a cylinder height H, and the radius of the cylinder R (h, H and R from 0.05 m to 2.5 m). In a single simulation only one parameter was changing. The base parameters (were one of the other was changing) were equal to $f = 1$ MHz, $h =$

2a, H = 2a, R = a. The impact of the change of those parameters on the value of the impedance (real, imaginary part, ratio imaginary to real) was investigated.

The relationship of the real and imaginary parts of the impedance for metal and tissue for different distances between coil and the cylinder are illustrated in Figs. 2a and 2b. The influence of the cylinder height is shown in Figures 3a-3b. Figure 3a presents the relation of the real parts of the impedance for metal and tissue for different cylinder height. The imaginary part of the impedance as a function of the cylinder height is shown in Figure 3b.

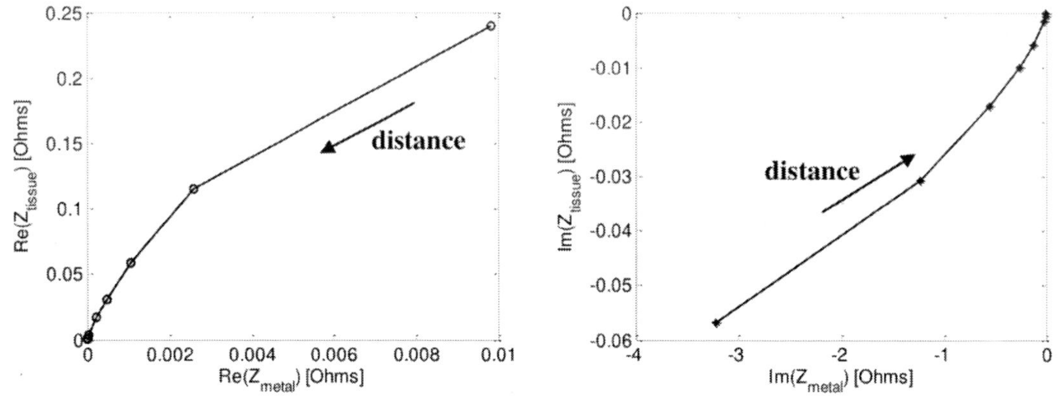

Figure 2. The real vs. real (a) and the imaginary vs. imaginary (b) part of the impedance. Both for various coil – cylinder distance h (simulation).

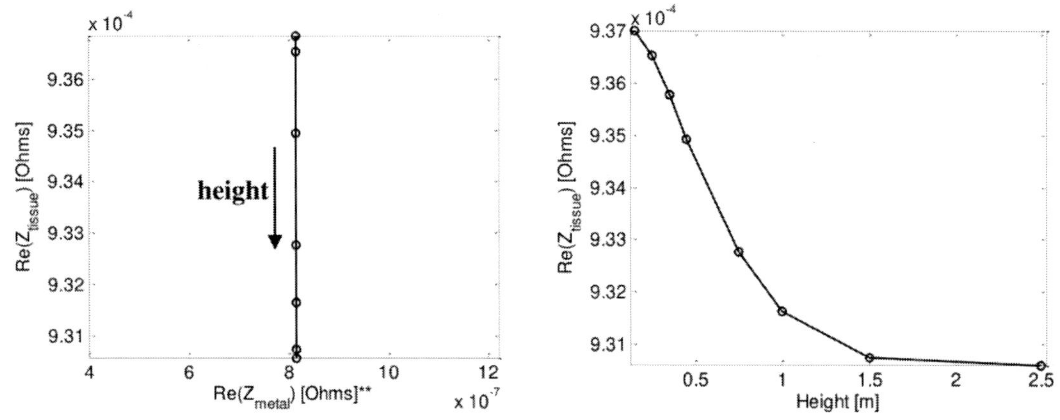

Figure 3. The real vs. real part of the impedance (a), and the real part of the impedance (b). Both for various cylinder heights H (simulation).

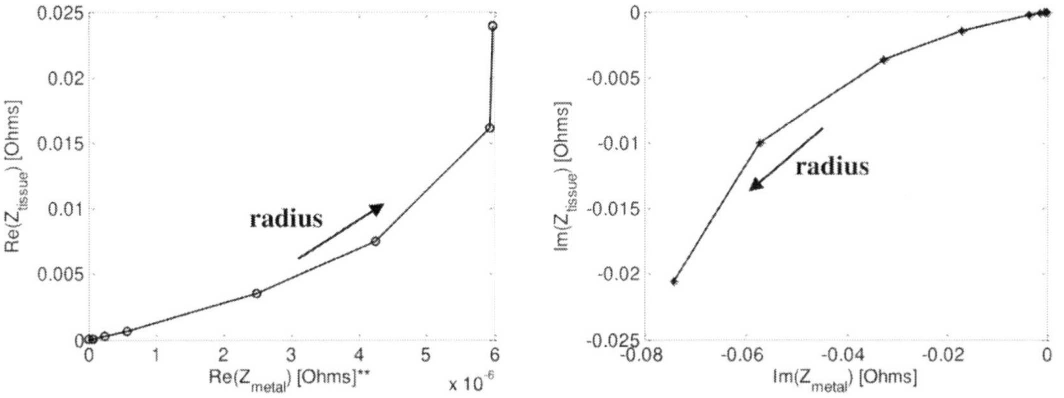

Figure 4. The real vs. real part of the impedance (a), and the imaginary vs. imaginary part of the impedance (b). Both for various cylinder radiuses *R* (simulation).

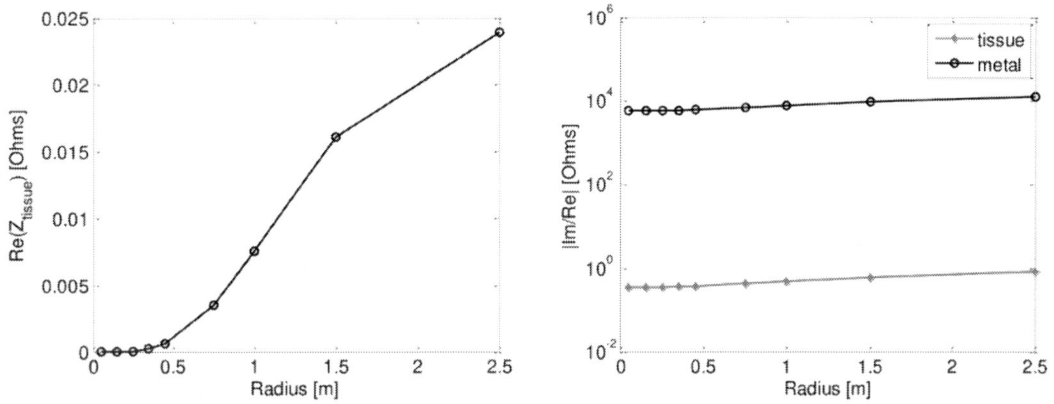

Figure 5. The real part of the impedance (a), and the ratio of the imaginary to the real part of the impedance for metal and tissue (b). Both for various cylinder radiuses *R* (simulation).

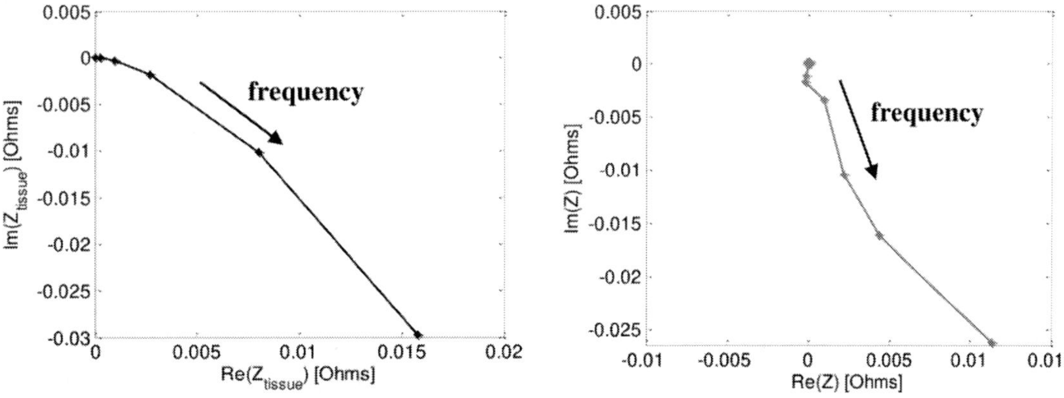

Figure 6. The Cole-Cole plot for tissue (simulation) (a), and human tissue (measurement) (b).

The influence of the cylinder radius on the measured impedance was tested, as shown in Figures 4a-4b. The relation of the real part of the impedance as a function of the cylinder radius is shown in

IOP Conf. Series: Materials Science and Engineering **104** (2015) 012037 doi:10.1088/1757-899X/104/1/012037

Figure 5a. Figure 5b presents comparison of the imaginary and real part of the impedance for considered materials. The relationship between the real and the imaginary part of the impedance for tissue as a function of frequency obtained from simulations is shown in Figure 6a. The error bars (Figure 7b) show the standard deviation.

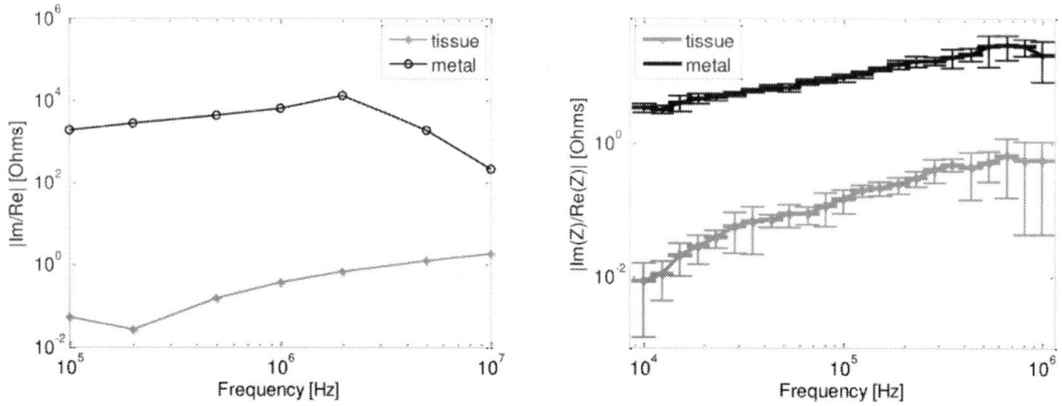

Figure 7. The ratio of the imaginary to the real part of the impedance from simulations (a), and from measurements with the standard deviation (b).

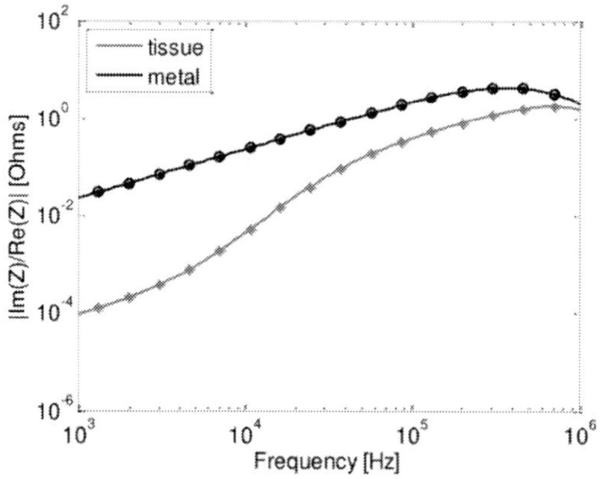

Figure 8. The ratio of the imaginary to the real part of the impedance for transformer model.

Similar plot for the human tissue obtained from measurement as seen in Figure 6b. The ratio of the imaginary and to the real parts of the impedance for metal and tissue obtained from simulations and measurements respectively as illustrated in Figures 7a and 7b. Figure 8 shows the same ratio as in Figure 7 for transformer model.

4. Discussion

Burrows [22] was the first, up to our knowledge, who introduced the use of reciprocity in eddy current probe analysis. Similar approach for biomedical applications was derived by Geselowitz [16]. Further on it was generalized for nonquasi-static sinusoidal sources and fluctuating geometry by Mortarelli in 1980 [15]. Because of the complexity of the problem one can consider the simplified assumptions: the change in the electrical field caused by introducing disturbance is so small that the field is the same as in the case without disturbance. This assumption is valid in the case of relatively small disturbance and/or small dielectric properties change introduced by the disturbance. This allows approximating the field E_b in (4) by E_a, thus simplified numerical calculations. Otherwise the numerical methods such as FEM had to be used to calculate the E_b. We have made similar approach in biomedical applications [13, 23], where these assumptions were valid and gave good results. However, the difference in

dielectric parameters between air and disturbance are so large that this simplification may not be valid. Because of this we have used another simplification, i.e. the disturbance expands in the half space $z<0$. The simulations show that the impedance change rapidly decreasing as a function of coil – cylinder distance and those are similar for the tissue and metal (Figure 2). The vector potential decay for metal is so fast that the height of the cylinder has no influence on impedance change. This is opposite to the tissue (Figure 3). However, even in that case, larger values of the height of the cylinder do not have much influence on the impedance. In the considered range of the cylinder radius, its influence on the impedance is similar in the case of metal and tissue (Figures. 4-5). The main difference is in the amplitude of the signal – it is larger for the real part in the case of tissue, while smaller for the imaginary part of the impedance.
The simulations of the impedance change show that our results are not inconsistent with those obtained by Cheng [19] for half space (our results are smaller due to smaller disturbance size).
The main difference in the impedance change for metal and living tissues is that for the metal the permittivity can be neglected comparing to conductivity. It means that sensitivity is multiplied by real number, which is only the enlargement of the value. In the case of living tissues the real and imaginary part of the complex conductivity are similar, i.e. sensitivity (complex number) is multiplied by another complex number so resulting real and imaginary part of the impedance is dependent of both parameters in more complex way.
The simulation results were confirmed by measurements (Figures 6). It follows from the simulation and measurements that metal and living tissue can be differentiate basing on the ratio of imaginary to real part of impedance change measurements (Figures 7-8 and Figure 5b). However the time of measurements was too long for practical applications but it was increased to obtain more reliable results.
From the presented study it is obvious that multifrequency measurement allows distinguishing between tissues and metal. Similarly the dependence of the real and imaginary parts of the impedance on the distance from the coil is different for tissues and metal. The size of the detected object influences the amplitude of the signal. In the case of the tissues the relative permittivity is quite large, so its influence on the reflected impedance is large and allows distinguishing tissues from the metal.

5. Conclusions

The presented simulations and measurements allow to conclude that it is possible to detect and differentiate tissues from metal objects using multifrequency measurements basing on the ratio of imaginary to real part of impedance change. The amplitude of measurement signal depends on the size, the location and the complex conductivity of the detected object. Thus it may be possible to detect object and determine its approximate location and size. However, due to complex signal dependence of these parameters more studies are required.

6. References

[1] Gao L, Parker K J, Lerner R M and Levinson S F 1996 **22/8** 959-977
[2] Gorczewska A 2015 *PhD Interdisciplinary Journal* No. 1
[3] Bujnowski A, Skalski L and Wtorek J 2012 *J Med Imag Health In* **2** 27-34
[4] Wilson D L and Geselowitz D B 1981 *IEEE Trans. Bio. Eng.* Vol. BME-28 No-**10**
[5] Teichmann D, Foussier J, Jia J, Leonhardt S and Walter M 2013 *IEEE Trans. Biomed. Eng.* 60 pp. 2-41 2152
[6] Wimmer R 2011 *Whole Body Interaction Human-Computer Interaction Series 2011*, pp 121-133
[7] Alanis A, Thai T, Dejean G, Gilad-Backrach R and Lymberopoulos D 2014 No. *MSR-TR-2014- 81*
[8] Norgia M and Svelto C 2007 *Instr. Meas. Tech. Conf. IMTC 2007*
[9] Sakthivel M and George B 2014 *IEEE Trans. Inst. Meas.*
[10] George B, Zangl H, Bretterklieber T and Brasseur G 2009 *Inter. Inst. Meas. Tech. Conf.*
[11] Scharfetter H, Casanas R and Rosell J 2003 *IEEE Trans. Bio. Eng.* Vol. BME-**50** No-7
[12] Gabriel C 1996 *Fin. Tech. Rep.*

[13] Wtorek J and Bujnowski A 2007 *ICEBI 2007 IFMBE Proceedings 17* pp. 280-283
[14] Liu X, Ng W M, Lee C K and Hui S Y 2008 *IEEE App. Pow. Elec. Conf. Exp. APEC 2008*
[15] Mortarelli J R 1980 *IEEE Trans. Bio. Eng.* Vol. BME-27 No-**11**
[16] Geselowitz D B 1971 *IEEE Trans. Biom. Eng.* Vol. BME-**18** pp. 38 – 41
[17] Ktistis C and Peyton A J 2007 *ICEBI 2007 IFMBE Proceedings* **17** pp. 464-467
[18] Auld A B, Moulder J C 1999 *Jour.of Non. Eval.* 18(1): pp. 3-36
[19] Cheng D 1965 *IEEE Trans. on Instrum. and Imp.* Vol. IM-14 No. **3**
[20] Kwok M C and Pepper M G 1991 *Med. & Biol. Eng. & Comput.* Vol. **29** pp. 136-140
[21] Pepper M G, Taylor D J E and Kwok M C 1991 *Med. & Biol. Eng. & Comput.* Vol. **29** pp. 141-148
[22] Burrows M L 1964 *Ph.D. dissertation* University of Michigan Ann Arbor MI
[23] Bujnowski A, Palinski A, Koscinski P, Skalski L, Skurczynska A and Wtorek J 2013 *Journal of Physics: Conference Series* 434 (2013) 012035

Acknowledgments

This work is partially supported by Statutory Funds of Faculty of Electronics, Telecommunications and Informatics, Gdansk University of Technology.

IOP Conf. Series: Materials Science and Engineering **104** (2015) 012038 doi:10.1088/1757-899X/104/1/012038

Multichannel Human Body Communication

Piotr Przystup, Adam Bujnowski, Jerzy Wtorek

Department of Biomedical Engineering, Faculty of Electronics, Telecommunication and Informatics, Gdansk University of Technology, Poland

E – mail: przystup@biomed.eti.pg.gda.pl; bujnows@biomed.eti.pg.gda.pl; jerzy.wtorek@eti.pg.gda.pl

Abstract. Human Body Communication is an attractive alternative for traditional wireless communication (Bluetooth, ZigBee) in case of Body Sensor Networks. Low power, high data rates and data security makes it ideal solution for medical applications. In this paper, signal attenuation for different frequencies, using FR4 electrodes, has been investigated. Performance of single and multichannel transmission with frequency modulation of analog signal has been tested. Experiment results show that HBC is a feasible solution for transmitting data between BSN nodes.

1. Introduction

Many diagnostic procedures based on long term monitoring require a number of biosignals. Typically sensors or electrodes are placed all over the patient's body and are connected with the central unit using wires. An example of such a procedure could be polysomnography, which involves all-night recording of electrocardiogram (ECG), electroencephalogram (EEG), electrooculogram (EOG), respiration signal and blood oxygen saturation. Each sensor is connected by wire with the central unit, which makes sleeping uncomfortable and unnatural. This could lead to incorrect diagnosis. Recording and examination of physiological signals and associated parameters is another example of possible application [1]. Each wire, that could be removed from the measurement system increases the patient's comfort during examination and the accuracy of the diagnostic procedure.

To reduce the number of wire connections, wireless data transmission could be used. Traditional technologies can be applied to gather biosignals, like Near Field Communication (NFC) [2], ZigBee [3] or Bluetooth [4]. Unfortunately, these methods have some drawbacks. ZigBee and Bluetooth are characterized by quite large range, way beyond the limits of patient's body, which can easily cause interference with other Body Sensor Networks (BSN) and leads to performance degradation [5]. Additionally, relatively high power consumption of Bluetooth and low data rate of ZigBee limit the possible uses of those technologies in BSN. On the other hand, NFC has low data rate and it's not suitable for application that requires constant connectivity between sensor nodes [6]. To overcome these issues, Human Body Communication (HBC) could be utilized. The main concept of HBC is to transmit the information via the human body through electrodes attached to the skin [7]. Relatively high conductivity of the human body, in comparison to air, makes the HBC an energy saving method, while data rates of 2Mb/s and higher make it an interesting alternative for wireless communication within BSN [8].

HBC can be categorized into two different types: electric field coupling (figure 1) [9], and electromagnetic coupling (figure 2) [10]. Using electric field coupling, only one electrode is attached to the patient's body on the receiver and transmitter side. High frequency signal is propagated through the body, while the return path is coupled by the near electric field. This implies that the signal quality is dependent on the surrounding environment. On the other hand, electric field coupling provides lower attenuation of the signal than the other method. Electromagnetic coupling requires two electrodes on each side of the communication channel. The human body is treated as a waveguide, with the high frequency electromagnetic waves generated by transmitter. Transmission quality is not affected by the surrounding environment, however signal attenuation is much higher.

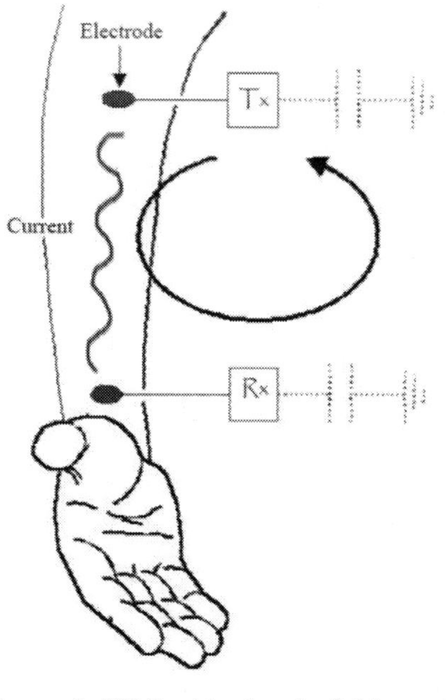

Figure 1. HBC with electric field coupling

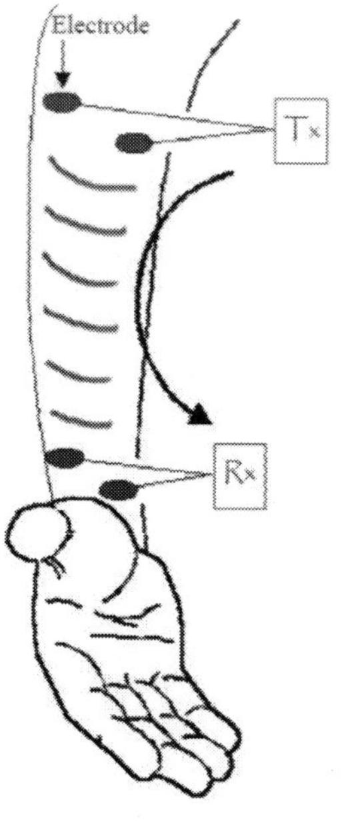

Figure 2. HBC with electromagnetic field coupling

The main goal of this study is to test the performance of the electric filed coupling HBC for analog signal transmission. Two scenarios will be tested: 1 transmitter / 1 receiver, and 2 transmitters / 2 receivers working simultaneously.

2. Methods and materials

Signals which are transmitted with HBC need to be modulated with a high frequency carrier. Two methods are available in order to transmit the analog signal. It can be directly connected to Voltage Controlled Oscillator (VCO) input. The receiver will demodulate the signal and pass it to the Analog to Digital Converter (ADC). Alternatively, the ADC could be placed on the transmitter side. Data would be modulated using Frequency Shift Keying (FSK) or similar method, and then transmitted. It could be saved, or displayed directly after demodulation. The first method is cost- and power-effective, since it requires only one ADC on the receiver side in central unit, while the second is more noise immune, since it transfers binary data. Transmitting analog signal has one more advantage – it does not introduce any delay to the signal which could be extremely important in case of biosignals recording, for example for pulse wave velocity measurement [11]. In this study, analog signal without ADC conversion will be sent.

2.1. Electrodes' placement and attenuation measurement

In order to evaluate the performance of HBC, three different types of electrodes have been tested. They were square pieces two-side FR4 laminate with dimensions 10 mm × 10 mm, 20 mm × 20 mm and 30 mm × 30 mm. One side of the electrode was connected to the signal input / output, while the other was connected to ground. Electrodes were attached to the patient's body in four different locations, presented in figure 3a-d, signal side towards the body. Distances between the transmitter and receiver were approximately 30 cm, 60 cm, 160 cm and 50 cm respectively.

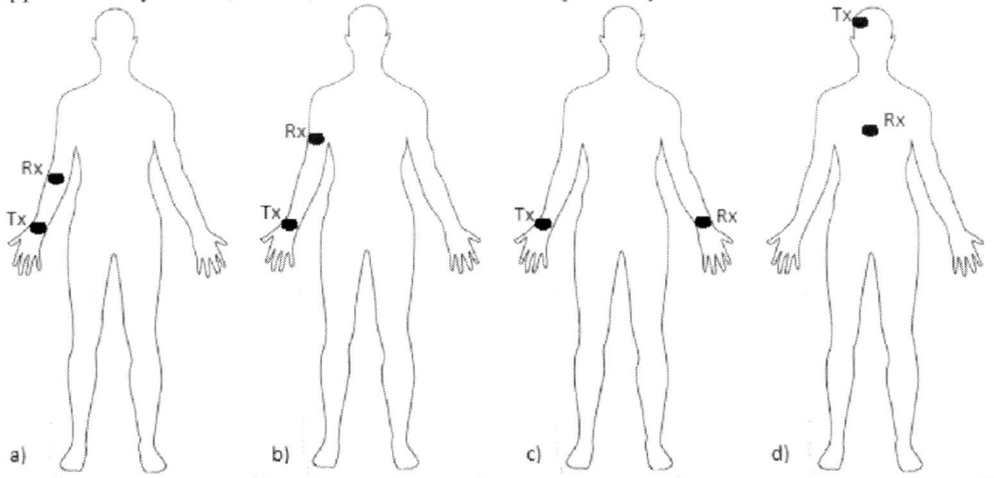

Figure 3. Electrodes' placement

Signal attenuation was measured for frequencies between 0,5 MHz and 20 MHz. Function generator was used as signal source, while the digital oscilloscope played the role of the receiver.

2.2. Signal transmission performance

For next experiments, the transmitter and receiver circuits were designed. Transmitter modulates the analog input signal with high frequency carrier. For frequency modulation, VCO was used. On the receiver side, signal was amplified and demodulated using a Phase Locked Loop (PLL) device. It contains a VCO and a phase comparator which is exclusive-or network. Demodulated analog signal was connected to the input of a 14-bit ADC.

Five sinus waves with frequencies 1 Hz, 5Hz, 10 Hz, 20 Hz and 40 Hz were modulated and transmitted. After demodulation, signals were sampled with a 14-bit ADC, 250 samples/s. For each of them, Signal to Noise Ratio (SNR) was calculated, defined as a ratio between the power of the input signal, and noise.

2.3. Multichannel transmission

Two transmitter circuits with different output frequency ranges were used. The first one generates an output signal in a range of 0.67 MHz – 0.79 MHz, while the output of the second was within 1,6 MHz – 1,86 MHz. Transmitters are battery powered. Since both of them worked simultaneously, an additional filter stage at the receiver side was necessary. Simple LC bandpass filters were designed for this purpose, with 3 dB frequency bandwidth matching the output frequency range of the modulators. (figure 4).

After input filtering the signal was amplified and demodulated using a PLL. The demodulated analog signal was connected to the input of the 14-bit ADC. Sinusoidal signals with different frequencies, between 1 Hz and 40 Hz were transmitted, using the designed devices. SNR was determined for both frequencies. For this experiment, transmitter circuits were placed on both wrists, while the receiver circuit was attached to the patient's chest (figure 5).

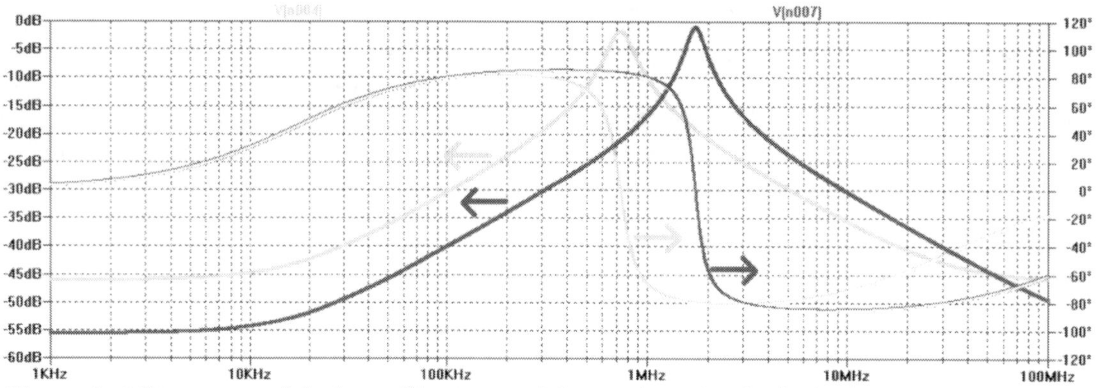

Figure 4. AC response of the input filter stage of the receiver circuit, for both used ranges (0.67-0.79 and 1.60-1.86 MHz)

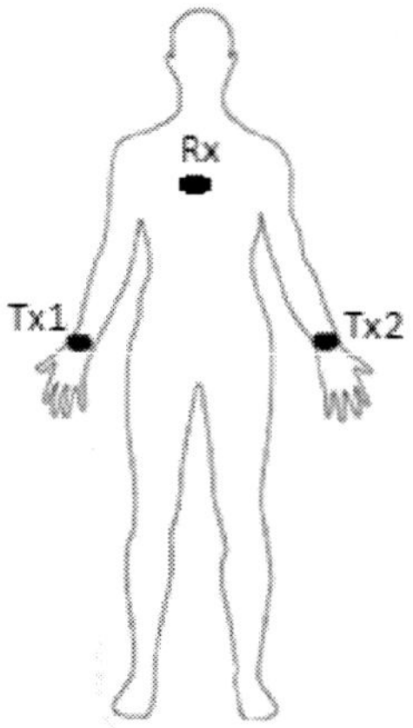

Figure 5. Electrodes' placement for multichannel configuration

3. Results

Signal attenuation for different electrode sizes and signal frequencies is shown in figure 6.

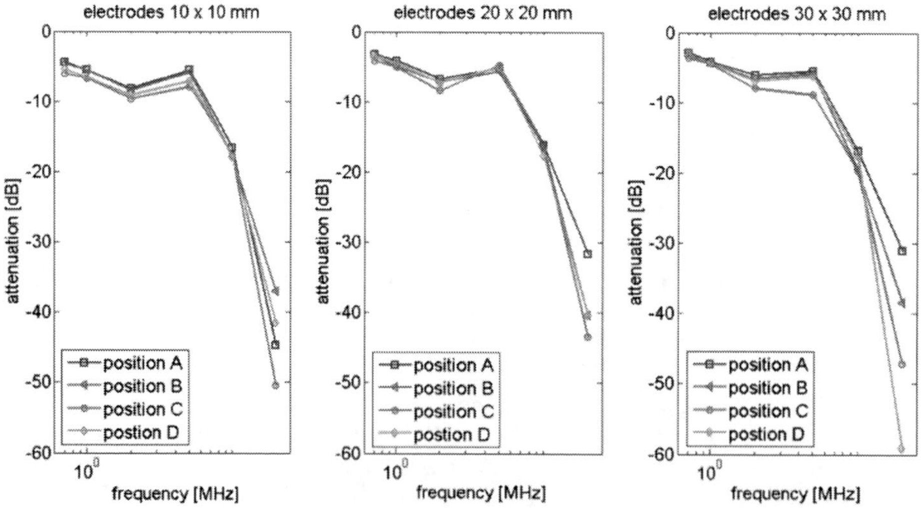

Figure 6. Signal attenuation

SNR for single-channel transmission as a function of input signal frequency is presented in figure 7. For this test, the worst-case scenario was used – transmitter and receiver were placed on the left and right wrist respectively (the longest distance to travel for the signal), and the signal was transmitted with the smallest electrodes (the greatest signal attenuation).

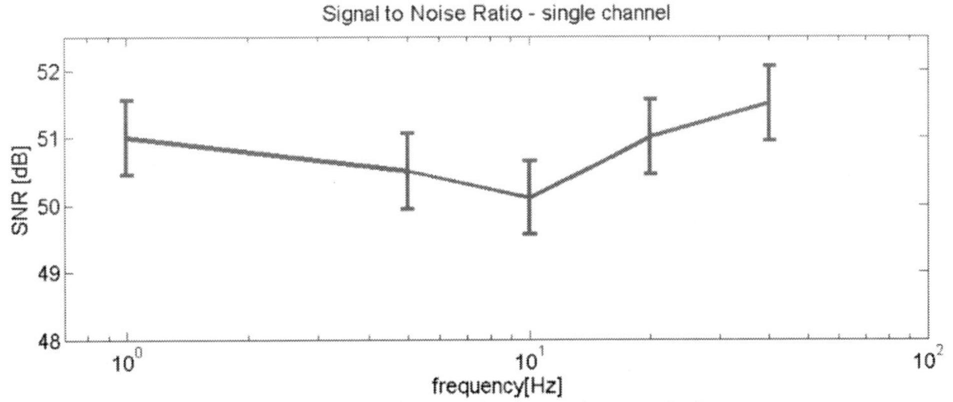

Figure 7. SNR for single channel transmission.

Finally, the multichannel configuration was tested. There was no significant difference in transmission performance between the two channels with different carrier frequency. Figure 8 presents SNR as a function of the input signal frequency for one of them.

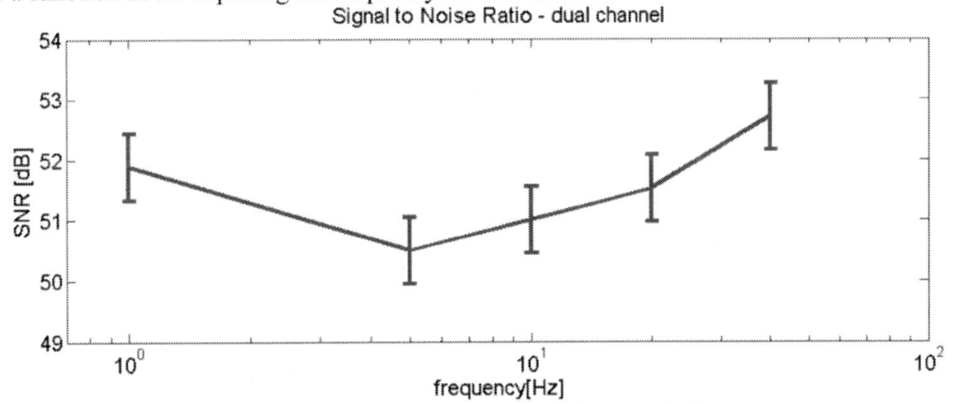

Figure 8. SNR for multichannel transmission

All experiments were conducted twice. First with no patient movements, later with the patient's arm moving up and down during transmission. No differences between those two sets of measurements were observed.

4. Discussion and Conclusion

In this paper, the performance of HBC has been investigated. High frequency signal attenuation measurements allow for choosing the optimum carrier frequency for the modulator circuit. The obtained results are slightly different from those presented in previous work [10], [12]. This can be caused by the differences in the size and materials of the electrodes used in different studies. In our case, the optimum band is below 5 MHz. The experiment has proved that the size of the electrode is not critical in this frequency range and could be as small as 1 cm^2.

Frequency modulation of the analog signal could be used for analog data transmission. Such approach could be extremely useful in case of biosignals transmission, where the delay between the actual and transmitted signal might be critical. Analog signal modulation is more noise-sensitive. Thus, low noise transmission channel is desirable. The presented solution is characterized by SNR as high as 50 dB. Channel performance in higher frequencies is limited by the sampling rate of the recording device [13].

Modern medical devices contain a number of sensors connected by wires with the central unit. Reduction of such connections will increase the patient's comfort, and make the examination procedure less invasive. This could lead to the increase in accuracy for some type of diagnostic methods, like sleep disorders detection. The solution is to reduce the number of sensors [14], or to use wireless data transmission. For each sensor, that has to be connected with the central unit without a wire, a pair of transmitter and receiver is necessary. Each pair needs to use a different carrier frequency. The presented experiment proves that it is possible to use more than one transmitter-receiver pair at a time. Moreover, no degradation in transmission channel performance was observed.

Patient's movements have negligible influence on signal transmission. This means that it is suitable to use HBC in long term patient monitoring.

Future research of HBC will cover the selection of the optimal size and material of the electrodes, as well as evaluation of HBC performance in all night recordings.

5. References

[1] J Siebert, J Wtorek and J Rogowski 1999 Stroke volume variability - Cardiovascular response to orthostatic maneuver in patients with coronary artery diseases *Annals of the new york academy of sciences,* vol. **873**, 182-190.

[2] K Yamashita, S Izumi, M Nakano, T Fujii, T Konishi, H Kawaguchi, H Kimura, K Marumoto, T Fuchikami, Y Fujimori, H Nakajima, T Shiga and M. Yoshimoto 2013 A 38 μA wearable biosignal monitoring system with near field communication in *New Circuits and Systems Conference (NEWCAS), 2013 IEEE 11th International* (Paris)

[3] S Sudin, F Aziz, N Mohd Hishamuddin, F Ahmad Saad, A Md Shakaff, A Zakaria and A Salleh 2014 Wearable Heart Rate Monitor Using Photoplethysmography for motion *Biomedical Engineering and Sciences (IECBES), 2014 IEEE* (Kuala Lumpur).

[4] J Lovelace, T Witt and F Beyette 2013 Bluetooth enabled elctroencephalograph (EEG) platform *Circuits and Systems (MWSCAS), 2013 IEEE 56th International Midwest Symposium* (Columbus).

[5] K Naik, D Wei, Y Su and N Shiratori 2005Analysis of packet interference and aggregated throughput in a cluster of Bluetooth piconets under different traffic conditions *Selected Areas in Communications, IEEE Journal* vol. **23** no. 6 1205-1218.

[6] E Strommer, J Kaartinen, J Parkka, A Ylisaukko-oja and I Korhonen 2006 Application of Near Field Communication for Health Monitoring in Daily Life *Engineering in Medicine and Biology Society EMBS '06. 28th Annual International Conference of the IEEE* (New York).

[7] T Zimmerman 1996 Personal Area Networks: Near-field intrabody communication *IBM Systems Journal* vol. **35** (3) 609-617.

[8] C-H Hyoung, J-H Hwang, S-W Kang, S-O Park and Y-T Kim 2015 A Feasibility Study on the Adoption of Human Body Communication for Medical Service *Circuits and Systems II: Express Briefs, IEEE Transactions* vol. **62**(2) pp. 169-173.

[9] R Xu, H Zhu and J Yuan 2011 Electric-Field Intrabody Communication Channel Modeling With Finite-Element Method *Biomedical Engineering, IEEE Transactions* vol. **58**(3) 705-712.

[10] M Wegmueller, M Oberle, N Felber, N Kuster and W Fichtner 2010 Signal Transmission by Galvanic Coupling Through the Human Body *Instrumentation and Measurement, IEEE Transactions* vol. **59**(4) 963-969.

[11] W Lee and S Cho 2015 Integrated All Electrical Pulse Wave Velocity and Respiration Sensors Using Bio-Impedance *Solid-State Circuits IEEE Journal of* vol. **50**(3) 776-785.

[12] N Zedong, L Tengfei, W Wenchen, G Feng and W Lei 2012 Experimental Characterization of human body communication in shield chamber *Biomedical and Health Informatics (BHI) IEEE-EMBS International Conference* (Hong Kong).

[13] P Przystup, A Bujnowski, J Ruminski and J Wtorek 2013 A multisensor detector of a sleep apnea for using at home *Human System Interaction (HSI) 6th International Conference* (Sopot).

[14] P Przystup, A Bujnowski, A Poliński, J Rumiński and J Wtorek 2014 Sleep apnea detection by means of analyzing electrocardiographic signal *Advances in Intelligent Systems and Computing,* vol. **300** 179-192.

Acknowledgments

This work has been partially supported by Statutory Funds of Electronics, Telecommunications and Informatics Faculty, Gdansk University of Technology.